Understanding Flowers and Flowering

Understanding Flowers and Flowering

An Integrated Approach

Second Edition

Beverley Glover

OXFORD
UNIVERSITY PRESS

OXFORD

UNIVERSITY PRESS

Great Clarendon Street, Oxford, OX2 6DP,
United Kingdom

Oxford University Press is a department of the University of Oxford.
It furthers the University's objective of excellence in research, scholarship,
and education by publishing worldwide. Oxford is a registered trade mark of
Oxford University Press in the UK and in certain other countries

First Edition published in 2007
Second Edition published in 2014
Impression: 1

Published in the United States of America by Oxford University Press
198 Madison Avenue, New York, NY 10016, United States of America

British Library Cataloguing in Publication Data
Data available

Library of Congress Control Number: 2013944236

ISBN 978–0–19–966159–6 (hbk.)
ISBN 978–0–19–966160–2 (pbk.)

Printed and bound by
CPI Group (UK) Ltd, Croydon, CR0 4YY

Preface

Flowers are the features of plants that most endear them to human beings. We grow flowers in our gardens, we display them in our homes, and they feature prominently in our artistic history. Scientists, too, have been fascinated by flowers, and there is a long tradition of botanical study of floral structure and floral ecology. However, it is only in the last few decades that the tools of molecular genetics have been applied to flower induction, development, and morphology. The data arising from these studies, when combined with painstaking observation and analysis of the interactions of plants with their pollinators, are beginning to provide the first truly integrated understanding of both how and why flowers take the forms we so admire.

Traditionally, flowers are studied from one of a number of viewpoints. Molecular biologists may study the genetic control of flower induction or flower development, usually focusing on a single model species. Evolutionary biologists may study how flowers evolved, the forms of the earliest flowers, or the morphology of flowers of ancient lineages alive today. Pollination biologists may study the interactions between plants and their pollinating animals, the natural selection exerted by those pollinators, and how these pressures affect plant population dynamics. However, it is becoming clear that these various disciplines each have enormous power to inform and shape the work of the others. An understanding of how flower colour is controlled biochemically and genetically, for example, can be of great benefit when studying how pollinator-imposed selection might have influenced the evolution of petal colour within a group of plants. Similarly, an understanding of how time of flowering influences competition with other species in the same community can be of great help in understanding why and how certain genetic components of the floral transition mechanism are conserved and others are less so.

I hope that this book might serve as a starting point for those interested in taking such an integrated approach to the study of flowers. I have written it, and revised it for this second edition, with the intention of helping to bridge the gaps between the different disciplines that work with flowers. My aim is to provide students and researchers studying one aspect of floral biology with an overview of other important aspects of flowers, both to help them to set their own work in context and to encourage them to consider experiments which might lead to greater integration of the field. In particular, I hope that this book will encourage dialogue between floral biologists of all varieties, with a long-term view to ensuring a continuing increase in interdisciplinary studies of flowers.

The book is divided into three main sections. Section I is introductory, giving some necessary background to the evolution of flowers and to the history of scientific thought on flowers and flowering. In this second edition the chapter on the evolution of flowers has been extensively revised and updated. Section II considers the molecular mechanisms that control flower induction, flower development, and floral mating type, providing coverage of the genetic material available for shaping by natural selection. This section is initially focused on a very few species of model plants, looking at the molecular similarities which unite all flowers. In the later chapters it considers the development and reproductive strategies of plants from a range of species, with a new chapter on the floral transition in diverse species, and extensive revision of the chapter on floral development in various groups. Section III extends this analysis much further, considering the explanations for the differences between flowers, rather than their similarities. This section moves between molecular explanations for flower morphology and the ecological consequences of that morphology, in an attempt to integrate what we know both of how and why different flowers take their different forms. A new chapter on the lability of floral form considers how floral traits change within phylogenetic contexts. Finally, the epilogue attempts to draw out some themes which persist throughout the book, suggesting possible future directions for the field.

Many people have contributed to the development of both the first and second editions of this book, and I am particularly grateful to all members of my own research laboratory, past and present, for enthusiastic support and helpful discussions at many points in the process. The second edition has benefited from the suggestions and advice of reviewers of the first edition, and I am particularly grateful to Doug Soltis, Elena Kramer, and Martin Ingrouille for constructive comments. John Parker, Caroline Dean, David Hanke, Cathie Martin, Jeff Ollerton, Rob Raguso, and Nick Waser read various sections of the book in detail, and I must thank them all for the time and care that they took and for their excellent suggestions and advice. Many people were kind enough to provide me with images for figures. While these are acknowledged in the relevant figure legends, I particularly thank Nick Waser, Enrico Coen, and the Cambridge University Botanic Garden for their great generosity in sharing images, Matthew Dorling and Heather Whitney for photographic assistance, Mike Webb for biochemical pathways, and Rosie Bridge for line drawings. For this second edition I owe a great debt to Alison Reed, whose brilliant drawings and photographs have greatly enhanced the figures throughout the book. Thanks also to Roy Barlow and Don Manning for their excellent cover design for the first edition, which has been adapted by the OUP team for edition 2. At Oxford University Press, Ian Sherman has provided steadfast support for this project, dealing with various changes to the schedule with calm good humour, while Helen Eaton, Christine Rode and Lucy Nash have kindly shepherded me through the production process. On a personal note I am still grateful to Jocelyn, Duncan, and Katie Taylor for lending me the space and quiet to really begin writing the first edition, rather than just thinking about it. In recent years, and particularly during the development of this second edition, I have relied heavily on Sam Brockington and Edwige Moyroud for the discussion of ideas and the development of new lines of thought, as well as for practical and personal support in day-to-day academic life. Thank you both. And finally, as with everything I do, the writing of a second edition has only been possible because of the patience and support of Stuart, Sam, and Katie—I do appreciate you all, really.

Contents

SECTION I

Introduction

CHAPTER 1

The evolution of flowers

The oldest fossil flower currently known is around 125 million years old. Flowers, and the plants that produce them (angiosperms or flowering plants), are relatively recent innovations in evolutionary terms. The first land plants, which did not possess flowers, arose around 470 million years ago, but fossil evidence indicates that only after another 340 million years did the angiosperms appear. However, following their appearance in the fossil record of the early Cretaceous period, the angiosperms spread geographically from their point of origin in the tropics and diversified dramatically to become the ecologically dominant plant group in the great majority of terrestrial habitats. This extraordinary geographic and morphological radiation took a mere 40 million years, and was even more extraordinary for the number of species it generated. The 250–400,000 species of extant angiosperms represent the most species-rich plant group by far, and are exceeded in numbers in the speciose animal kingdom only by the arthropods. Furthermore, the differences in growth habit, morphology, and life history within the angiosperms are vast, leading Darwin to describe the speed and scale of this recent radiation as 'an abominable mystery'. It is not possible to provide here a full analysis of the extensive literature on the origins and radiation of angiosperms and their flowers. The aim of this chapter is to provide an overview of the key issues surrounding the origin of the flowering plants and their flowers, and to conclude with an introduction to the major groups of flowering plants, many of which will be referred to in later chapters.

1.1 The origin of flowering plants

Angiosperms are defined by a number of features, of which possession of a flower is only one. Some of these traits are shown in Fig. 1.1. They have fully protected and enclosed ovules with two layers of protective integuments surrounding them, enclosed in a carpel within which the seed eventually develops. Their wood contains true, continuous vessels, as well as the more widespread tracheids in which water has to cross a membrane between individual cells. Their phloem consists of sieve tube elements and the unique companion cells, both derived from the same mother cell. Angiosperms have distinctive pollen, with columnar structures providing support for the surface layer. In addition, only angiosperms undergo double fertilization, whereby two genetically identical sperm cells are released into the ovary with one fertilizing the egg and the other fusing with the central cell to form the endosperm. Traditionally, fossil evidence was all that was available to probe the origins of the angiosperms, but, more recently, molecular data obtained from extant species have been used to inform the debate. The following two sections consider insights from these two types of evidence into the age and nature of the earliest angiosperms, and their relationships with other seed plants.

1.1.1 Fossil evidence for angiosperm origins

Fossil evidence dates the origin of the angiosperms to the early Cretaceous period, with the oldest fossil flowers (125 million years ago), angiosperm fruits (121 million years ago), angiosperm pollen (130 million years ago), and angiosperm leaves (120 million years ago) all supporting this conclusion (Hughes 1994; Krassilov and Dobruskina 1995; Brenner 1996; Friis *et al*. 1999, 2001; Sun *et al*. 2002). The oldest fossils suggest that the first angiosperms were aquatic plants. For example, *Archaefructus*, a fossil dated to

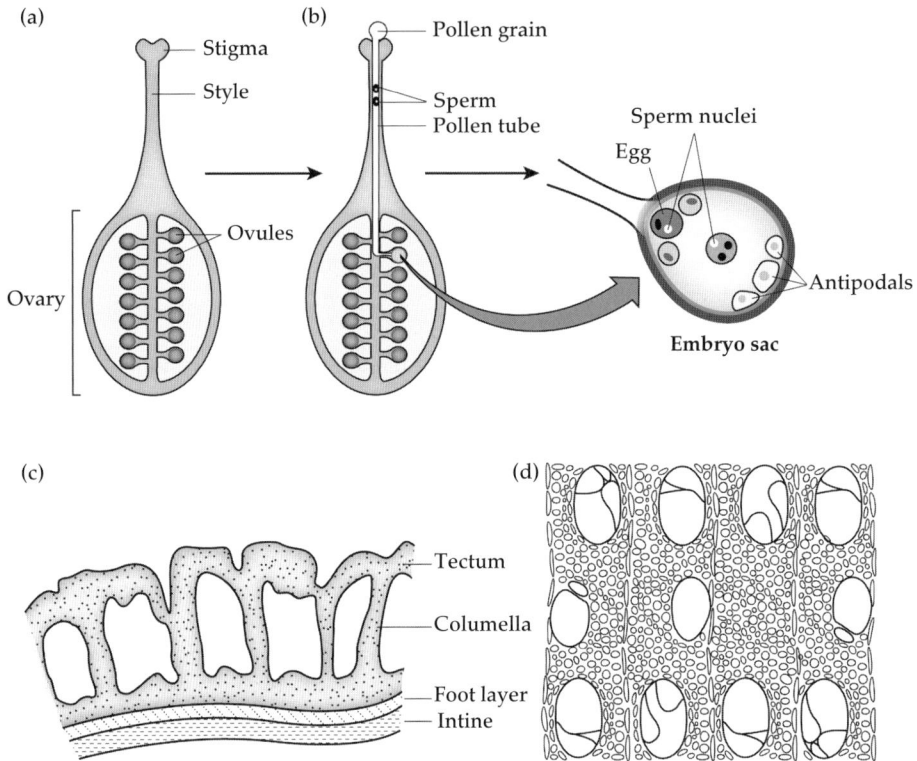

Figure 1.1 Some defining features of angiosperms. (a) Enclosed ovules, enfolded within the carpel. (b) Double fertilization, with two sperm nuclei arriving in the pollen tube. One fertilizes the egg cell and the other fertilizes the central cell with its two nuclei, generating the triploid endosperm. (c) Columnar pollen exine, shown in cross section. (d) Wood with true vessels as well as the narrower tracheids found in gymnosperms.

124.6 million years ago, has the long thin stems and highly dissected compound leaves that are typical of aquatic species. Perhaps even more convincingly, it is found with fossilized fish mixed in with the plant tissue (Sun *et al.* 2002). Similarly, Friis *et al.* (2001) identified a fossil flower from deposits up to 125 million years old as a member of the Nymphaeales (the modern water lilies), on the basis of its unique centrally protruding floral axis and its distinctive seeds with wavy cell walls in their seed coats.

The recent identification of *Leefructus mirus*, a fossil with features diagnostic of the eudicot order Ranunculales, in deposits dated at 122–125 million years old, suggests that the angiosperms might be older than paleontologists had previously thought (Sun *et al.* 2011). The eudicots are a more recently derived group of the angiosperms (see Section 1.6 and Fig. 1.5 for an overview of relationships within

the angiosperms). If the eudicots were already present in the early Cretaceous period, it is likely that the first angiosperms arose considerably earlier. Molecular analyses also date the origin of the angiosperms a little earlier than 130 million years ago (see Section 1.1.2 below).

The fossil record suggests that the angiosperms rapidly diversified from their aquatic origins to occupy understorey and early successional niches on dry land, with this first diversification occurring during the early Cretaceous period (Friis *et al.* 2005). Their subsequent radiation over the course of the Cretaceous quickly led to the adoption of late successional ecological positions, presumably as their size and woodiness increased. The scarcity of angiosperm wood in the early Cretaceous fossil record, along with the apparently small size of early angiosperm seed and leaves, supports the conclusion that the early angiosperms were small herbaceous species with a weedy lifestyle, whether on land or in fresh water. Some authors have speculated that this herbaceous habit followed an early loss of woodiness, since most gymnosperms are woody and most of the earliest diverging groups of extant angiosperms are also woody (Willis and McElwain 2002).

The first angiosperms appear to have originated in tropical regions, only spreading to higher latitudes after 20–30 million years (Barrett and Willis 2001). The earliest angiosperm fossils (of pollen) have been found in modern-day Israel and Morocco, land that lay just north of the equator in the early Cretaceous period. The subsequent spread of the angiosperms into higher latitudes was very rapid, accompanied by an increasing dominance of those areas already occupied (as measured by the relative proportions of angiosperm and other pollen types retained in the fossil record) (Willis and McElwain 2002).

The identity and morphology of the last common ancestor of the angiosperms and other land plant groups have long been a source of debate. The most closely related extant group is the gymnosperms, traditionally viewed as a cluster of three or four divisions. Of these divisions, the conifers (Coniferophytes) are familiar, and dominate many high-latitude forests. The cycads (Cycadophytes) are currently less prominent, but fossil records indicate that they were once ecologically highly significant. The Ginkgophytes are currently represented by only one surviving species, *Ginkgo biloba*, a commonly grown tree in parks and gardens. The remaining group, the Gnetophytes, was, until recently, considered likely to be the closest relative

of the angiosperms, on the basis of both fossil and morphological evidence (Crane *et al.* 1995). There are currently three extant genera of gnetophytes, namely *Gnetum*, *Ephedra*, and *Welwitschia*, which share a range of morphological similarities with some angiosperms (see Fig. 1.2). In particular, leaf morphology and venation in *Gnetum* closely resemble that of angiosperms, their xylem does contain vessels, and some gnetophytes produce reproductive structures containing both male and female parts. Double fertilization, often considered to be a defining feature of angiosperms, has also been documented in both *Ephedra* and *Gnetum* (Friedman 1990). However, recent molecular studies have refuted this hypothesis (see below), and have even questioned the status of the Gnetophytes as a division, suggesting that they are part of the Coniferophytes (Qiu *et al.* 1999; Chaw *et al.* 2000). These recent studies confirm the monophyletic nature of the extant gymnosperms, indicating that none of them provide a clear link to the angiosperms.

The fossil record informs the debate on the relationship of angiosperms to other land plant lineages by providing details of extinct groups. Two extinct gymnosperm groups in particular, the Bennettitales and the Mesozoic 'seed ferns' (such as *Caytonia*), have frequently been proposed as ancestors or close relatives of the angiosperms, and several studies based on fossil and extant morphology linked the Bennettitales, the Gnetophytes, and the angiosperms into a clade known as the anthophytes (Crane 1985; Doyle and Donoghue 1986). These conclusions were based on the distinctive morphologies of the extinct plants, analysed by painstaking

Figure 1.2 The Gnetophytes. (a) *Ephedra distachya* subsp. *monostachya* (male). Photo by Le.Loup.Gris (Wikimedia Commons). (b) *Welwitschia mirabilis* (male). Photo by Franzfoto (Wikimedia Commons). (c) *Gnetum latifolium* var. *funiculare*. Photo by Vinayaraj (Wikimedia Commons). See also Plate 1.

microscopical observation of fossils. *Caytonia*, for instance, has ovules almost entirely enclosed within cupules, bearing some similarity to angiosperm carpels. The Bennettitales have many morphological features in common with angiosperms, most notably the production of a bisexual reproductive shoot surrounded by sterile (possibly perianth-like) organs. Because these fossil taxa cannot be incorporated into molecular phylogenies, attempts to integrate them into our understanding of seed plant relationships requires a combined molecular/morphological approach. Such approaches often produce conflicting results, but it is certainly likely that some of these extinct lineages represent branches of the land plant phylogeny that diverged along the branch leading to angiosperms.

1.1.2 Molecular evidence for angiosperm origins

Molecular evidence for the origins of angiosperms is based on the comparison of DNA sequences or fingerprints from extant species. In addition, use of a molecular clock, which calculates the age of divergence of two sequences according to the number of differences between them, can provide estimates of the date of evolutionary events. Recent molecular dating studies have used methods to allow for divergent evolutionary rates, particularly those arising following branching of a lineage, when different branches might experience very different evolutionary rates according to whether they adopt a woody (long generation time) or herbaceous (short generation time) habit (Smith and Donoghue 2008).

Early molecular dating studies indicated an origin of the angiosperms considerably earlier than the 130 million years ago that the fossil record suggests. Studies indicated variously that the angiosperms arose in the late Carboniferous period, 290 million years ago (Kenrick 1999; Qiu *et al.* 1999), and that the monocotyledonous angiosperms diverged from the dicotyledonous species 250–200 million years (Wolfe *et al.* 1989) or 300 million years ago (Martin *et al.* 1989). However, improvements in dating technology are mainly responsible for the convergence of recent reports on an age for the angiosperms of 140–180 million years (Sanderson *et al.* 2004; Bell *et al.* 2005a). The most recent detailed study used a 'relaxed clock' method that allowed different lineages to experience

different evolutionary rates, and concluded that the angiosperms originated 167–199 million years ago (Bell *et al.* 2010). The same study dated the origin of key angiosperm clades, with the eudicots appearing around 130 million years ago, the Rosids arising 108–121 million years ago, and the Asterids appearing 101–119 million years ago (see Section 1.6). These dates are highly compatible with the fossil record, suggesting only that the very earliest angiosperms have not yet been retrieved from fossils. The dating of the eudicots at 130 million years old is particularly interesting, given the recent discovery of the eudicot fossil *Leefructus mirus* from deposits 122–125 million years old (Sun *et al.* 2011).

If the angiosperms originated 167–199 million years ago, the absence of angiosperm fossils from deposits earlier than 130 million years can be readily explained in a number of ways. It is possible that the fossil record does not contain the very earliest angiosperms because they were not woody, were relatively rare within their communities, or were predominantly found in dry or alpine environments not conducive to fossil formation. Some authors have suggested that their early Cretaceous diversification actually represents a migration event from a previous habitat less suited for fossil formation.

The molecular phylogenies of Qiu *et al.* (1999) and Chaw *et al.* (2000) have also shed light on questions of relatedness between gymnosperm groups and the early angiosperms (see Figs. 1.2 and 1.3).

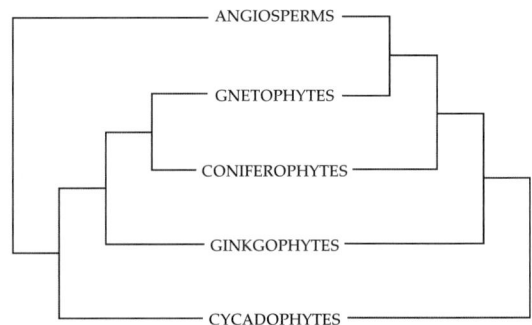

Figure 1.3 Molecular evidence and morphological evidence suggest different relationships between the angiosperms and the various groups of extant gymnosperms. Molecular data (left) place the angiosperms as sister group to the monophyletic gymnosperms, whereas traditional morphological analysis (right) placed the angiosperms and the gnetophytes as sister taxa.

To the astonishment of many botanists, the extant gnetophytes resolved in a single clade with the conifers, refuting the suggestion that their ancestors were the precursors of the angiosperms. Instead, the extant gymnosperms have been shown to be monophyletic and to have diverged from the total land plant lineage around 300 million years ago. They can therefore be thought of as the sister group to angiosperms, and do not necessarily provide much relevant information about the likely morphology of the ancestral angiosperm. Indeed, the derived nature of many extant gymnosperm reproductive structures, combined with the partial nature of the fossils of extinct gymnosperms, makes it very hard to draw conclusions even about the nature of ancestral gymnosperm reproductive structures.

The phylogeny of Qiu *et al.* (1999) provides useful information on the relationships of early diverging angiosperm clades, which can be used to infer features of ancestral angiosperm reproductive structures. The Nymphaeales (water lilies) are indeed ancestral to many other plant groups, but are not the sister group of all the other angiosperms. This position was awarded to *Amborella trichopoda*, the only extant species of the Amborellales, and a native of New Caledonia (see Section 1.6; Fig. 1.4). *Amborella* is a small weedy shrub, supporting the idea that the basal angiosperms were understorey or early successional species, but indicating that

Figure 1.4 The flower of *Amborella trichopoda*. Photograph kindly supplied by Sangtae Kim and Pam Soltis (University of Florida). See also Plate 2.

they were probably woody, not herbaceous, despite the relative scarcity of early fossilized angiosperm wood. Following the publication of the phylogeny of Qiu *et al.* (1999), the position of *Amborella* as the most basal extant angiosperm was controversial for a while. Goremykin *et al.* (2003) analysed 61 protein-coding genes common to 13 fully sequenced land plant chloroplast genomes, which placed the origin of *Amborella* later than the origin of the monocots. However, the controversy was short-lived and the position of *Amborella* as sister to the other angiosperms has been confirmed by numerous other studies (see detailed discussion in Stefanovic *et al.* 2004; Soltis and Soltis 2004; Soltis *et al.* 2004; Martin *et al.* 2005; Lockhart and Penny 2005). This controversy highlights some of the important considerations involved in designing approaches to reconstruct phylogeny, and also emphasizes the point that any phylogenetic tree can only be viewed as the current best hypothesis based on available data.

1.2 Seed plant reproductive structures

The reproductive structures of most plant lineages prior to the angiosperms, including most of the gymnosperms, were unisexual. The evolution of seeds freed plants to reproduce in the absence of a film of external water (previously necessary to allow fragile free-swimming male gametes to fuse with a static egg). In seed plants the female gametophyte is surrounded by parental sporophyte tissue, usually derived from bracts. These enfolding bracts act to protect the ovule and may also serve as a protective coat when the seed is dispersed. In gymnosperms the ovule is not completely enclosed in sporophyte tissue, but is protected within a chamber, into which wind-blown pollen is drawn after being trapped by a drop of secreted liquid.

In the cycads, dioecy is the rule, with female plants producing clusters of ovules on the edges of modified leaves called megasporophylls. The whole female reproductive structure forms an ovulate cone. Male plants produce cones of differently specialized leaves, the scale-like microsporophylls, arranged in a spiral phyllotaxis to produce a cone. These microsporophylls possess a pollen sac on their abaxial surface, each of which produces numerous pollen grains. A similar system operates in *Ginkgo*, although the ovules are born in pairs on stems rather than as a cone. In contrast, most conifers are monoecious, producing male and female cones on the same individual plant, although some taxa within the group are dioecious (such as the junipers and yews). The Gnetophytes are usually dioecious, although some species of the genus *Ephedra* are monoecious. Analysis of cone structure has revealed that the pollen cone of conifers is a condensed branch with the microsporophylls representing the modified leaves along the branch. However, the seed cone of conifers, and both cone types in the Gnetophytes, are thought to be derived from a condensed branch with branches. In this scenario the central stem produces bracts with modified stems in their axils—the modified stems are the scales of the cone (Judd *et al.* 2007).

1.3 The first flowers

Despite the absence of a clear picture of the last common ancestor of angiosperms and gymnosperms, there are clearly some key innovations that arose in the angiosperm lineage and gave rise to flowers, such as the combining of male and female reproductive structures within a small space on the same shoot and the production of perianth organs (Theissen *et al.* 2002). Indeed, it is these innovations that give us the modern definition of a flower (*the bisexual reproductive shoot of an angiosperm, in which the reproductive organs are surrounded by whorls of sterile organs*). Analysis of the development of these innovations allows us to build a general picture of the evolution of the first flowers.

1.3.1 A bisexual reproductive shoot

The typical angiosperm flower is hermaphrodite, and this is believed to be the ancestral condition for flowers. Although there are many examples of derived unisexual flowers—either on hermaphroditic plants (monoecy) or on unisexual plants (dioecy)—the majority of angiosperm flowers produce both male and female reproductive organs. This is in marked contrast to the reproductive structures of most other plant lineages, where unisexual reproductive structures are the norm. In the gymnosperms, as described above, the two types of

unisexual reproductive cones occur either together on a single plant or each on a separate plant. The combination of male and female organs on a single shoot is an angiosperm innovation. The development of a single shoot containing both male and female reproductive organs was therefore a key event in the evolution of flowers.

Discussion of the development of a bisexual reproductive shoot is complicated by the lack of clear homology between gymnosperm cones and angiosperm flowers. Since the cones of cycads and the pollen cones of conifers are condensed branches with the scales representing leaves, they could be described as homologous to a flower (with the reproductive and perianth organs derived from leaves). However, the compound cones of the Gnetophytes and female conifers are clearly not homologous to flowers. In these cones the scales are modified stems, and so the structure is much more like an angiosperm inflorescence, where each flower is formed from a separate axillary meristem. Whatever the homology relationships, the evolution of the flower required the development of a bisexual shoot, which may then have required much subsequent reduction to form the flowers we see today. Assuming that the ancestor of angiosperms and gymnosperms produced its male and female reproductive structures on separate shoots, each composed of a spiral of organs, then the development of a bisexual shoot (and later flower) requires either the development of female organs at the top of the male shoot or the development of male organs at the base of the female shoot. All angiosperm flowers contain an outer whorl of male reproductive organs and an inner whorl of female reproductive organs, suggesting that this evolution of bisexual flowers occurred only once (Cronk 2001). A single exception to this rule (with male reproductive structures inside a whorl of female ones) has been identified in the inside-out flower of *Lacandonia schismatica* (Pandanales), as a result either of homeotic organ conversion or of reduction of the inflorescence to resemble a flower (for a review, see Garray-Arroyo *et al.* 2012).

An adaptive explanation for the combination of male and female reproductive organs in the same shoot was proposed by Frohlich (2002), who observed that ectopic sterile ovules are sometimes produced by the male reproductive cones of the gnetophyte *Welwitschia*, and appear to attract pollinating insects to the male branches by exuding the same droplets of liquid that are secreted by fertile female ovules (Endress 1996). This would provide a selective advantage in terms of male fitness (pollen export) to a male shoot with additional female characters. From there it is easy to imagine the full feminization of the ectopic ovules into fertile structures. Bisporangiate cones are also occasionally produced by other gymnosperms, and the presence of ovule droplets is commonly reported in these cones (Rudall *et al*. 2011). Flores-Renteria *et al*. (2011) even observed that the bisporangiate cones of *Pinus johannis* were fully fertile, unlike the sterile ovules in male cones reported in *Welwitschia*. Other authors have noted that bisexual shoots facilitate selfing, which is potentially of great importance to plants (like the early angiosperms) invading new habitats.

Several models have been proposed to explain the evolution of the bisexual shoot from a molecular genetic perspective. These focus on two sets of genes—those controlling the conversion of the meristem to the reproductive form (particularly *LEAFY*; see Chapters 7 and 9), and those controlling the production of male organs in the angiosperm flower (B function genes; see Chapter 10). This section will outline these models, but the reader might find the models clearer when they have read later chapters on floral meristem and floral organ development.

The first model to take a genetic approach to the evolution of bisexual flowers was the Mostly Male Theory. In this model the bisexual shoot evolved by the production of ectopic ovules in the centre of a male cone, the cone retaining a 'mostly male' identity (Frohlich 2002). This hypothesis was based on analysis of sequence and expression patterns of genes predicted to regulate gymnosperm cone development, which suggested that an angiosperm gene required for flower production (*LEAFY*) is most closely related to a gymnosperm gene involved in male (but not female) cone development (Frohlich and Parker 2000). The *PrFLL* gene of *Pinus radiata* (and the equivalent gene in *Welwitschia* and *Ginkgo*) is very similar to the angiosperm *LEAFY* gene, which encodes a protein crucial in determining the floral nature of the meristem. In *Pinus*

radiata, *PrFLL* is expressed in male cones only, and a duplicate gene, *NEEDLY*, which has been lost from the angiosperm lineage, is expressed only in female cones. Frohlich interpreted these observations as implying that the angiosperm flower is descended from an ancestor of the male gymnosperm cone. In the ancestral angiosperm lineage the genes responsible for ovule development were recruited to the control of *PrFLL*, resulting in the production of ectopic ovules, and eventually the evolution of carpels. However, more recent data have largely disproved the exclusively male expression of *PrFLL* and the exclusively female expression of *NEEDLY* in gymnosperms. First, Shindo *et al*. (2001) showed that the *PrFLL*-like gene of *Gnetum parvifolium* is expressed in the female reproductive structure, rather than the male one. Dornelas and Rodriguez (2005) similarly showed that the *PrFLL*-like gene of *Pinus caribaea* is expressed in female, but not male, cones. Vazquez-Lobo *et al*. (2007) reported expression patterns for the *PrFLL*-like and *NEEDLY* genes from *Picea abies*, *Podocarpus recihii* and *Taxus globosa*, finding that both genes are expressed in both male and female reproductive structures of all three species. A recent comparison of transcriptome profiles of the male and female cones of *Ginkgo biloba*, the cycad *Zamia fisheri* and the gnetophyte *Welwitschia mirablis* with the transcriptomes of flowers of *Arabidopsis thaliana* (Brassicales) and *Oryza sativa* (Poales) detected no significant differences between the proportion of gymnosperm orthologous genes that were expressed both in the male cone and in the angiosperm flowers, and the proportion of gymnosperm orthologous genes that were expressed both in the female cone and in the angiosperm flowers (Tavares *et al*. 2010). From this analysis the authors concluded that the angiosperm flower was not a 'mostly male' structure, as its transcript content did not more closely resemble a male cone than a female cone.

If the bisexual shoot did not evolve by the ectopic development of ovules on a male cone (or vice versa), it must have evolved by the conversion of organs at the axis of a male cone into the female form, or by the conversion of the organs at the base of a female cone into the male form. These homeotic models do not imply that the flower is predominantly male or female, but that conversion of one organ type into the other results in a more nearly equal shoot. They are therefore more compatible with the transcriptomic analysis of Tavares *et al.* (2010). Theissen *et al.* (2002) described these ideas as the 'Out of Male' hypothesis and the 'Out of Female' hypothesis, depending on the shoot type in which the conversion occurred. Homeotic conversion of organ types is not usually attributed to meristematic function genes like *LEAFY*, but to the genes controlling floral organ development that are discussed in Chapter 10. The key set of genes to this discussion are called the B function genes, and in the angiosperm flower they specify the development of the stamens (male reproductive structures) in conjunction with the expression of C function genes. They also specify petal development, but that is incidental to this discussion. The C function genes alone specify carpel development (female reproductive structure development). Both B and C function genes encode transcription factors called MADS box proteins, which activate transcription of the downstream genes involved in organ identity (see Chapters 10 and 11 for further details). Genes related to angiosperm B function genes have been isolated from conifers, and from *Gnetum*, and are expressed in the male cones, but not in the female cones (Sundstrom and Engstrom 2002; Winter *et al.* 2002). Similarly, genes related to angiosperm C function genes have been isolated from *Gnetum*, *Cycas*, and *Picea*, and shown to be expressed in both the female cones and the male cones, indicating that C function is active in all reproductive structures in gymnosperms (Rutledge *et al.* 1998; Tandre *et al.* 1998; Zhang *et al.* 2004). The true roles of these genes in gymnosperms remain hypothetical, as no functional tests have yet been possible. However, on the basis of expression patterns alone, it appears that expression of the C function genes, in angiosperms and in gymnosperms, confers a reproductive identity on an organ. It is expression of the B function gene that determines which sort of reproductive structure is produced. Expression of B function genes as well as C function genes results in the development of a male reproductive structure, a pollen cone. Expression of C function genes alone results in the development of a female reproductive structure, a seed cone. These observations provide the basis for the Out of Male and Out of Female

hypotheses. Loss of B function gene expression from the upper regions of a male axis would result in the production of female organs within whorls of male organs. Similarly, gain of B function gene expression in the lower regions of a female axis would result in the production of male organs outside whorls of female organs (Theissen *et al.* 2002). The current lack of functional tools with which to explore gymnosperm cone development limits our ability to test these various hypotheses.

1.3.2 Evolution of the perianth

The typical angiosperm flower produces two whorls of sterile protective perianth organs—often, but not always, differentiated into protective green sepals and attractive petals. The production of these perianth organs was another angiosperm innovation. However, it is not necessary to assume that the first angiosperm flower possessed a perianth. Indeed, some authors have argued to the contrary, proposing a perianth-less structure containing both male and female reproductive organs as the first flower (Theissen *et al.* 2002). Some fossil evidence does seem to support this hypothesis, including the morphology of *Archaefructus* and other early fossil angiosperms (Sun *et al.* 1998). Other authors have argued that the specialized habitats of early fossil angiosperms, particularly the aquatic ones, might have resulted in specialized floral morphology and even secondary loss of a perianth. However, it is clear that at some point the bisexual reproductive shoot was surrounded by sterile organs to generate the familiar form of the angiosperm flower.

The development of the perianth may have occurred in a number of ways. One hypothesis proposes that the perianth developed by loss of reproductive function of the outermost male organs, a model that has molecular support from studies of ranunculid flower development (Kramer and Irish 1999). Alternatively, bract-like structures may have been modified to produce a perianth, shifting position on the shoot to develop immediately outside the reproductive organs. Both mechanisms have some support, and some authors postulate that the perianth evolved multiple times. However, a review of the available evidence (Specht and Bartlett 2009) suggests that the ancestral flower did have a perianth (that is, there was a single ancestral origin), and that this perianth was lost repeatedly in multiple lineages, including some of the earliest diverging groups (see Section 1.6), such as the Hydatellaceae (Nymphaeales) and the Chloranthaceae (Chloranthales). Given the great morphological similarities between sepals and leaves, most authors now agree that the sepals (and thus the ancestral perianth) are derived from bracts. During subsequent evolution of the angiosperm flower a differentiated perianth arose in multiple lineages. The inner organs of this differentiated perianth, the petals, are thought to have evolved from both sepals and stamens in different lineages (see Section 1.3.3).

1.3.3 Early 'flower' morphology

The first angiosperm flowers with perianth organs almost certainly did not have separate sepals and petals, but an undifferentiated and unspecialized perianth. Analysis of the fossil record and of extant basal angiosperms suggests that the anthers of these flowers made a lot of pollen, while the ovules may not yet have been entirely enclosed by carpel tissue, in the same way that the carpels of *Amborella* are not fully closed (Dilcher 2000; Endress and Igersheim 2000). The carpel tissue that had developed would have enfolded individual groups of ovules, with the multiple fused carpels of many modern flowers still to evolve.

The presence of small unisexual flowers with few flower parts as well as large bisexual flowers with many flower parts as fossils in deposits of similar ages has led to a range of suggestions as to the likely morphology of the first flowers (Endress 1987). The larger fossil flowers are reminiscent of those of modern-day *Magnolia* or waterlily, and both the Magnoliales and the Nymphaeales are early diverging groups. However, the smaller fossil flowers resemble those of the Piperales (containing the spice peppers). Interpretation of these fossil data is further complicated by the possibility that even early fossil flowers had diverged from the ancestral morphology. For example, adaptation to an aquatic lifestyle might change flower morphology considerably, and many of the earliest angiosperm fossils show clear evidence of an aquatic habit (Friis *et al.* 2001; Sun *et al.* 2002).

It was noted in Section 1.1.1 that similar flower-like structures also evolved in the now extinct Bennettitales, a group of gymnosperms that were widespread between 248 and 140 million years ago (Willis and McElwain 2002). Some species of this group combined both male and female reproductive organs on a single axis, and appear to have enclosed those organs within a circle of bracts, superficially like angiosperm flowers. There is debate as to whether these bracts opened, like the perianth organs of flowers, or whether structural constraints would have prevented this. It is also unclear whether the bracts contained pigmentation, believed to be a significant factor in the success of angiosperm flowers.

1.3.4 Evolution of the differentiated perianth

The perianth of most angiosperms is clearly divided into protective green sepals and attractive colourful petals. However, this differentiation of roles is a derived feature that appears to have been gained and/or lost in multiple different angiosperm lineages. The early diverging angiosperm groups often exhibit a perianth that shows gradual transition from sepaloid outer organs to petaloid inner organs, with no clear division point (discussed by Ronse de Craene *et al.* 2003). This morphology is facilitated by their perianth phyllotaxy, which in many species is spiral and continuous, rather than discretely whorled. This gradual transition of perianth organs is shown by *Amborella* (Buzgo *et al.* 2004), the Nymphaeales (Warner *et al.* 2008), and some other early diverging groups.

Differentiation of the perianth can in theory occur either by modification of inner sepals to produce petals, or by production of sterile stamens that evolve morphologically to produce petals. Both mechanisms have been shown to have occured in different lineages, with a sepal-derived origin of petals usually postulated in early-diverging angiosperm groups and the monocots, but a stamen-derived origin of petals assumed for the eudicots. Some studies have assumed that the petals of the core eudicots (see Section 1.6) evolved only once, from stamens, although recent reports have suggested that this is an over-simplification of a group which has itself experienced repeated petal loss and gain (Ronse de Craene 2007; Brockington *et al.* 2012). The molecular genetic basis of these various evolutions of the differentiated perianth probably involves similar homeotic mutations to those that generated the bisexual reproductive shoot. The molecular basis of perianth morphology will be discussed for several examples in Chapter 11.

1.4 Floral diversification

The first angiosperm flowers were borne by a lineage that subsequently underwent a dramatic radiation, including an astonishing diversification of floral form. It has been hypothesized that this process was initiated by the association of male and female organs on the same axis, which increased the effectiveness of animal pollination (previously less significant than wind pollination for most plant species). An alternative hypothesis is that the appearance of coloured pigments in the new flowers had a dramatic effect on their attractiveness to animals, again establishing biotic pollination as the norm. Studies of the colour vision of a range of arthropods have inferred that the Cambrian ancestors of today's insects possessed the same trichromatic colour perception system as that used by modern pollinating insects to discriminate between flowers (Chittka 1996). If insects were pre-adapted for viewing flower colour 500 million years ago, then the appearance of coloured pigments in the perianth organs of the first flowers may have rapidly led to a strong association between flowers and animal pollinators. Increased animal pollination is likely to have resulted in increased outcrossing, in part through the dual service provided by an animal in both removing pollen from the anthers and applying non-self pollen to the stigma of a single flower in a single visit. This increase in outcrossing will have increased the amount of recombination between plant genomes and therefore the speed of angiosperm evolution (Dilcher 2000). Outcrossing can be both advantageous and disadvantageous to plants, having consequences for the genetic make-up of populations and effective population size (Barrett 2002; Charlesworth 2006; see also Chapter 13). However, in many situations, selection will strongly favour adaptations that maintain increased outcrossing, resulting in the appearance and maintenance of a suite of novel colours, shapes, structures, scents, and rewards. At the same time, animal pollination is more likely than abiotic pollination to result in reproductive isolation, particularly where specific pollinating animals are attracted only to a subset of floral attributes but the wind is random in its dispersal of pollen. Such reproductive isolation may have facilitated speciation and divergence of form, further increasing the speed of angiosperm radiation. Finally, full closure of the carpel and the ability to recognize and reject self pollen through biological means may have evolved in concert to reinforce outcrossing (Dilcher 2000).

That rapid evolution coincided with the recruitment of animal pollinators is not in doubt, with the advantages of outcrossing and the ability to exchange genetic information between widely spaced individuals representing major benefits in terms of increased genetic diversity and thus 'evolvability'. What is less clear is the extent to which angiosperm and flower radiation can be directly attributed to a subsequent coevolution between angiosperms and pollinating insects. This attractive hypothesis has informed much of pollination ecology for decades, and can best be summarized by viewing flowers with a perianth as an angiosperm's way of manipulating an insect into carrying its pollen around for it. Once the flower has acquired the ability to manipulate one insect, it may radiate (genetic variability permitting) into as many different forms as there are insects with different preferences to manipulate.

There is some fossil evidence to support the hypothesis that angiosperm radiation was the result of angiosperm–insect coevolution. Fossils of early Cretaceous flowers show some features indicative of animal pollination, including a larger size than necessary for wind pollination (Crane *et al.* 1995). The first Lepidoptera (butterflies and moths) appear in the fossil record at the same time as the first angiosperms, and some groups of the Hymenoptera (bees, wasps, and ants) also appear to have evolved at the same time. However, there is little correlation between the origin or speciation of other insect groups and the origin and speciation of the angiosperms (Willis and McElwain 2002). This suggests that angiosperm–insect coevolution may have played an important role in the great diversification of some groups and floral forms, but that a significant part of this radiation must instead be attributed to the short lifespans and high rates of outcrossing of the new plants, and their consequent rapid acquisition of new habitats and new niches.

1.5 Morphological diversity of the flower

The radiation of the angiosperms has resulted in a tremendous variety of floral form. This variety encompasses size, shape, symmetry, organ number, whorl number, phyllotaxis, organ elaboration, colour, pigmentation pattern, texture, scent, and reward availability. The ecological and evolutionary explanations for this variety are discussed in Section III of this book, while its molecular genetic basis is discussed within Sections II and III (notably Chapters 11, 15, 16, 17, and 18). However, it is worth noting at this point that this diversity is not evenly distributed across the angiosperm phylogeny. While the eudicot group contains by far the greatest species diversity, estimated at around 70% of angiosperm species, its floral diversity is restricted to the details rather than the body plan of the whole flower. Eudicot flowers generally have a whorled phyllotaxis, differentiated perianth, and floral organs in groups of five (although there are deviations from this basic pattern). Their diversity is at the level of size, symmetry, organ elaboration, colour, texture, and scent. The monocots also have whorled phyllotaxis, with floral organs in groups of three. They usually produce an undifferentiated perianth with two whorls of brightly coloured tepals, or else have dramatically modified wind-pollinated flowers such as those of the grasses (Poales). It is in the early diverging angiosperm groups, where species diversity is only a few per cent of the angiosperm total, that the greatest diversity in floral body plan occurs. As noted earlier, species in these groups have spiral, whorled, or transitional phyllotaxis. They show great variation in organ number, as well as diversity of colour, scent, and floral elaboration. There is variation in the shape of their ovules and carpels, and in the degree to which the carpels are fully closed. It is likely that the relatively more constrained form of the eudicot flower has facilitated greater species diversification by establishing a successful and flexible floral form that is subject to variable selection pressures from different pollinators.

1.6 An introduction to angiosperm phylogeny

Throughout this book, reference will be made to many different angiosperm species. In many cases the aim will be to allow the reader to make comparisons between species, or to draw evolutionary conclusions. To allow ready access to a skeleton angiosperm phylogeny, this section and the associated Fig. 1.5 provide an overview of current thinking on relationships of different groups of flowering plants. This summary is based on the work of the Angiosperm Phylogeny Group (APG), and the phylogeny published as APG III (Angiosperm Phylogeny Group 2009). The APG is an informal collaboration of plant systematists from several different countries, who work together to update and publish a holistic picture of the whole angiosperm phylogeny, often incorporating detailed data on specific groups contributed by other plant systematists. Their phylogeny has come to be regarded as the most reliable reference point for angiosperm relationships.

The phylogeny presented in Fig. 1.5 is based on orders of plants. Within each of these orders are variable numbers of families. Throughout this book, species will be described at first mention in a chapter, with their order in brackets following the species name. The two exceptions to this rule are the molecular genetic models *Arabidopsis thaliana* (Brassicales) and *Antirrhinum majus* (Lamiales), which are discussed in almost every chapter and are introduced in detail at first mention. Readers who are interested in more detailed phylogeny can find family relationships within each of the orders on the Angiosperm Phylogeny Website (<www.mobot.org/mobot/research/apweb>), along with updates to the phylogeny as they occur.

At the base of the angiosperm phylogeny are a set of three sequentially branching orders, known together as the early diverging angiosperms. The Amborellales, at the base of the tree, contains only *Amborella trichopoda*, described above (see Section 1.1.2) and now much studied as the earliest diverging extant angiosperm. The Nymphaeales contains the waterlilies, and the Austrobaileyales is a group of mainly tropical shrubs. After these three orders, the phylogeny branches into three points. The group known as the Magnoliids

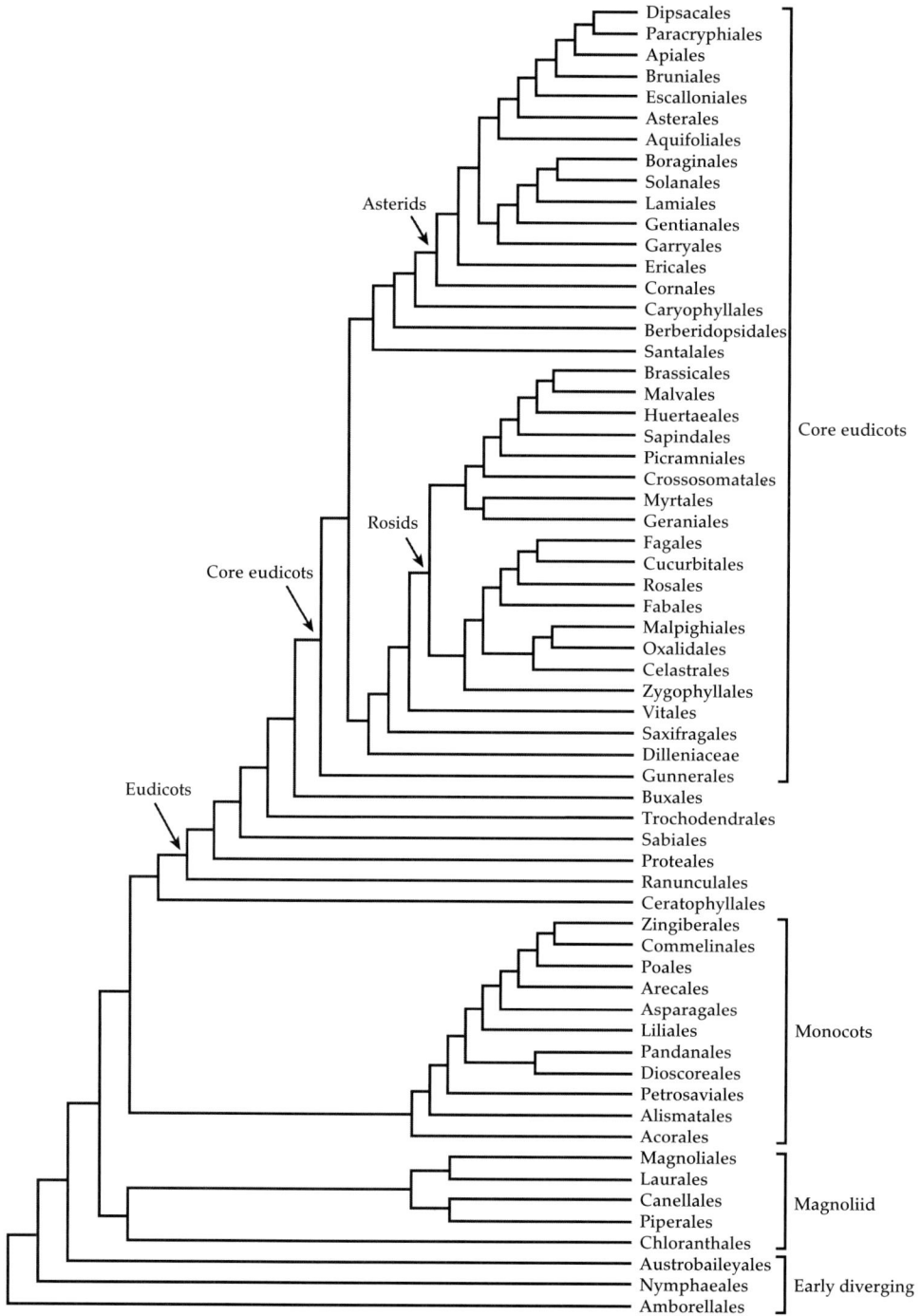

Figure 1.5 A summary of angiosperm phylogeny, based on APG III (Angiosperm Phylogeny Group 2009).

contains the Magnoliales, Laurales, Canellales, Piperales, and possibly the Chloranthales (which are sometimes excluded from the rest of the group). Many of these orders contain flowers that appear primitive, or show features in common with the early diverging angiosperms, but it is not clear whether these traits are ancestral or secondarily derived. The Monocots are a very important group, containing 11 orders, including the economically important Poales (the grasses). Within this order are many important model and crop species, including maize, rice, and wheat. The Liliales contains many horticulturally important species, such as lily, tulip, and freesia, while the Asparagales contains the enormously speciose orchid family. The third branch contains the Ceratophyllales and the Eudicots. The Eudicots are the most species-rich section of the tree, representing around 70% of angiosperm species diversity. Five orders of early diverging eudicots, including the Ranunculales (buttercup family), precede the branching of the Core Eudicots. The Core Eudicots contains many orders, some branching independently but most contained within the Rosids or the Asterids. Important groups within the Rosids include the Fabales (containing key model and crop legumes, such as pea, soybean, and *Medicago truncatula*), the Malvales (cotton), and the Brassicales (vegetable brassicas and the main genetic model, *Arabidopsis thaliana*). The Asterids also contains important orders, notably the Asterales (the species-rich daisies, including sunflower), the Solanales (including potato, tomato, and aubergine), and the Lamiales (containing the genetic model *Antirrhinum majus*).

For a thorough treatment of the diversity of the flowering plants in a phylogenetic context I recommend the books of Soltis *et al.* (2005) and Judd *et al.* (2007).

CHAPTER 2

Historical interpretations of flower induction and flower development

The modern analysis of flowers and flowering rests on a wealth of literature concerned with the description and interpretation of plant form. Chief among these works is Goethe's foliar theory, which proposes that all aerial plant organs are analogous to a single organ, which he calls the leaf. The foliar theory has underpinned all work on flower development, including modern molecular genetic analyses, as well as providing a frame of reference for evolutionary studies. This chapter describes Goethe's theory with reference to the flower, and its use and expansion by twentieth-century botanists. An analysis of the differences and similarities between vegetative and floral organs is presented, providing a framework for a section that interprets the foliar theory in an evolutionary context. The second half of the chapter moves on to consider historical interpretations of the transition from the vegetative state to the flowering state, describing the mechanisms proposed by early plant physiologists to explain this transition. Again, this overview is designed to provide context for further chapters on the current state of knowledge with regard to this floral transition.

2.1 The foliar theory of the flower

The scientific study of plant morphology, and in particular the study of comparative morphology, both of different organs within a plant and of the same organ from different plants, was effectively the invention of the great German philosopher Johann Wolfgang von Goethe. In 1790, Goethe published his seminal essay on the metamorphosis of plants (translated and discussed by Arber 1946), in which he proposed that all plant organs could be thought of as equivalent or analogous to a single type organ, which he called the leaf. However, it is important to establish that he was not proposing a developmental or evolutionary concept, with other plant organs being descended from a leaf. Indeed, it might have helped to separate his thinking on equivalence from later evolutionary thinking on relatedness if he had used a different name for his hypothetical type organ. Nageli (1884) suggested 'phyllome' as a more suitable name, a suggestion with which later morphological botanists such as Arber (1937) strongly concurred. Because Goethe's theory implied no developmental or evolutionary progression, it is effectively reciprocal. Thus it is just as easy to describe leaves as analogous to petals as it is to describe petals as analogous to leaves. In essence, Goethe's view was that reproduction (and the development of reproductive organs) ought to be treated as a function of the entire plant, and that the flowering shoot could not be understood except in relation to the vegetative shoot of the same plant (Arber 1937). This holistic approach was a new development in a field that had previously relied on analysis of individual components of organisms. Its genius lay in the detail with which it compared different parts of the same plant, combined with the simplicity of its approach. This approach involved no speculation about the meaning behind the analogy of different plant organs. It provided, however, a data set, free of preconceptions, which was then adopted by later evolutionary and developmental biologists and has provided remarkable support for their theories.

Figure 2.1 Analogous floral and reproductive organs in tobacco (*Nicotiana tabacum*, Solanales). (a) The inflorescence shoot (left) is analogous to the vegetative shoot (right). (b) The bract (below) is analogous to the leaf (above). (c) The sepal (below) is analogous to the leaf (above). (d) The petal (below) is analogous to the leaf (above). (e) The stamen (below) is analogous to the leaf (above).

Goethe's analysis of the equivalence of the different plant organs to a basic vegetative unit was summarized by Arber (1937), from which the following analysis of each of the reproductive structures is derived and expanded. Pairs of analogous structures are shown in Fig. 2.1.

2.1.1 Inflorescence shoot

The inflorescence shoot consists of the shoot apical meristem (SAM), the stem which is laid down from the rib meristematic region of the SAM, and the bracts, flowers, and secondary branches that are produced laterally from that stem. Its similarities to the vegetative stem from which it derives are clear. There is little difficulty in accepting such similarities, as the two are merely different phases in the life cycle of the SAM, which itself undergoes no major reorganization of structure in the transition from the vegetative to the flowering stage, although many changes in patterns of gene expression within that structure have been observed. The SAM usually remains indeterminate in the inflorescence shoot, allowing indeterminate growth similar to that shown previously by the vegetative shoot. The inflorescence shoot's direction of growth and its function—the production and support of lateral organs—are also the same as those of the vegetative shoot.

Differences between the inflorescence shoot and the vegetative shoot are primarily related to the organs that they produce, and these will be considered individually below. However, within the shoot itself a number of changes may occur. First, in some species, phyllotaxy in the inflorescence shoot is altered, with organs being produced in a pattern different from that of the vegetative shoot. Secondly, the distance between lateral organs may vary. In many species the inflorescence shoot is characterized by internodes shorter than those of the vegetative shoot. Finally, the details of cellular differentiation within the epidermis of the stem itself may be altered. For example, the vegetative stem of *Antirrhinum majus* (snapdragon, Lamiales) is entirely hairless, but the inflorescence stem is covered with short multicellular trichomes with glandular heads.

2.1.2 Floral shoot

The floral shoot consists of the floral meristem and the lateral organs derived from it. It should not be considered analogous to the main vegetative shoot (derived directly from the SAM) but to the axillary shoots which develop from the flanks of the SAM. In the same way, the floral meristems develop on the flanks of the inflorescence meristem. The floral shoot usually arises in the axil of a leaf-like organ, the bract, whereas the axillary vegetative shoot always arises in the axil of a leaf. So the architecture of the floral shoot, and its positioning with respect to the main axis of plant growth, is very similar to that of a vegetative axillary shoot. The function of the floral shoot is also analogous—the production and support of lateral organs.

However, the differences between the floral shoot and an axillary vegetative shoot are not solely restricted to the organs that they produce, very different though those may seem. There are again three significant differences between the floral shoot and a vegetative shoot. First, phyllotaxy is almost always altered in the floral shoot, with whorled phyllotaxy being the norm in flower development (with notable exceptions), irrespective of the usual phyllotaxy adopted by the leaves of the plant. Secondly, and more significantly, there is an enormous difference in the extent to which the shoot elongates. The axis of a vegetative shoot will almost always elongate, so that the leaves are separated by well-defined internodes. In contrast, the floral axis very rarely elongates at all after the production of the first sepal, producing organs so closely placed that fusions both within and between different whorls of floral organs are common. This difference was noted by Goethe, and has been described as the primary divergence between floral and vegetative shoots (Arber 1937). However, there are examples of primary vegetative shoots adopting a similar form—many species (e.g. in the Brassicales) adopt a rosette form in the juvenile phase, with almost no internodal elongation at all. The loss of internodal expansion

in the floral shoot can therefore be thought of as a reversion by the axis to a more juvenile growth form. Such retention of juvenile characteristics in mature organisms is termed neoteny, and has been hypothesized to be responsible for many apparent saltational leaps in the animal fossil record, including the divergence of humans from apes. It is also well known in plants, where it can be responsible for significant morphological evolution—for example, in floral form (Box *et al.* 2008; Box and Glover 2010). A third, and equally significant, difference between the flowering shoot and a vegetative shoot is the determinacy of the meristem and consequently of the shoot. The early morphologists were concerned with the fate of the meristematic cells within a flower, believing that consumption of the meristem by the terminal organs (the carpels) would indicate a significant difference from the activity of the axillary shoot. However, careful analysis has shown that many species retain an apex to the floral axis, clearly distinct from the floral organs (see Fig. 2.2; Arber 1937). Instead of consumption by the carpels, then, the floral meristem is simply determinate and ceases to produce more organs after the flower is

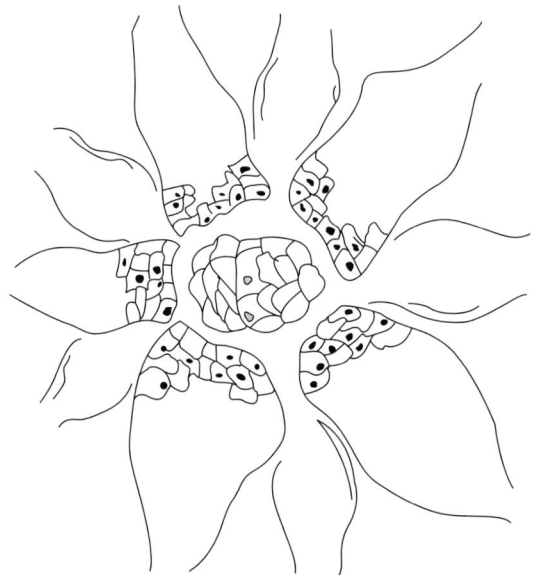

Figure 2.2 The shoot apical meristem still present between the carpels in a flower of *Ranunculus acris* (Ranunculales) (redrawn from Arber 1937).

complete. It is likely that this determinacy, unusual in a plant meristem, is secondarily imposed. Genetic analyses have revealed that floral meristem determinacy can be broken by mutation of a single gene (called *AGAMOUS* in *Arabidopsis thaliana*; Coen and Meyerowitz 1991) and the meristem returned to an indeterminate state (see Fig. 2.3). The analogy between floral and vegetative meristems is therefore not broken by the determinacy of the floral meristem, a secondarily imposed character controlled by a single gene.

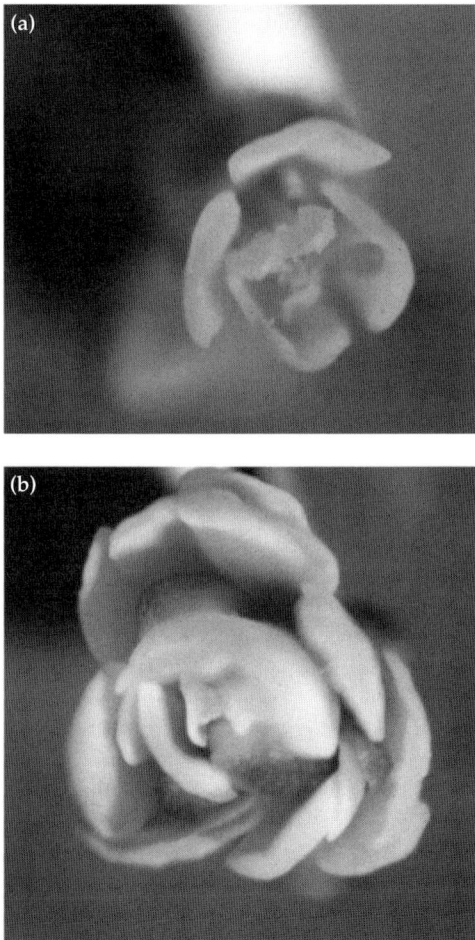

Figure 2.3 The indeterminate nature of the *AGAMOUS* mutant flower (b), which produces flowers within flowers, confirms that determinacy has been imposed upon the wild type floral meristem (a) by the action of this single gene. Photographs kindly supplied by Ian Furner (University of Cambridge).

2.1.3 Bracts

If the evidence suggests that the vegetative and reproductive shoots are analogous, the various appendages borne by these shoots are increasingly difficult to relate to one another. The bract is the easiest case, being positionally and morphologically often barely distinguishable from a leaf. Bracts, like leaves, are lateral organs produced by the SAM and which subtend a meristem. They are also in possession of dorsoventrality, are usually laminate in form, and are frequently green and photosynthetic.

The differences between bracts and leaves are usually limited to details of size, shape, pigmentation, and cellular differentiation, although there are some species in which bracts are entirely absent. In many species the bracts are smaller than the foliage leaves, perhaps reflecting the transition of the plant from an organism primarily concerned with photosynthetic production to an organism concerned with investing the products of that photosynthesis in reproduction. In some plants the bracts have become modified to attract pollinators to the flowers, and may therefore contain anthocyanin or other pigments, either as well as or at the expense of chlorophyll. Similarly, plants that use their bracts in this way, such as some species of *Cornus* (Cornales), may develop light-focusing papillae on their bracts (Kay *et al*. 1981), while other species may show different trichome distributions on their bracts relative to their leaves.

2.1.4 Sepals

The first whorl of perianth organs, containing the sepals, is often morphologically very similar to a whorl of leaves. The sepals are produced as lateral appendages from the floral meristem, and are usually laminate in form and green in colour. Exceptions to this occur in species where both whorls of perianth organs are specialized for pollinator attraction, either as two very similar whorls, such as in the Liliales, or as two very different whorls, such as in *Fuchsia* (Myrtales). In both cases the two whorls of petaloid organs are sometimes known as tepals. Apart from this particular difference, related to adoption of a novel function, the sepals usually differ from leaves only in their size, often being much smaller.

2.1.5 Petals

The second whorl of perianth organs can, in species with a simple undifferentiated perianth, be morphologically very similar to leaves, too. However, in many species, particularly in some of the eudicots with zygomorphic (or bilaterally symmetrical) flowers, the petals can be morphologically very distinct both from leaves and from one another. However great the differences in morphology between petals and leaves may seem, they are nonetheless analogous in terms of their position as lateral appendages from a meristem. Other similarities include dorsoventrality and a laminar form. Both petals and leaves can clearly be seen to distinguish their dorsal surface from their ventral surface. The leaves of most angiosperm species differentiate stomata at a considerably higher frequency on their ventral epidermis than on their dorsal epidermis. Internally, the palisade mesophyll (the columnar cells that provide the extended surface area for photosynthesis) is located on the dorsal side of the leaf. Similarly, many petals develop specialized epidermal cells on their dorsal surfaces, and may even be differentially pigmented or patterned on their dorsal and ventral surfaces. The laminar form, produced by growth perpendicular to an initial axis, is also common to both leaves and petals.

If the similarities between petals and leaves are in major architectural and developmental characteristics, the differences are largely in the detail of function. Petals are not usually green and photosynthetic, but instead may contain pigments that make them stand out against vegetation. This can be interpreted as a secondary adaptation to their specific function as pollinator attractants. Similarly, petals usually have no palisade mesophyll (since they do not photosynthesize), and often develop elaborate papillate epidermal cells to enhance their apparent colour and texture. These differences, like changes in size and shape, do not reflect major differences between petals and leaves, but rather minor modifications of a similar ground plan, and therefore do not perturb Goethe's vision of the analogous nature of leaves and petals.

2.1.6 Stamens

The outer whorl of reproductive organs, the stamens, was also considered analogous to leaves by Goethe. The similarities of stamens to leaves lie in the presence of chlorophyll and in their growth form, which consists of elongation in a single plane (although usually with little or no laminar growth). Their differences may seem more striking, but can again be attributed to minor adaptations to changing function. The absence of a laminar form may at first seem significant. However, analysis of mutants with perturbations in leaf development has shown that, in order for a leaf to develop a lamina, it must first have dorsoventrality (Waites and Hudson 2001). Put simply, a leaf cannot grow sideways unless it can detect which way is up, which way is down, and thus which way is sideways. There is no need for a stamen to have dorsoventrality, as it is simply a filament supporting the pollen-containing locules. In the absence of dorsoventrality a laminar form cannot develop, so the lack of a lamina can be interpreted as a consequence of the loss of dorsoventrality. That leaves the locules themselves as the only significant difference between stamens and leaves.

A further argument in favour of the analogy between leaves and stamens is the interrelationship between petals and stamens. A variety of authors, beginning with Goethe himself, observed significant similarities between stamens and petals. It had also been noted that the absence of stamens in species which produce some female-only flowers was often associated with a reduction in petal development. Modern molecular genetic analyses have confirmed the association between petals and stamens, with both requiring the activity of 'B-function' genes (see Chapter 10) for their development (Coen and Meyerowitz 1991). It follows, then, that if petals are analogous to leaves, and petals and stamens are variations on the same theme, then stamens must also be analogous to leaves.

2.1.7 Carpels

The female reproductive structures have traditionally presented the greatest difficulty to people intent on interpreting the flower in a foliar context.

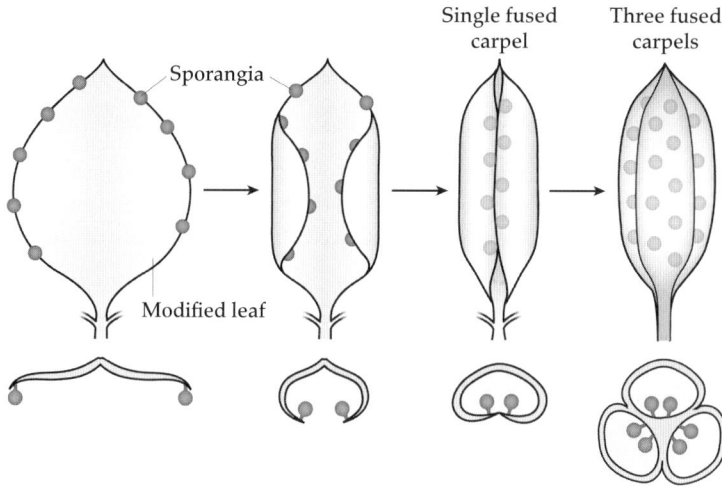

Figure 2.4 Development of the angiosperm carpel. A modified leaf with sporangia at the edges folds up to make a single fused carpel, several of which may fuse together in a single gynoecium.

Goethe himself had some difficulties with the carpel, and it was not until microscopical techniques advanced considerably that it was possible to see how the development of each carpel was analogous to the development of a leaf. Each carpel is essentially composed of a folded up leaf-like structure, with ovules developing on its adaxial side (see Fig. 2.4). In most species, multiple carpels fuse together to form a single compound pistil. Ovules aside, the only differences between carpels and leaves are the development of specialized stigmatic cells and the patterning of the vasculature. The stigmatic cells, and the stylar tissue beneath them, are clearly adaptations to the function of the carpel, and can be interpreted as secondary consequences of the novel function. The significance of the patterning of carpel vasculature has been the subject of much debate in the literature. The essential problem is that leaves have a dominant mid-vein with secondary vasculature branching outwards from it, whereas carpels have dominant lateral veins with secondary veins branching inwards from them. Although it is not clear how the distribution of vasculature could have altered, it has generally come to be accepted that this pattern reflects the lateral positions of the ovules, and the need to supply them with nutrients, and is thus likely to be a secondary consequence of ovule development.

Overall, then, comparative morphology shows us that Goethe's foliar theory of the flower was both innovative and insightful. All of the organs that make up a flower look as if they could be produced with only a few variations using the same basic developmental programmes that make leaves. Flowers, in fact, seem to be nothing more than bunches of slightly strange leaves bearing spore-producing structures.

2.2 The foliar theory in an evolutionary context

The observations of Goethe on the analogous nature of leaves and floral organs paved the way for later evolutionary interpretations of flower development. Where organs can be seen to represent variations on a theme, it is only a short step to blur the boundaries between them and imagine intermediate forms. This 'fuzzy' approach to plant morphology (Rutishauser and Isler 2001) fits perfectly with the idea, propounded by Darwin, that organisms were formed by gradual transitions between types. Goethe's work provided important data that supported the theory of evolution by natural selection, suggesting that the various organs of the flower evolved by gradual changes, as a result of mutations, from an ancestral leaf-like structure. As evolutionary theory became accepted as part of mainstream biology, Goethe's ideas were seamlessly incorporated into an evolutionary context, with his cautious 'analogous' converted to a more radical 'related to' or 'derived from.'

The idea that floral organs are only a few mutations different from leaves has been the foundation stone of modern molecular genetic analysis of flower development. Genetic analysis in the 1980s and 1990s identified lines of plants with mutations that caused interconversion of floral organs. Combination of these mutations into a single plant line resulted in a plant that produced four whorls of leaves in place of the four whorls of floral organs, finally providing the molecular 'proof' for Goethe's foliar theory (Coen and Meyerowitz 1991). These mutant lines, and the genes in question, are discussed in more detail in Chapter 10. However, their interpretation, and indeed much of the impetus for their identification, rely heavily on the historical work of Goethe himself.

2.3 The transition to flowering

The history of scientific work on flower development may belong to comparative morphologists, but the history of work on the transition to flowering lies with plant physiologists. The transition to flowering consists of a major phase change in the plant's life cycle. A plant that has previously produced only leaves and axillary meristems from its apical meristem must now switch to producing bracts and floral meristems, along with any changes to the general axis, phyllotaxy, and pattern of growth. This switch can be thought of as a 'decision' on the part of the plant—it is not a random event, but occurs in response to one or several of a large set of potential stimuli, which vary from species to species. The transition to flowering can be broken down into two distinct processes, induction and evocation. Floral induction is the commitment of the plant to start producing flowers, which often happens as a result of processes occurring in the leaves. Once induced, a plant is set on the path to flowering, even though no flowers have yet been produced. The induced plant then evokes the production of flowers by triggering changes in gene expression in the apical meristem. Induction and evocation are usually considered separately for the simple reason that they often occur in distinct organs. It is possible to induce many plants to flower by providing appropriate stimuli to a single leaf, and it is possible to evoke an uninduced plant to

flower by grafting a single induced leaf from another plant on to it. It is clear from this physical separation of processes that floral evocation requires the production of a signal by the induced organs, which is transported to the apical meristem and activates flower production. The mechanisms of induction and evocation, and the nature of the signal between them, have long been the subjects of physiological analysis.

2.4 Developmental explanations of floral induction

The variables that interact to determine when a plant is induced to flower can include developmental stimuli and environmental stimuli. For the majority of plants in temperate climates the environment is a key factor in the decision to flower, as producing vulnerable flowers and seeds in the depths of winter, for example, would not be a successful strategy. For this reason, environmental stimuli are considered in much greater depth than developmental ones, both in this chapter and in the literature as a whole. Developmental factors are of more importance to plants living in environments with little annual variation. It is unusual, for instance, to find a tropical forest species with a floral transition determined by factors such as day length or temperature, which are more or less constant in that environment.

The developmental factors that may influence the floral transition generally refer to the age and health of the plant. Some trees, for example, can only flower after a certain number of years of growth. This ensures that the tree has had time to build up a store of photosynthate with which to support any developing seeds and fruit. Similarly, some herbaceous and weedy species flower after the production of a certain number of leaves, again allowing for the development of sufficient photosynthetic tissue to support the nutritional needs of non-photosynthetic flowers, seed, and fruit (King 1997). Although the human observer might interpret these responses as implying a 'counting' mechanism within the plant, it is more likely that each leaf produces a certain amount of a stimulatory substance, and that the 'right' number of leaves are required to produce sufficient stimulus to cross a threshold, beyond which flowering responses are activated. The early literature on flower induction is primarily concerned with the search for this stimulatory substance, as discussed in Section 2.6.

It is likely that all plants, even those with a strong environmental stimulation of flowering, use developmental cues to a certain extent. The production of organs that do not contribute to the carbon balance of the plant is an expensive business, and one not undertaken unless the plant is sufficiently large and in sufficiently good health. The only exception to this general rule is that, in extreme cases of stress, flower induction may occur to ensure that seed is produced before a plant dies. This can often be observed in badly treated garden and house plants!

2.5 Environmental explanations of floral induction

The early plant physiologists established very quickly that the environmental variables which seemed to influence flowering were day length, temperature, and, to a lesser extent, water stress.

2.5.1 Day length

The most commonly used environmental variable is the length of time for which light is available, the day length. Almost all plants outside equatorial regions have flowering responses at least partially triggered by day length.

Plants that use light availability to tell them when to flower are described as photoperiodic. Photoperiodic plants use day length, and the change in day length as the seasons progress, to predict which season and therefore which weather conditions will come next. In the same way, if one had no access to a calendar, it would be more reliable to predict when autumn would arrive by measuring the shortening days than it would be to base everything on the weather. A late hot spell in September, for instance, could hide how far advanced autumn was, but day length is an entirely reliable indicator.

The fact that plants use day length to determine when to flower implies that there is an optimal time for flowering to occur. This optimal time is different between species, and within species it may differ in relation to geographical distribution. The optimal time to flower depends on a large number of factors, and is a reflection of the selective pressures that the plant experiences. It is essential that flowers are produced in sufficient time to set seed before winter weather conditions kill off developing tissues. The vulnerability of the developing seed and the harshness of the local climate will interact to determine how late any individual plant can afford to leave flowering. At the same time, if seed or fruit is to be dispersed by animals, it is essential for that seed or fruit to be mature while the animals are foraging. Again, this will depend on the animals used, whether they migrate, hibernate, or feed all winter, and how the local climate conditions affect the animals themselves. Flowers themselves may also rely on animals for pollination, and if this is the case, flowers must be produced when the appropriate pollinators are active. Finally, competition between flowers of different species for pollinator attention has been hypothesized to drive some plant species to flower at a time slightly suboptimal with regard to the above factors, simply to avoid competition and reduced levels of pollination (see Chapter 19).

So photoperiodic plants are plants that use day length to predict when their optimal flowering time is approaching, and flower accordingly. In fact, there are two types of photoperiodic plants—long day plants and short day plants. This is a very important distinction in photoperiodic research, and one that the early plant physiologists had to establish in order to avoid much later confusion. A long day plant is a plant that flowers when the days are longer than a certain minimum length. The minimum length itself will vary greatly between different species, but the key point is that long day plants flower only when the days exceed this length. In a temperate climate this means that the plant is likely to flower in late spring or early summer. Many common European garden and crop plants are long day plants, including species such as wheat (Poales), lettuce (Asterales), and sugar beet (Caryophyllales). Short day plants, on the other hand, are plants that require the days to be shorter than a certain maximum length in order to flower. Again, the critical length itself can vary widely, but it is the flowering when days are shorter than this that defines a short day plant. Short day plants are usually found in latitudes lower than those of long day plants. In a temperate climate they are represented

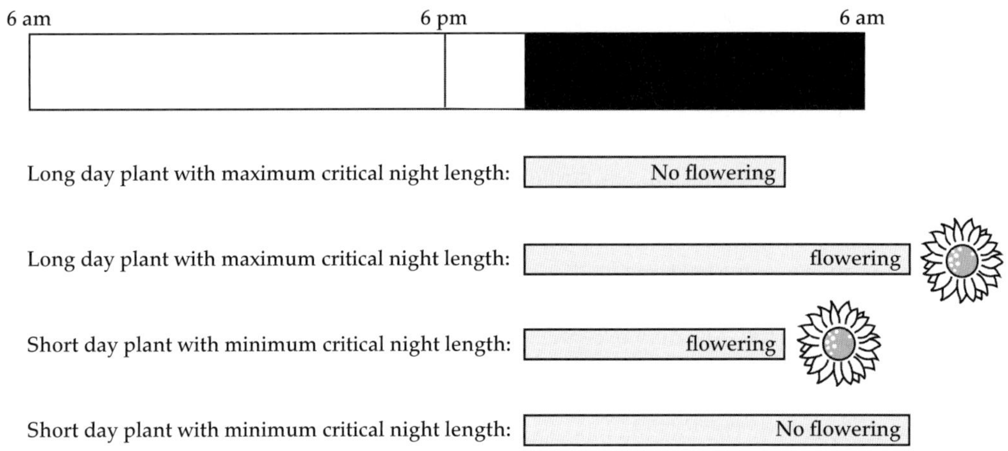

Figure 2.5 Responses of long and short day plants with varying critical night lengths. The long day plant will only flower in the light regime shown if its critical night length is longer than the period of darkness. The short day plant will only flower if its critical night length is shorter than the period of darkness.

by species that flower very early in spring, such as the snowdrop and the crocus (both Asparagales), and by autumn-flowering species, such as chrysanthemum (Asterales). (Figure 2.5 shows the responses of long and short day plants to an example light regime; note that it refers to critical night length, not day length, as explained below.)

Having established that plants appeared to flower in response to day length, it then took a number of simple experiments to show that plants actually appeared to be measuring the length of the night, not the day. This was shown by interrupting the 'day' experienced by plants in controlled growth conditions, by turning the lights off for a variety of periods. These interruptions had no effect whatsoever on flowering. However, interrupting the 'night' by turning the lights on for short periods had dramatic effects on flowering. The effect of flashes of light during the night was to persuade the plants that the nights were much shorter than they really were, and to induce early flowering in long day plants but to repress flowering in short day plants. These experiments also showed that red and far red light were the most effective at disturbing the plants' flowering responses (Borthwick *et al.* 1952). We now know that these 'night breaks' affected flowering by interfering with the normal cycling of the circadian clock (see Chapter 5), causing a mismatch between the timings of signals derived from the clock and signals derived from the light regime.

It was a short step from establishing that plants sensed the light regime to investigating where in the plant this occurred. Perhaps surprisingly, it has been shown quite conclusively that day length is measured in the leaves of the plant, not in the meristem, even though it is the meristem that actually undergoes the change from a vegetative to a reproductive state. Again, very simple physiological experiments provided this important breakthrough. Plants were kept in non-inductive conditions, with only single organs exposed to the correct day length. This could be achieved by placing a box around the organ in question, with a light in that box under separate control from the main lights of the room. These experiments showed that providing a single leaf with the appropriate day length was enough to cause flowering to occur. However, exposing just the apex of the plant had no effect at all. Although it may seem counter-intuitive to measure day length in the leaves and then send a signal to the meristem to cause flowering to occur, in fact there are a number of good reasons why leaves are the best place to sense the light regime. Leaves have a much bigger surface area than the apical meristem, and are adapted to maximize light capture. They also act as a fail-safe mechanism—measuring light in many different leaves at once means that if some leaves are shaded and do not perceive the correct day length, other leaves are likely not to be. The plant thus has multiple chances to perceive day length, and multiple chances to flower.

These experiments provided a baseline for modern research into the floral transition. The early physiologists had established that most plants flowered in response to the light regime, that they needed either days longer than a certain minimum (long day plants) or days shorter than a certain maximum (short day plants), that they appeared to measure night length rather than day length, and that they did this measuring in their leaves. Modern research (discussed in Chapter 5) has focused on the mechanism by which light is perceived, the clock that allows plants to measure day length, and the signalling that connects the two.

2.5.2 Temperature

The second most commonly used environmental stimulus to regulate flowering time is temperature. However, temperature signals are generally used in addition to photoperiod, and most often by plants that live for several years and only flower in some of them. Temperature provides information on the passing of years rather than seasons. Henbane, *Hyoscyamus niger* (Solanales), for example, is a long day plant, but will not flower at all unless it has been exposed to cold. It is biennial and only flowers in its second year. Putting the photoperiod and cold requirement together, we can see that henbane, which germinates in the spring, cannot flower in its first summer, because it has not been induced by cold. It cannot flower in winter, or the next spring, despite having been induced by cold, because it is in the wrong day length. Only in the second summer, when the cold induction has happened and the day length is right, will the plant flower. The degree of cold needed to induce flowering and the amount of time it is needed for vary a lot between species according to the environments in which they occur. Treating a plant with cold to induce flowering is called vernalization, and it is used extensively in agriculture, where crops may be sown in the autumn but will not flower until the following spring, induced by the winter cold.

Like the sensing of photoperiod, temperature sensing has been shown to take place in the leaves of the plant, although there is also evidence that apical meristems can be induced to flower by the chilling of them alone, through the repression of meristem-expressed genes (Searle *et al.* 2006).

2.5.3 Water stress

Early physiological experiments also showed that various kinds of stress could induce flowering. The most effective of these is water stress. Droughting plants, or even cutting off their roots, is frequently able to induce flowering, as has been shown for example in *Citrus latifolia* (Sapindales) (Southwick and Davenport 1986). This seems to ensure that a plant which is unlikely to survive does at least set seed for the next generation. It is a phenomenon which has been studied less extensively than light- or temperature-induced flowering. One explanation for this is that drought stress is not generally conducive to the production of much seed or seed of very good quality, so this form of flower induction is unlikely to be of use in agriculture or horticulture.

2.6 The florigen problem

Once the early plant physiologists had established that it was leaves that sensed the light regime, they quickly realized that there must be a signal which travelled from the induced leaves to the apical meristem to activate the floral transition. The search for this signal was one of the most active areas of plant research for many years. Although a single protein (FT; see Chapters 5 and 7) has now been identified as the key signal moving between leaf and meristem to induce flowering in Arabidopsis, we also now know that multiple signals interact to evoke the flowering response in the meristem, and that the process is much more complex than anything previously imagined.

2.6.1 Florigen

It was initially hypothesized that the signal sent by the induced leaf to the apical meristem was likely to be a plant growth regulator (otherwise known as a plant hormone). There were a number of pieces of

evidence to support this suggestion. Most compel-
lingly, it could be shown that the signal travelled
at about the same speed as the phloem stream (by
analysing time from leaf induction to changes in
enzyme activity in the meristem). Ring-girdling experiments were then conducted, removing only the
phloem from a ring around the stem of the plant
(and thus blocking the phloem at that point). Ring-
girdled plants no longer flowered, unless a leaf had
been induced above the ring. This served to confirm
the idea that the signal travelled in the phloem, the
location of much of plant growth regulator transport (Zeevaart 1976).

Grafting experiments suggested to the early plant
physiologists that the same signal could induce
flowering in all species. They therefore concluded
that the signal represented a novel plant growth
regulator, present in all plants, and they named it
'florigen' (Chailakhyan 1936, cited in Hoffmann-
Benning *et al.* 2002). The evidence in support of
florigen was initially very convincing. Numerous
reports were published in which a single induced
leaf of species A was grafted on to the stem of a plant
of species B (which was kept under non-inductive
conditions), resulting in the appearance of flowers
on plant B. The number of species to which this
applied was large, and led to a concerted effort to
identify the mysterious florigen. However, no novel
plant growth regulator could be extracted from the
phloem of these plants. And as research continued,
it became clear that the same signal did not promote
flowering in all species. One of the final nails in the
florigen coffin was the discovery that grafting an induced leaf from tobacco (Solanales) on to the stem
of a potato plant (also Solanales) resulted in the production not of potato flowers but of potato tubers
(discussed by Martinez-Garcia *et al.* 2001).

2.6.2 The anti-florigen theory

An alternative hypothesis to explain the transmission of a signal from the leaves to the meristem is
known as the anti-florigen theory. This proposes

that non-induced leaves emit a signal that travels to
the apical meristem and prevents flowering. When
the leaves are induced, it is the absence of this inhibitory message which allows the transition of
the meristem. Initially, the anti-florigen hypothesis
assumed a specific inhibitory signal that only prevented flowering, and was common to all species. It
is now generally accepted that the anti-florigen hypothesis is a useful model which could apply to the
release of any growth regulator or other signalling
molecule, and which might vary in different species
(Colasanti and Sundaresan 2000).

2.6.3 The source/sink resource allocation model

Another model to explain the evocation of flowering by induced leaves is the source/sink resource
allocation model, which states that the meristem
of a non-induced plant is not sufficiently supplied
with the appropriate nutrients to change into a
flowering meristem. Induction of the leaves changes the source/sink properties of various organs
of the plant, resulting in a rush of nutrients to the
meristem, which can then undergo the transition to
flowering. Although there is no specific evidence
to support source/sink relations as the major determinant of flowering, it is certainly the case that
extra energy, along with sufficient nitrogen and
phosphate for all the RNA and protein synthesis
associated with flowering, is required at the apex.
The source/sink properties of plant organs clearly
do change during development, and it is very likely
that there are many such changes at the transition to
flowering, some of which are probably critical to the
correct development of flowers (Heyer *et al.* 2004).

The various ideas and models that have arisen
from the work of the early plant physiologists have
been as important in understanding the data arising from modern molecular genetic analyses of the
floral transition as the ideas of Goethe have been in
understanding floral development. In Section II of
this book, we look at the current state of knowledge
of these processes.

Induction of Flowering

CHAPTER 3

Flower induction in *Arabidopsis thaliana*

Section II of this book explores our current state of knowledge concerning the induction, evocation, and development of flowers. In the twenty-first century, research into these processes is almost exclusively conducted through molecular genetic and genomic approaches. The molecular genetic approach uses the backwards-seeming logic of searching for a plant with a mutation that prevents a process from occurring properly. By analysing the process in the mutant plant, and by identifying and analysing the mutated gene, it is possible to build up a picture of how a normal plant and a normal gene work. When molecular genetic analysis becomes the usual way of exploring a biological process, there is a tendency for the majority of work to focus on one or a few species. This is partly because information already gained in a species informs the next set of experiments, but also partly because there are some species which are simply better suited for molecular genetic analysis than others. Thus the fruit fly, *Drosophila melanogaster*, is by far the best-studied animal, simply because a number of features of its biology make it an excellent model system. Similarly, the weedy annual plant *Arabidopsis thaliana* (Brassicales) has become the most extensively studied plant species. Once model species have been established, genome sequencing, and more recently next-generation re-sequencing of multiple lines or varieties of the model species, allow intensive genomic characterization as well. This chapter introduces Arabidopsis and the reasons why it has become the model of choice for the study of floral induction. It then goes on to look at the description and characterization of mutants with perturbations in floral induction, mutants

which have allowed the development of models explaining the different ways in which the flowering response can be induced.

3.1 *Arabidopsis thaliana* as a model system for the study of flowering

3.1.1 Why Arabidopsis?

Arabidopsis thaliana is commonly referred to simply as Arabidopsis, although its traditional English names are thale cress and mouse ear cress. It is a weed native across Eurasia and East Africa, and a member of the Brassicaceae and Brassicales, a family and order in the Rosid clade of the Eudicots. Arabidopsis was selected as a model species by plant geneticists as early as the 1950s, but it was not until the 1980s that it became the system of choice for the majority of plant scientists (Redei 1973). By December 2000, when its complete genome sequence was published (The Arabidopsis Genome Initiative 2000), Arabidopsis had become so much the most usable plant model species that to work on anything else was highly unusual. In the post-genomic era, Arabidopsis continues to be the main focus, with many facilities and services available that allow experiments which would be unthinkable with any other species. In particular, advances in next-generation sequencing technology have allowed the sequencing of the genomes of multiple Arabidopsis lines, with an aim of having 1001 different Arabidopsis genomes comparable by the time this edition is in print (<http://1001genomes.org>).

In the wild, Arabidopsis tolerates a very wide range of conditions. It is commonly found as a

Understanding Flowers and Flowering. Second Edition. Beverley Glover.
© Beverley Glover 2014. Published 2014 by Oxford University Press.

garden weed, in disturbed fields, on railway lines, and on walls. This unfussy approach to habitat was one of the first reasons why Arabidopsis was identified as a useful plant model—it is perfectly capable of completing its life cycle on sterile, chemically defined media in a Petri dish. Other advantages include its small size and fast life cycle. The mature plant can be grown within 1 cm² of space, and in soil does not usually spread more than 8–10 cm in diameter and 12–15 cm in height (see Fig. 3.1). Arabidopsis has a very short generation time—just 6 to 8 weeks—allowing several generations of a mutant to be studied within a relatively short period. It is self-fertile, which is very important for genetic analysis as it allows the simple generation of homozygotes. However, Arabidopsis can also be cross-pollinated to generate double and triple mutants. A single plant can produce over 1000 seeds, making segregation analysis straightforward. Its small genome is another great advantage. It has a haploid DNA complement of around 130,000 kilobase pairs—only 15 times that of the bacterium *Escherichia coli*, 5 times that of yeast, and about half the amount that *Drosophila* has. Mice, humans, and most other flowering plants have 30 to 40 times as much DNA as Arabidopsis, making their molecular analysis much more unwieldy (Redei 1973).

Being such a widespread plant, Arabidopsis is not entirely homogeneous. In its many different habitats it is found as a range of natural accessions (sometimes called ecotypes), each with significant differences in development, morphology, and behaviour. Even the commonly used laboratory lines have thousands of polymorphisms (Putterill *et al.* 2004), which may be morphological, easily identified by the naked eye, or molecular, requiring the use of DNA fingerprinting techniques to identify them. In both cases they provide useful tools for the mapping of genes of interest, as we shall see below. More recently, the natural variation present in the different ecotypes has been the focus of studies to explore the nature of plant adaptation to different environments (e.g. Todesco *et al.* 2010).

A final advantage of Arabidopsis as a model system is the remarkable ease with which transgenes can be introduced into its genome. Most plant species have not yet been tested for ease of transformation, and many that have were found to be resistant. Although it is relatively easy to introduce transgenes into some species, such as the garden snapdragon *Antirrhinum majus* (Lamiales), it is then incredibly difficult to regenerate new plants from their transgenic cells (Heidmann *et al.* 1998). However, an annually increasing number of plants are being reported as 'transformable', and the techniques used all rely on tissue culture

Figure 3.1 *Arabidopsis thaliana.*
(a) Seedlings growing on an agar plate.
(b) A mature vegetative plant. (c) During the reproductive phase of the life cycle.

for regeneration. In general, plants are transformed using the crown gall-inducing bacterium *Agrobacterium tumefaciens*, which has the ability to transfer a portion of a plasmid into the genome of a host plant cell. In modern genetic engineering, a modified version of both the bacterium and its plasmid are used. The transgene of interest is inserted into the modified plasmid, which is then transferred back into the bacterial cell (Zupan *et al*. 2000). A piece of plant tissue, typically a leaf disc or a section of root or hypocotyl, is infected with the *Agrobacterium*, and gene transfer occurs. It is then necessary to culture the transformed plant tissue on a variety of chemical media, to kill any remaining bacterial cells and potentially invading fungi, while supplying the nutrients and plant growth regulators that will allow regeneration of a new plant. Even in species that are amenable to this process, the tissue culture can take many months, and is labour intensive. However, Arabidopsis is routinely transformed using a much simpler and faster procedure. Flowering Arabidopsis plants are dipped into a culture of *Agrobacterium* containing the transgene in the modified plasmid. After a few minutes the plants are removed and returned to the greenhouse. The *Agrobacterium* inserts the transgene into the ovules within the flowers. When the plants set seed over the next few weeks, a proportion of those seeds will contain the transgene, as a result of its presence in the egg from which they developed. Seed from the dipped plants is sown on a selective medium, which allows the identification of transgenic plants, and they are simply picked off and used in further experiments. The speed and simplicity of this procedure has added greatly to the range and extent of experiments which can be conducted using Arabidopsis (Clough and Bent 1998).

3.1.2 Generating mutants in Arabidopsis

The best genetic approach to a developmental process is to find mutants which prevent that process from occurring normally. The rationale behind this method is that the study of the impaired function would give one insight into the genes and proteins involved in the normal process. Generating mutants in Arabidopsis is a simple process, and is usually done in one of three different ways. EMS mutagenesis uses ethyl methane sulfonate (EMS) to

induce chemical damage to DNA. Typically, seed are soaked in EMS prior to germination. The chemical induces single base changes in the DNA, known as point mutations, by alkylating bases, primarily guanine. Alkylation causes mispairing of bases, with alkylated guanine pairing to thymine rather than cytosine, thus introducing single base pair changes when DNA is copied. A second approach uses X-rays to cause deletions of pieces of DNA or large-scale rearrangements of sections of chromosome. Thirdly, it is possible to use transformation technologies to insert foreign DNA randomly into the genome. Insertion of foreign DNA into a gene causes mutation of that gene, and has the added advantage of providing a tag or handle which can later be used for molecular cloning of the mutated gene. This approach, called T-DNA or insertion mutagenesis, is the method of choice for large-scale mutagenesis programmes. However, individual research groups focusing on a single developmental process might prefer EMS mutagenesis because it is simpler and generates new mutants more quickly.

These seed which are mutagenized, for example by soaking in EMS, are called the M0 generation (M for mutant). They germinate and the plants that they produce are most likely to be heterozygous for any mutation. This is simply because it is vanishingly unlikely that damage to DNA will occur at precisely the same position on two homologous chromosomes. This heterozygous generation is called the M1 generation. Since most mutations are recessive, M1 plants usually look perfectly normal. M1 plants are allowed to self-pollinate and set seed, and the seed grows into the plants of the M2 generation. In the M2 generation there will be segregation of the mutation, so one in four plants will be wild type, two in four will be heterozygous, and the final one in four will be homozygous for the mutation. Assuming this mutation causes a visible phenotype, the mutant plants will be easy to pick out from the rest (see Fig. 3.2).

3.1.3 Complementation analysis: finding out how many genes have been mutagenized

When a mutagenesis programme goes well, a large number of mutants with interesting phenotypes might appear in the M2 generation. Usually only

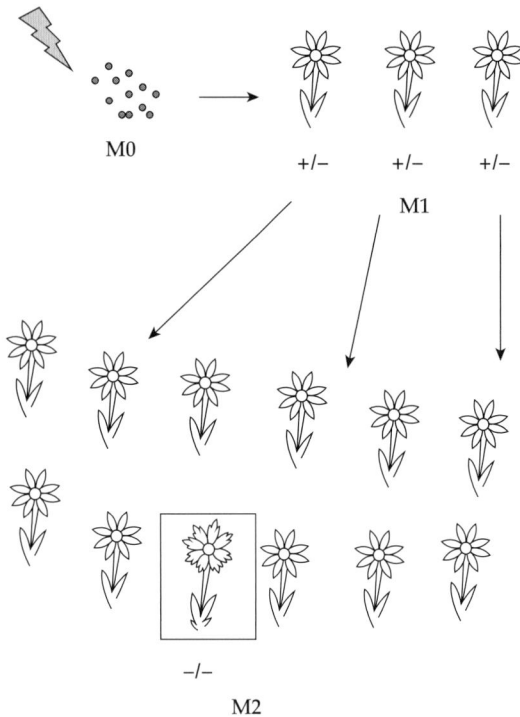

Figure 3.2 Mutagenesis of seed (M0) results in the growth of heterozygous plants, in which recessive mutations are masked by dominant wild type alleles (M1). Homozygous mutants are selected in the next generation (M2).

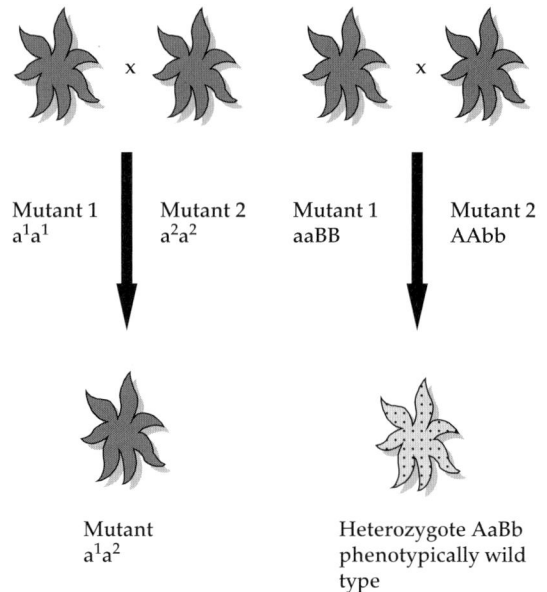

Figure 3.3 Complementation analysis determines whether two phenotypically similar mutants contain mutations in the same (left) or different (right) genes.

those mutants with a phenotype relevant to the process of interest are selected—so the 17 mutants (for example) which have problems with the induction of flowering are picked, but the dwarf plants and the yellow plants and the plants with strange-shaped leaves are ignored. However, before any analysis of these mutants is carried out, it is important to establish how many genes they represent. In theory, 17 mutants with the same phenotype might mean 17 different genes have been mutagenized, a different one in each plant, or that a single gene has been mutagenized 17 times. Or, more likely, one has managed something in between. The process of sorting out how many genes have been mutagenized is called complementation analysis.

Complementation analysis begins by making a cross between two mutant lines (illustrated in Fig. 3.3). Thus pollen from mutant 1 is used to pollinate flowers of mutant 2. Remember that both

parents were homozygous for their mutation. So if the mutation in mutant 1 and the mutation in mutant 2 are in the same gene, A, the progeny of this cross will have one mutated allele of gene A from each parent. They will still not have a functional version of gene A, and will still look mutant. However, if the mutation in mutant 1 and the mutation in mutant 2 are in different genes, A and B, the story is very different. In this case, mutant 1 will be homozygous for a mutation in gene A, but homozygous for a wild type allele of gene B. Mutant 2 will be homozygous for a mutant allele at gene B, but also homozygous for a wild type allele at gene A. So in this case the progeny of the cross will have one wild type allele of gene A (from mutant 2) and one mutant allele (from mutant 1), and one wild type allele of gene B (from mutant 1) and one mutant allele (from mutant 2). These plants will be heterozygous at both loci, so will have a functional copy of each gene. Since the functional copies are usually dominant, and the mutant copies recessive, these plants will appear wild type. Very simply, then, if the progeny of a cross between two mutant plants is also mutant, then the two mutations are in the same gene. However, if the progeny of a cross between two mutants is wild type, then the two mutations are in different genes as the mutants are able to *complement* one another.

3.1.4 Cloning genes following mutagenesis

Once mutants have been identified and characterized, it is useful to isolate the mutated gene and investigate its normal function. There are a number of different ways of going about this. If the mutant was generated by insertion mutagenesis, then it is possible to use the transgene as a tag with which to clone the gene. A variety of PCR-based techniques (such as TAIL-PCR, iPCR, and Genome Walking) allow the amplification and isolation of a piece of DNA for which the sequence of one flanking region (i.e. the transgene) is known.

If the mutation was generated using EMS then it is necessary to map the position of the mutation within the genome (see Fig. 3.4). Once that has been done, it is a simple matter to analyse the DNA sequence within that region using the whole genome sequence available on public databases. Mutations are mapped on to the genome using the many polymorphisms, both morphological and molecular, that exist between different laboratory lines of Arabidopsis. Mapping populations are readily available, containing many polymorphisms at known map positions. These are crossed to the mutant lines of interest, and the resulting F1 generation is self-pollinated to produce a segregating F2 population. Within this population the new mutation will segregate with some of the polymorphisms, but completely independently from others. This indicates that the mutation is on the same chromosome as those polymorphisms with which it segregates. Fine mapping of the mutation involves investigating how frequently there is crossing over between the mutated genes and other polymorphisms on the same chromosome. This approach allows one to pinpoint the location of the newly mutated gene between known polymorphisms. Investigation of the genome sequence within that area may indicate the presence of several genes, but it is usually a simple matter to find out which is responsible for the mutant phenotype. The usual approach is to take wild type copies of each of the candidate genes and insert them into the mutant plant, using the transformation protocol described earlier. The one that restores a normal phenotype to the mutant plant is the one which has been mutated.

Recent advances in next-generation sequencing technologies are raising the possibility of identifying mutations by simply sequencing whole or partial genomes of mutant and wild type siblings to identify the points of difference, allowing mapping and cloning in a single step.

Once cloned, the gene of interest can be analysed in a number of ways, including investigating its expression pattern within the plant, the consequences of changing that expression pattern using transgenic methods, and identifying other proteins with which its protein product interacts.

3.1.5 Reverse genetic approaches in Arabidopsis

The genetic approach described above has been responsible for enormous advances in understanding many processes, including floral induction. In recent years, and particularly following the publication of the genome sequence, it has become increasingly possible to use a reverse genetic approach as a counterpoint to the traditional genetic method. The reverse genetic approach begins with a gene, perhaps one with a sequence similar to another gene that has an interesting function, and progresses towards an understanding of its function. Many thousands of insertion lines containing foreign DNA tags have been generated by a range of research groups and consortia, and the DNA flanking the tags sequenced and made available on the Internet, making it a simple matter to search for tags which have inserted in or near a gene of interest (see <http://signal.salk.edu/cgi-bin/tdnaexpress>). The lines containing these tags can then be ordered and the plants grown for phenotypic characterization and further experimental analysis.

3.2 Mechanisms of gene silencing

Before we go on to discuss the specific approaches used to study flowering time in Arabidopsis, it is necessary to introduce a particular set of mechanisms of gene regulation. Throughout the chapters

(a) Stage 1

Unknown chromosome position for locus ∗

×

F1
self
F2

Mapping population with markers
on each chromosome

All markers combine with ∗ except ◇,
so ∗ is on chromosome 3

(b) Stage 2

Locus 1

× → F1 →self F2

1 in 5 through chiasma formation.
∗ far from ◇

Locus 2

× → F1 →self F2

1 in 1000 through chiasma formation.
∗ is close to ◇

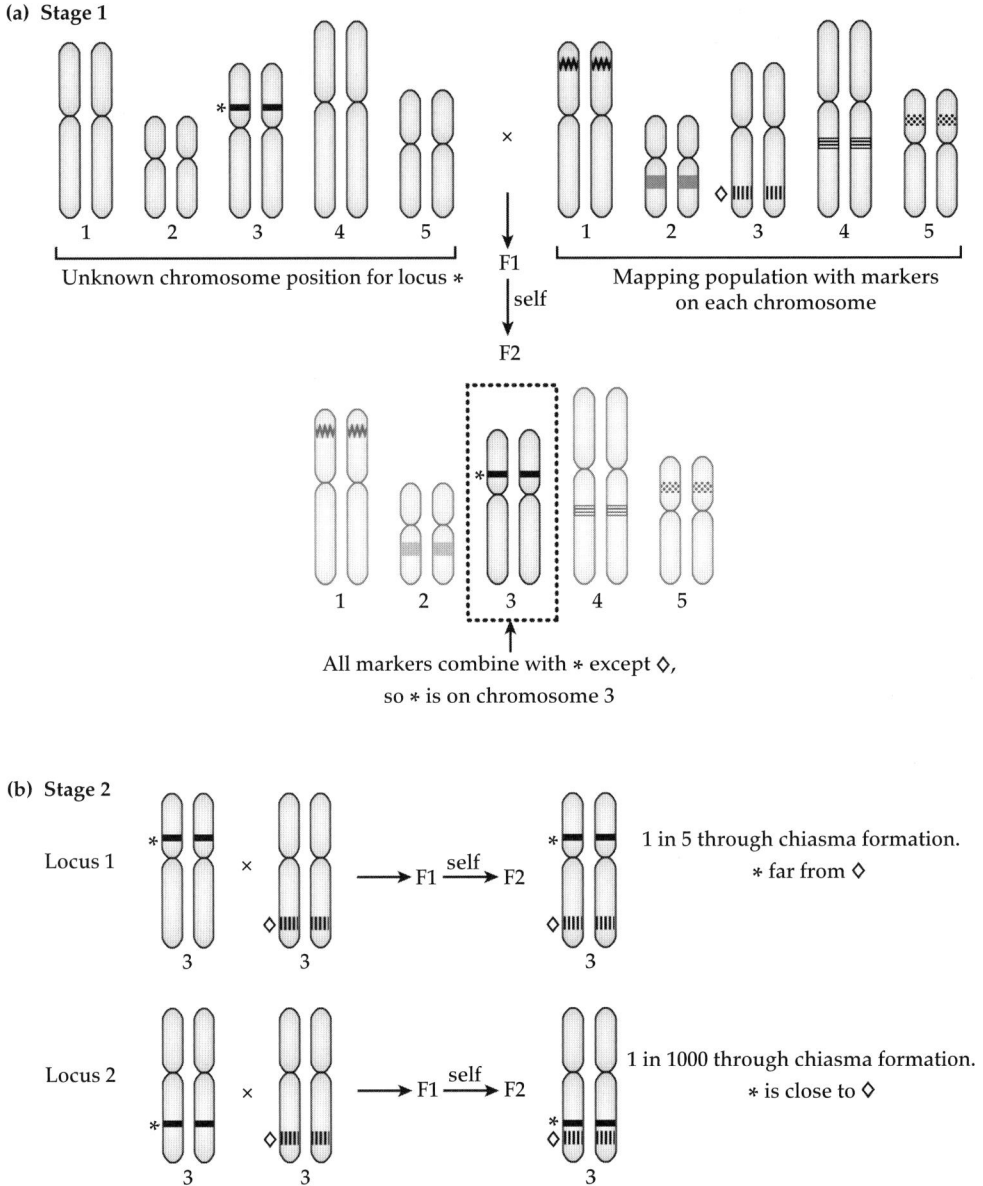

Figure 3.4 Mapping allows the position of a mutated genetic locus to be determined. (a) In stage 1, crosses between the mutant and a mapping population with polymorphisms on each chromosome allow the identification of which chromosome the mutation is on. (b) In stage 2, fine mapping allows the position of the mutation on the chromosome to be determined.

that follow I shall assume a basic understanding of gene function, and of the regulation of gene expression through transcription factors interacting with RNA polymerase and the promoter sequences usually found upstream of gene coding regions. The

resulting RNA is translated to protein, in the central dogma of molecular biology: DNA → RNA → protein. In cases where particular genes are behaving in unusual ways the text will explain the novel process. However, it has become increasingly clear in the last decade that much gene regulation occurs through more complex mechanisms, operating either at the level of the chromatin itself, or through small RNA molecules that interfere with the basic processes of transcription and translation. These mechanisms are briefly introduced here, to save repetition in the following chapters.

3.2.1 Gene silencing involving small RNA molecules

The cells of plants, and other eukaryotes, contain systems that enable them to detect and destroy foreign nucleic acids introduced by viruses. These systems have also been recruited to the control of endogenous gene regulation, where they provide an impressive degree of fine-tuning, allowing the evolution of complex developmental pathways. The basic defence system involves a group of RNA molecules called small interfering RNAs (siRNAs), first discovered not in Arabidopsis but in petunia and tobacco (both Solanales) in 1999 (Hamilton and Baulcombe 1999). These molecules are usually double-stranded pieces of RNA, 21–24 nucleotides in length with 2 nucleotide overhangs at either end. They are produced by cleavage of larger double-stranded RNA molecules found in viral genomes or as a result of transgene insertion. The larger molecules are cleaved by one or more of a small family of RNaseIII enzymes called Dicer(s). The siRNAs separate into single strands which base pair with complementary messenger RNA molecules (usually derived from the invading virus or transgene, but sometimes endogenous mRNAs transcribed from the host genome are also targeted). The siRNAs are associated with a complex of proteins called the RNA-induced silencing complex (RISC), which contains one of the family of Argonaute proteins. Argonaute cleaves the bound messenger RNA strand, preventing its translation and the production of a functional protein. In this way, foreign nucleic acid molecules are destroyed, along with the transcripts produced from them by the host cell machinery (for a review, see Mello and Conte 2004).

The same basic machinery is sometimes used in the regulation of endogenous gene expression in all eukaryotes. There are slight differences in mechanism between plants and animals, and this description focuses on the plant system. In the Arabidopsis genome around 300 loci that code for microRNAs (miRNAs) have so far been identified. These DNA sequences are transcribed and produce an RNA molecule which contains two regions of complementary sequence, reading in opposite directions. The molecule therefore folds to generate a hairpin or stem loop structure, with the complementary arms base paired and the unique sequence forming a loop between them. This structure is usually around 70 nucleotides long. A specialized member of the Dicer RNase protein family then cuts away the loop, leaving a double-stranded RNA molecule around 22 nucleotides long. This separates into single strands, which associate with the RISC complex and target complementary messenger RNAs for degradation. The elegance of this system lies in the degree of control it gives a cell over the amounts of protein produced by individual genes that are targeted by miRNAs. First, the usual mechanisms can control the amount of mRNA transcribed from the gene, and therefore the amount of protein produced. However, transcriptional control of the targeting miRNA locus then allows variable amounts of miRNA to be produced too, giving very fine control over the final amount of non-cleaved mRNA remaining to be translated into functional protein (for a review, see Voinnet 2009).

3.2.2 Chromatin modifications and gene regulation

A variety of structural modifications to the DNA in the genome can also have significant effects on the regulation of individual genes. One of the most frequently referred to in the next few chapters is DNA methylation. This is the addition of a methyl group (CH_3) to a cytosine or adenine base in the DNA. The conversion of a cytosine to a 5-methylcytosine molecule does not affect its base pairing, but does seem in general to reduce transcription. The mechanisms by which methylation reduces transcription are not yet understood, but it is clear that the more methylation occurs in a region of the genome, the more silent that region appears to be. Methyl groups are added to the DNA through the activities of a diverse family of proteins called DNA methyltransferases. In plants a number of genes have been shown to encode these enzymes, including *DRM2*, which is responsible for adding methyl groups at previously unmethylated sites in the genome. Methylation of the DNA in some sequences can be heritable through cell divisions, and several DNA methyltransferases have been shown to be responsible for semiconservative replication of methyl marks (reviewed by Furner and Matzke 2010). As with RNA-mediated gene regulation, regulation of the activity of the genes encoding these enzymes can influence the transcription of many structurally unrelated genes. It is not yet clear how the position of novel methylation in the genome is determined. It is likely that a number of mechanisms are at work, with each responding to a variety of different cues. One mechanism links RNA-dependent regulation to methylation, as it has been shown that siRNAs and miRNAs can direct methylation at the genomic locations responsible for double-stranded RNA synthesis and to regions homologous to them. This mechanism appears to be important in silencing transposons and viral insertions in the genome (reviewed by Furner and Matzke 2010).

Structural changes to the histone proteins around which DNA is wound to generate chromatin can also influence gene regulation. A variety of chemical modifications are possible, but the best studied is acetylation and deacetylation of lysine residues in the histone tails (the N-termini of the histone proteins). As with DNA methylation, these processes involve the activities of enzymes, which are themselves encoded by genes and hence under transcriptional control. Acetylation by histone acetyltransferase enzymes introduces an extra negative charge, loosening the link between the DNA molecule and histone, allowing the cell's transcriptional machinery freer access to the DNA. In contrast, deacetylation by histone deacetylase will increase silencing of that region of the genome (reviewed by Chen and Tian 2007). It is also possible for histones to be methylated (reviewed by Liu *et al*. 2010a). Methylation of certain lysine residues in histone molecules acts like DNA methylation to reduce transcription, but methylation of other lysine residues appears to enhance transcription of associated DNA (Xu *et al*. 2008). All of these effects vary according to the amino acid position on the histone molecule that is affected, which histone in the complex is involved, and what else is happening in neighbouring regions of the genome.

3.3 Flowering-time mutants

The molecular genetic approaches outlined above have generated enormous insights into the ways in which flowering is induced in Arabidopsis. In the chapters that follow we shall focus on the results of this analysis, looking at the pathways that have been proposed to induce and repress flowering and the genes and proteins that work within them. The rest of this chapter focuses on the original mutants that began this process of discovery, and the ways in which they are categorized. Over 80 genes have been shown to influence floral induction to some degree, although many have effects on other developmental processes as well.

3.3.1 Arabidopsis is a vernalization-sensitive facultative long day plant

Arabidopsis is a useful model for the study of the floral transition, because it uses several different pathways and thus provides insight into a wide range of processes. To begin with, Arabidopsis is clearly a plant that has developmental control of its floral transition. Even when grown under optimal environmental conditions for flowering, Arabidopsis remains vegetative until it has produced a rosette of leaves—it will not flower immediately after germinating. In fact, this observation indicates that Arabidopsis has an autonomous pathway which represses the flowering response until a certain developmental stage has been reached. Secondly, Arabidopsis is a long day plant. If grown in a growth room with 16 hours of daylight and 8 hours of darkness every 24 hours, Arabidopsis plants will flower within 2 to 3 weeks of germination. However, if grown in a growth room at the same temperature but with 12 hours of daylight and 12 hours of darkness, the plants will show no sign of flowering after the same time period has passed. They clearly measure day length (or night length) and respond to photoperiod by flowering only when the days are longer than a certain critical minimum. However, this photoperiodic response is not absolute. A plant entirely limited by photoperiod would never flower in the growth room with 12-hour days. Arabidopsis, on the other hand, continues to grow vegetatively in the short day room for several weeks after its counterparts in the long day room have flowered, and then it flowers itself. For this reason it is described as a facultative long day plant—long days promote flowering and cause it to occur early, but in their absence the plant will still flower eventually (Jack 2004). This shows that, besides its photoperiodic induction pathway, Arabidopsis must also contain an autonomous pathway which induces flowering eventually, irrespective of environmental variables. Finally, Arabidopsis is also sensitive to vernalization. Accessions from parts of the world with cold winters germinate in the autumn and remain vegetative until the next spring. Any remaining long days in their autumn growth season do not induce flowering, because they require between 1 and 3 months of cold temperature to become responsive to day length. Accessions from warmer regions usually germinate in spring or summer and flower in the same year that they germinate. This most probably occurs through lack of this vernalization requirement, but it could also result from especially great sensitivity to vernalization, ensuring that the

plants flower early in their first spring, before full summer with its attendant droughts arrives (Stinchcombe *et al.* 2004). By extending molecular analysis into a range of accessions, it has been possible to use this one model species to investigate developmentally controlled, photoperiod-sensitive, and cold-induced flowering responses.

3.3.2 Classification of flowering-time mutants

When mutant plants are first described, they are classified and categorized according to their phenotype alone. Later, as molecular information on gene function is accrued, it sometimes becomes apparent that the initial classification was unhelpful. However, since only phenotypic data are available in the first instance, it continues to make sense to classify according to those data. Arabidopsis mutants in floral induction are traditionally described as either late flowering or early flowering (illustrated in Fig. 3.5). These classifications refer to their time (or developmental stage) of flowering in different growth rooms. A late flowering Arabidopsis mutant is one whose behaviour in a long day growth room is similar to that of a wild type plant in a short day room—it flowers later than a wild type plant would

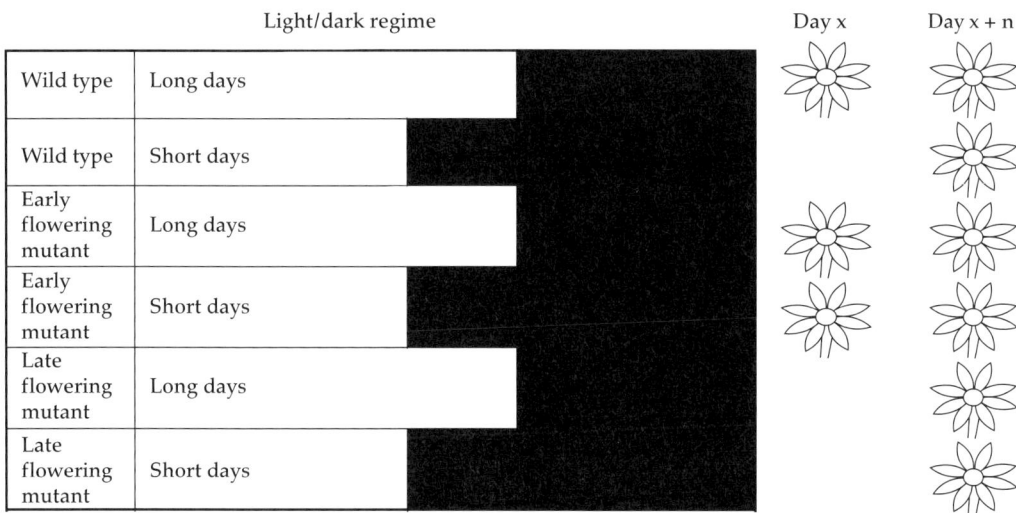

Figure 3.5 Flowering-time mutants are described according to how they behave under long and short day conditions. Wild type plants flower at an early time point under long day conditions, but at a later time point under short day conditions. Early flowering mutants flower at the early time point in both long and short days. Late flowering mutants flower at the later time point in both long and short days.

do in the long day room. An early flowering Arabidopsis mutant, on the other hand, is one whose behaviour in a short day growth room is similar to that of a wild type plant in a long day room—it flowers earlier than a wild type plant would do in the short day room (Coupland 1995).

Because time of year, age of seed, and growth room temperature can all affect the rate of Arabidopsis growth and development, researchers often measure flowering time in terms of number of leaves rather than in number of days (Koornneef *et al.* 1991). Using this scheme, a wild type plant might flower after producing 6 leaves in a long day room, but not until it had produced 12 leaves in a short day room. A late flowering mutant is a plant which would not flower until it had produced 12 leaves *in the long day room.* An early flowering mutant, in contrast, is a plant which would flower after producing only 6 leaves *in the short day room.*

These classifications only refer to time of floral induction in a particular growth room—the late flowering mutant is defined by its behaviour in long days, whereas the early flowering mutant is defined by its behaviour in short days. In fact, these categories can be broken down further, as shown in Table 3.1. Note that the number of leaves is simply an indication of relative flowering time. The absolute number of leaves formed before flowering depends on the individual mutant and may also be variable between plants carrying the same mutation.

This further classification of flowering-time mutants is determined by their behaviour in the *other* day length—late flowering mutants are divided up by their behaviour in short days, and early flowering mutants are divided up by their behaviour in long days (Koornneef 1991).

Table 3.1 Approximate number of leaves produced before flowering by wild type and four mutant categories of Arabidopsis.

	Long days	Short days
Wild type	6 leaves	12 leaves
Late flowering class 1	12 leaves	18 leaves
Late flowering class 2	12 leaves	12 leaves
Early flowering class 1	4 leaves	6 leaves
Early flowering class 2	6 leaves	6 leaves

Late flowering mutants in class 1 show the typical late flowering in long days and produce more rosette leaves than wild type plants in this situation. Their response to short days is an even more dramatic delay in flowering, producing even more rosette leaves before being induced to flower. These mutants are clearly still responsive to photoperiod. Their ability to measure day length is nicely demonstrated by the fact that they flower earlier in long days than in short days. However, compared with wild type plants they are always very slow to flower. It seems, then, that these mutants are lacking a promoter of flowering not related to photoperiod. They are perturbed in an autonomous promotion pathway, one which is always required to initiate flowering.

Late flowering mutants in class 2 also show the typical late flowering in long days, producing around 12 leaves before flowering. However, when placed in the short day room these mutants still flower after producing 12 leaves. In contrast to class 1 mutants, class 2 mutants do not discriminate between different day lengths, but behave as though in short days regardless of how long the days are. This phenotype suggests an inability to measure day length, with the mutants always convinced that the days are short. The mutated genes in these plants might include those encoding the proteins which perceive light or measure time, so that no promotive long day signal is ever perceived, even if it is present.

Early flowering mutants can also be divided into two distinct classes. Early flowering mutants in class 1 show the typical response of flowering too early in the short day room—producing flowers with only around 6 leaves. If these plants are moved into the long day room, they flower even earlier—after making as few as 3 or 4 leaves. These mutants, like those in late flowering class 1, are clearly still sensitive to photoperiod, as they flower later in short days than they do in long days. However, they flower earlier than they are 'supposed' to, whatever the day length. These mutants are lacking something which represses flowering until the plant has reached a certain developmental stage. So this mutant class should make it possible to identify the genes that provide a developmental brake to flowering, giving the plant a chance to build up a certain amount of photosynthetic tissue before it starts to expend energy on producing flowers and seed.

Early flowering mutants in class 2 also show the typical pattern of flowering with only 6 leaves in short days. However, unlike mutants in class 1, they will also flower in the long day room only after they have produced 6 leaves. These mutants appear to be unable to respond to different photoperiods, and perceive all days as being long days. In the same way as mutants in late flowering class 2, these mutants have some problem with the perception of light or time. However, whereas that problem made mutants in late flowering class 2 behave as though all days were short, mutants in early flowering class 2 behave as though all days were long. In fact, these mutations may turn out to be in similar genes, with the differences in phenotype related to whether the mutation makes the gene more or less active.

3.3.3 Epistatic analysis places flowering-time genes into pathways

The phenotypes of flowering-time mutants give some clues as to which pathways they might operate in—those regulated by light, developmental cues, or temperature. This is confirmed by conducting epistatic analysis to test whether two mutated genes operate in the same pathway or different ones. If they function in the same pathway then epistatic analysis also indicates in which order they are required within that pathway.

Epistatic analysis begins by producing a double mutant (aabb) by crossing two different mutant plants (aaBB and AAbb) and selfing the heterozygote progeny (AaBb). There are a number of possible phenotypes which might show up in this double mutant plant. If the double mutant (aabb) has the same phenotype as one of the parents (say aaBB), the gene A is epistatic to the gene B. That means that gene A and gene B work in the same pathway, with gene A acting before gene B. The logic behind this is that if the two genes work in the same pathway, and if the pathway is blocked at the earlier point (A), then the phenotypic effect of mutating a gene further down the pathway (B) will be invisible. Alternatively, the double mutant (aabb) may have the same phenotype as parent plant AAbb—indicating

that gene B works before gene A in the pathway. It is also possible that the double mutant has the phenotypes of both parents. In this case we say that there is no epistasy between the genes, which work in different pathways and do not interact. Epistatic analysis is most easily imagined with morphological mutants, where crossing a mutant with no roots to a mutant with short roots would generate a double mutant with no roots—showing that one needs to make a root before one worries about its length. On the other hand, crossing a mutant with no roots to a mutant with purple leaves is likely to generate a mutant with no roots and purple leaves—because the pathways are unrelated. However, the process can be used to investigate more complex processes such as floral induction. In these cases, phenotype is not defined simply as late or early flowering, but factors such as the expression patterns of other known genes within the mutants are taken into account as well (Coupland 1995).

3.3.4 Arabidopsis contains one repression-of-flowering pathway and four promotion-of-flowering pathways

Epistatic analysis of the sort described above, using mutants including those above and other lines with perturbed responses to cold, has shown that at least five genetic pathways regulate flowering time in Arabidopsis (Simpson and Dean 2002). There is a single endogenous repression pathway which acts to prevent flowering. Its job is to delay flowering until the plant has reached a certain developmental stage and has the photosynthate to support developing flowers and seed. There are then four promotion pathways. Having multiple promotion pathways gives the plant plasticity, because it means that weak promotion in one pathway can be compensated for by strong promotion in another pathway, ensuring that flowering always happens. Two of these promotion pathways respond to environmental signals while two of them are under endogenous control. The photoperiodic promotion pathway relies on photoreceptors and a circadian clock to transduce information about day length. The vernalization promotion pathway is activated by exposure to low temperatures. The ability of plants like Arabidopsis to flower under non-inductive conditions indicates that plants must contain an autonomous promotion pathway, which ensures flowering occurs if all else fails. The fourth promotion pathway in Arabidopsis is the gibberellin pathway, which acts independently of the other pathways and may serve to activate flowering in response to certain developmental triggers. These different pathways are discussed in the next four chapters.

3.3.5 Integrating the flowering-time pathways

The various flowering-time pathways can be thought of as converging on a set of genes, which are known as the flowering-time integrators. The identification of these genes, whose function or expression is regulated by more than one of the input pathways, was described by Simpson and Dean (2002) as 'a key step forward in our understanding of flowering time control'. The inhibition pathway culminates in repression of the flowering-time integrators by the product of a gene called *FLC*. The autonomous promotion pathway results in downregulation of *FLC*, and thus allows activation of the flowering-time integrators. The vernalization promotion pathway also operates through downregulation of *FLC*, allowing induction of the flowering-time integrator genes. The photoperiodic promotion pathway bypasses FLC, and directly activates the flowering-time integrator genes. Gibberellin is also able to directly induce expression of flowering-time integrators, bypassing the FLC block. It can be seen, then, that the pathways all converge on the activity of a small subset of flowering-time integrator genes, either directly or through FLC. At any one time point these genes will be experiencing some degree of inhibition and some degree of activation. When the activation signals are stronger than the inhibition, the flowering-time integrators will be expressed and the floral transition will occur.

CHAPTER 4

The autonomous pathways for floral inhibition and induction

The autonomous pathways for floral inhibition and induction are central to the entire process of floral transition in Arabidopsis. The floral inhibition pathway is the central regulator of flowering, acting throughout the plant's life to ensure that flowering does not occur until the appropriate environmental, developmental, and physiological cues have been received. Without this inhibition pathway, Arabidopsis plants could flower immediately on germinating, before sufficient photosynthate had been produced to support the developing flowers and seeds. The inhibition pathway, as we shall discuss below, essentially operates through a single protein repressing the transcription of genes necessary for the floral transition to occur at the shoot apical meristem. Other proteins interact to enhance this basic repression mechanism. On the other hand, the autonomous induction pathway is equally essential to the life history of a weedy annual. Unlike long-lived perennial plants, Arabidopsis does not have the option of delaying flowering indefinitely until perfect environmental conditions are perceived. Instead, the autonomous induction pathway acts antagonistically to the inhibition pathway to ensure that the plant flowers eventually, irrespective of a lack of inductive environmental signals. These two pathways are the central players in the control of Arabidopsis flowering, around which the environmentally induced pathways that we shall discuss in later chapters revolve.

4.1 The floral inhibition pathway

The floral inhibition (or repression) pathway is the central regulator of flowering time in Arabidopsis.

The repression caused by this pathway is constitutive, and can only be broken by signals that either downregulate or bypass it. In Chapters 5 and 6 we shall consider how environmental signals achieve just that, but we must begin with a discussion of how the inhibition pathway itself works. The floral transition requires activation of a set of genes in the shoot apical meristem which convert the meristem from the vegetative state to the inflorescence state. The identities and functions of these floral meristem identity genes will be discussed in Chapter 9. In particular, it is essential that three genes described as flowering-time integrators are expressed (a concept introduced towards the end of Chapter 3). These genes, *FT*, *LEAFY* (*LFY*), and *AGAMOUS-LIKE 20* (*AGL20*) (also known as *SUPPRESSION OF OVEREXPRESSION OF CONSTANS 1* (*SOC1*)), activate the floral meristem identity genes, and it is thought that all pathways affecting flowering time are integrated at these three genes (He and Amasino 2005). The three genes themselves will be discussed in more detail in Chapter 7. The inhibition pathway works by repressing transcription of the flowering-time integrators, thus ensuring that the floral meristem identity genes are not activated and that the meristem remains vegetative (see Fig. 4.1). Repression of the flowering-time integrators is achieved primarily through the action of the FLOWERING LOCUS C (FLC) protein, reinforced by the activity of FRIGIDA (FRI). Dominant alleles at either locus result in plants that are extremely late flowering. Loss of function of either gene reduces the repression of flowering and causes an early flowering phenotype. In order for the plant to flower, signals promoting flowering must overcome repression of

Understanding Flowers and Flowering. Second Edition. Beverley Glover.
© Beverley Glover 2014. Published 2014 by Oxford University Press.

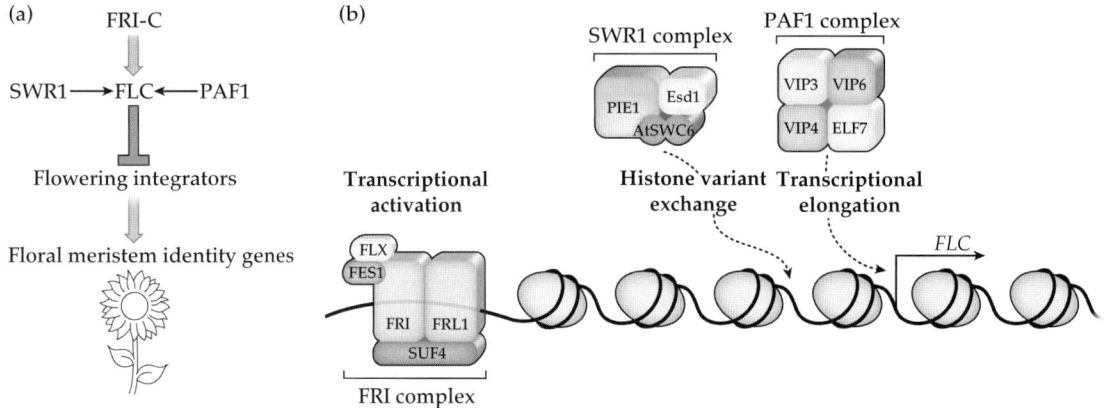

Figure 4.1 The floral inhibition pathway. (a) Expression of the flowering-time integrators, and thus the floral meristem identity genes, is prevented by FLC, itself enhanced by FRI-C and the SWR1 and PAF1 complexes. (b) The FRI complex, SWR1 complex, and PAF1 complex enhance FLC activity in different ways. FRI-C activates transcription, SWR1 modifies histone content, and PAF1 methylates histones and elongates transcripts. Redrawn from Yun *et al.* (2011).

the flowering-time integrators by FLC. This can be achieved in either of two ways. Proteins in the autonomous and vernalization promotion pathways promote flowering by downregulating *FLC* expression. These proteins directly target the *FLC* gene, and, through a number of routes, reduce its expression level to the point where it no longer represses the flowering-time integrators. The alternative approach is taken by proteins in the photoperiod and gibberellin-based promotion pathways. These proteins act to directly activate the flowering-time integrators, simply bypassing repression by FLC. This sets up a conflict at the flowering-time integrators, with numerous promotive signals, plus an FLC-based inhibitory signal which varies in strength, converging on them and controlling their expression levels and subsequent functions.

4.1.1 FLC

The *FLC* gene was isolated using positional cloning, and encodes a MADS box protein, a member of a large family of transcription factors found in all eukaryote genomes (Michaels and Amasino 1999). The central role of FLC is to repress the expression of the flowering-time integrators, and it was always thought likely that the protein would act as a negative regulator of transcription, binding

to promoter sequences but failing to activate transcription. Initial reports focused on establishing the central role of FLC in floral repression. Loss of *FLC* function causes early flowering, while ectopic expression of FLC from the cauliflower mosaic virus 35S (CaMV 35S) promoter caused plants to flower extremely late, with some failing to flower at all, senescing and dying after producing over 80 rosette leaves but never bolting (Michaels and Amasino 1999). Early experiments showed that the level of *FLC* transcript in a range of backgrounds correlated directly with the time those plants took to initiate flowering (Sheldon *et al.* 2000b). Later work confirmed that *FLC* transcript abundance correlated directly with FLC protein abundance across a range of genetic, developmental, and environmental variables, indicating that *FLC* regulation operates at the transcriptional level (Rouse *et al.* 2002).

More recently, analysis of FLC activity has centred around two aspects: the regulation of *FLC* expression both by proteins within the floral repression pathway and by proteins in floral promotion pathways, and the regulation by FLC of genes such as the flowering-time integrators. The activities of floral repressors that enhance FLC function are discussed below, while the regulation of *FLC* expression by floral promotion pathways is discussed in Section 4.2, as well as in Chapters 5 and 6.

Chapter 7 is primarily concerned with the activities of the flowering-time integrators. However, it is worth mentioning here the way in which FLC represses the activity of those flowering-time integrators. Hepworth *et al.* (2002) focused on the role of FLC in repression of *SOC1*, and identified a key region of 351 base pairs in the *SOC1* promoter. This region was essential for repression of *SOC1* expression by FLC, but also necessary for activation of *SOC1* expression by CONSTANS (CO, a promoter of flowering operating in the photoperiod pathway; see Chapter 5). FLC was shown to bind to this 351-base-pair region *in vitro*, probably at a CarG box site identified within the sequence. CarG boxes are the binding motifs usually recognized by MADS box transcription factors. Hepworth *et al.* (2002) proposed that FLC and CO compete for binding of similar regions within the *SOC1* promoter, and that it is only when insufficient FLC protein is present that sufficient CO-driven activation can occur to cause *SOC1* expression and subsequent function.

Searle *et al.* (2006) and Helliwell *et al.* (2006) extended this analysis and demonstrated that FLC also binds to the promoter of *FD*, the protein product (a bZIP transcription factor) of which is associated with FT in the shoot apical meristem at the floral transition, and to a CArG box in the first intron of *FT* itself. This binding represses transcription of these key flowering pathway integrators and delays the floral transition. Taking a transgenic approach with targeted expression in different tissues, Searle *et al.* (2006) showed that ectopic expression of *FLC* in leaves reduces expression of *FT* and *SOC1* and delays flowering accordingly, and that ectopic expression of *FLC* in the shoot apical meristem reduces expression of *FD* and *SOC1*, and similarly delays flowering.

An additional MADS box protein, SHORT VEGETATIVE PHASE (SVP), has been shown to interact with FLC and enhance its binding of promoter sequences (Lee *et al.* 2007; Li *et al.* 2008). Lee *et al.* (2007) showed that ectopic expression of *FLC* (and its enhancer *FRI*) could not delay flowering in an *svp* mutant background, although they strongly delayed flowering in wild type. Li *et al.* (2008) confirmed that SVP and FLC interact physically *in vivo*. These results suggest that FLC acts with SVP in the repression of floral integrators. It is possible that the FLC–SVP complex is the central repressor of flowering, rather than FLC alone.

4.1.2 FRI and FRIGIDA LIKE 1 (FRL1)

The main enhancer of FLC activity, FRI, was identified in 2000 by Johanson *et al.* It had long been known that the requirement for vernalization in Arabidopsis was attributable to a single locus, *FRI*. Where a dominant *FRI* allele is present, an Arabidopsis ecotype will flower only after a period of cold treatment (i.e. it is late flowering without cold), but where a recessive *fri* allele is present (as is the case with all lab lines) the Arabidopsis ecotype does not require vernalization and flowers early. *FRI* was isolated by map-based cloning, and found to encode a 609-amino-acid protein with little sequence similarity to proteins from other species. The protein contains two coiled-coil domains, one near the N terminus and one near the C terminus. Analysis of *FRI* loci in Arabidopsis ecotypes confirmed that most early flowering varieties contained mutations within *FRI* (Johanson *et al.* 2000). This result was confirmed by the later analysis of Le Corre *et al.* (2002), who sequenced *FRI* from 25 Western European ecotypes of Arabidopsis and concluded that recent positive selection for early flowering had occurred within a set of isolated populations, resulting in variation at the *FRI* locus. It had previously been shown that FRI is necessary for *FLC* expression (Michaels and Amasino 1999), but in 2001 Michaels and Amasino also demonstrated that, in the absence of a functional *FLC* allele, FRI has no effect on flowering time in long or short days. The role of FRI as an enhancer of FLC activity was thus clearly established.

Since mutagenesis has traditionally been conducted in lab lines of Arabidopsis, which all contain non-functional *fri* alleles, it is only since the discovery of the role of FRI that investigation into proteins which interact with it has begun. Michaels *et al.* (2004) conducted a mutant screen for early flowering plants in an Arabidopsis line containing a dominant FRI allele in a Columbia background. They identified the *frl1* mutant, and confirmed that it had no phenotypic effect in a normal Columbia (i.e. *fri* mutant) background. The *FRL1* gene was isolated using a T-DNA insertion as a tag, and found to encode a protein 23% identical to FRI but with no large blocks of conserved sequence. Analysis of FRL1 function in a number of backgrounds confirmed that its only role is to facilitate the FRI-mediated expression of *FLC*. In an *frl1* mutant background, FRI has no effect on *FLC* transcription. Analysis of the Arabidopsis genome sequence revealed a further five genes with significant sequence similarity to *FRI* and *FRL1*, suggesting that *FLC* regulation may be more complex than was originally thought (Michaels *et al.* 2004).

These observations were brought together by Choi *et al.* (2011), who showed that FRI and FRL1 act together as part of a larger complex. The FRI protein is the central scaffold for this complex, which is known as FRI-C. FRL1 stabilizes the complex, along with another protein, FES1 (FRIGIDA ESSENTIAL 1). FES1 and FLX (FLC EXPRESSOR) act within the complex to activate transcription, while SUF4 (SUPPRESSOR OF FRIGIDA 4) binds directly to the FLC promoter. The FRI-C complex also recruits chromatin modification factors to the FLC locus, ensuring continued active transcription of the *FLC* mRNA (Choi *et al.* 2011).

4.1.3 EARLY IN SHORT DAYS 1 (ESD1), PHOTOPERIOD INDEPENDENT EARLY FLOWERING 1 (PIE1) and the SWR1 complex

The *esd1* mutant causes early flowering, as a result of a reduction of *FLC* expression (Martin-Trillo *et al.* 2006). Analysis of *esd1* in conjunction with other mutants suggests that the ESD1 protein is a strong inducer of *FLC* expression, but may also act on other genes involved in the repression of flowering, as the *esd1/fca* double mutant is earlier flowering than the *fca* mutant alone. Map-based cloning revealed that *ESD1* encodes ACTIN-RELATED PROTEIN 6 (AtARP6), which is required for modification of *FLC* chromatin. These modifications, which include methylation and acetylation of histones, are necessary to achieve high levels of *FLC* expression and subsequent repression of flowering (Martin-Trillo *et al.* 2006).

PIE1 was first described from a screen to identify new early flowering mutants (Noh and Amasino 2003). The *pie1* mutant flowers early in short days

compared with wild type, and earlier still in long days, confirming its role in a non-photoperiod-dependent floral repression pathway. The gene was isolated using a T-DNA insertion in the mutant as a tag, and the very large (2055-amino-acid) protein it encodes found to resemble chromatin remodelling proteins from vertebrates with roles in the activation of gene expression. Two nuclear localization signals make it very likely that the protein is nuclear localized and plays a role in chromatin remodelling and gene expression. On the basis of this characterization, the authors investigated a potential role for PIE1 in activation of *FLC* expression. They discovered that PIE1 does regulate *FLC* expression in the shoot apical meristem, but not in other tissues, such as the root. This function is consistent with the expression pattern of *PIE1*, which is restricted to the shoot apical meristem. It appears that PIE1 is involved in the activation of *FLC* expression by FRI, and delays flowering time through this enhancement of FLC activity (Noh and Amasino 2003).

The similar roles of ESD1 and PIE1 are explained by their joint function in a protein complex resembling the yeast SWR1 complex (Lázaro *et al.* 2008). This complex modifies the histone content of chromatin, allowing transcription to occur. In Arabidopsis, ESD1, PIE1, and SWC6 physically interact and are necessary for the histone acetylation and methylation required for high levels of *FLC* expression. This SWR1 complex is proposed to be recruited to the *FLC* promoter by the FRI-C complex (Choi *et al.* 2011).

4.1.4 VERNALIZATION INDEPENDENCE genes and the PAF1 complex

Another protein complex, also first defined in yeast, has also been shown to interact with *FLC*. In 2002, Zhang and van Nocker reported the results of a screen for mutants which flowered early and had reduced levels of *FLC* expression. They called these new mutant lines the *vernalization independence* lines (*vip*), and they represent a class of mutants with perturbations in genes involved in *FLC* upregulation. The *VIP4* gene was isolated first, and found to encode an extremely hydrophilic protein with similarity to animal and yeast proteins involved in forming multiprotein complexes. *VIP4* is expressed throughout the plant, but is not downregulated after vernalization, confirming that it plays a role in an autonomous enhancement of *FLC* (and thus repression of flowering) pathway (Zhang and van Nocker 2002).

The following year the *VIP3* gene was isolated, by positional cloning, and found to encode a protein consisting almost entirely of repeated WD (tryptophan and aspartate) motifs. Such a structure is indicative of a protein which acts as a scaffold for the assembly of multiprotein complexes. Ectopic expression of *VIP3* from the CaMV 35S promoter was insufficient to activate *FLC* expression, even though *FLC* transcript levels are reduced in the *vip3* mutant. This suggested that VIP3 acts with other proteins to activate *FLC* expression, perhaps in a complex for which it provides the base (Zhang *et al.* 2003).

Oh *et al.* (2004) reported that *VIP6* encodes a protein with significant similarity to a yeast protein, CTR9, which forms part of the polymerase II-associated factor 1 (PAF1) complex. The PAF1 complex comprises five proteins which interact with RNA polymerase II to regulate gene expression through the methylation of histones and the elongation of transcripts. In the same year, He *et al.* (2004) also described a mutation in a gene they called *EARLY FLOWERING 8 (ELF8)*. Loss of function of *ELF8* resulted in loss of *FLC* transcript and early flowering. The *ELF8* locus is in fact the same locus as the *VIP6* locus, and both groups noted the similarity of the protein product to the CTR9 component of the yeast PAF1 complex. Indeed, Oh *et al.* (2004) also noted that the protein encoded by *VIP4*, previously described in yeast as forming multiprotein complexes, is now also known to associate within the PAF1 complex. Another component of this complex is encoded by the *EARLY FLOWERING 7 (ELF7)* locus. The *elf7* mutant flowers early in both long and short days and contains no detectable *FLC* transcript (He *et al.* 2004). The *ELF7* gene encodes a protein with great similarity to the PAF1 component of the yeast PAF1 complex. These data provide considerable support for the presence in Arabidopsis of a multiprotein complex involved in regulating transcription of *FLC* by methylating histones to ensure continued elongation of newly forming mRNA transcripts.

4.2 The autonomous induction pathway

The autonomous induction (or promotion) pathway was first formally described by Maarten Koornneef and colleagues, who conducted a mutant screen for flowering-time loci and then used epistatic analysis to describe the pathways within which these loci operated (Koornneef *et al.* 1991, 1998). Genes that confer autonomous promotion of flowering in Arabidopsis were identified by the isolation of late flowering mutants that retain photoperiod sensitivity and respond strongly to vernalization. That is, the mutants flowered later than wild type plants in long days, but later still in short days, and their flowering could be promoted by cold treatment. Such late flowering mutants were clearly not deficient in photoperiod-induced flowering, nor in vernalization-induced flowering, but in some general regulator of floral induction. Koornneef's original analysis identified four genes in this autonomous flowering pathway—*FCA*, *FY*, *FPA*, and *FVE*. However, a variety of other genes are now also considered to act in this pathway, including *LUMINIDEPENDENS (LD)* (Lee *et al.* 1994b), *FLK* (Lim *et al.* 2004), and *FLD* (Sanda and Amasino 1996). The proteins encoded by all of these genes have been shown to autonomously promote flowering in the same way—by repressing the FLC-based repression of flowering in the floral inhibition pathway (see Fig. 4.2). Indeed, this repression appears to be the result of direct inhibition of FLC activity in all cases, as each protein operates, albeit in different ways, to prevent the accumulation of *FLC* mRNA in shoot apical meristem cells (Simpson 2004). Each mutant has increased levels of *FLC* transcript when compared with wild type plants (Michaels and Amasino 1999).

4.2.1 FCA

The best studied of the genes operating in the autonomous induction pathway is *FCA*. The *FCA* gene was isolated by positional cloning and encodes a 747-amino-acid protein which interacts with RNA (Macknight *et al.* 1997). Specifically, FCA is classed as a plant-specific RNA-recognition motif-type RNA-binding-domain-containing protein, or RRM for short. In fact the protein contains

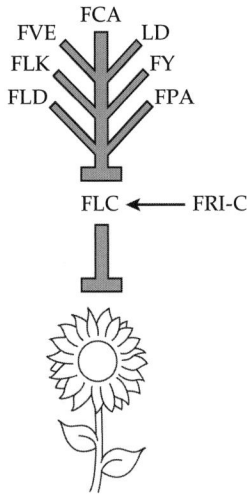

Figure 4.2 Diagram of the autonomous induction pathway. Proteins acting in the autonomous induction pathway repress FLC, eventually overcoming its repression of flowering.

two RRM domains, both in the N terminal region, and a C terminal WW domain, a region which is usually, and in this case has been confirmed to be, involved in protein–protein interactions. The WW domain is essential for FCA function (Simpson et al. 2003). The *FCA* transcript is found throughout the plant, although it is most abundant in areas of active cell division, including the root and shoot apical meristems (Simpson 2004). Within those cells the FCA protein is located in the nucleus, consistent with a role in RNA binding. Macknight et al. (1997) were able to confirm that FCA does bind RNA *in vitro*, with a preference for U- or G-rich sequences, but will not bind DNA.

FCA has been shown to downregulate *FLC* by interfering with RNA processing. The FCA protein promotes early polyadenylation of an *FLC* antisense transcript. This leads to downregulation of *FLC* sense transcription, apparently through changes to the histone methylation status of the locus (Liu et al. 2010b).

The *FCA* transcript has alternative polyadenylation sequences, which play an important role in its function (see below). This alternative polyadenylation is conserved in *FCA*-like sequences in *Brassica napus* (Brassicales) and *Pisum sativum* (Fabales)

(Macknight et al. 2002). Perhaps surprisingly, *FCA* transcript is found throughout the plant, and is present in particularly high levels not only in shoot meristems but also in the root apical meristem and root lateral meristems. Consistent with this expression pattern, *fca* mutant plants have shorter roots with 20% fewer lateral roots than wild type plants (Macknight et al. 2002). These data suggest that FCA plays other roles in plant development, besides that in promoting flowering. Recent reports suggest that FCA acts as a more general regulator of RNA 3′ processing, modifying transcript length and preventing intergenic transcription throughout the genome (Sonmez et al. 2011). FCA has long been known to activate responses over distances of some cells, as it has been shown not to act cell-autonomously. Furner et al. (1996) demonstrated that *fca* mutant sectors in a wild type plant flowered at the same time as wild type.

4.2.2 FY

The molecular identity of FY was established in 2003 as a result of a search for proteins that interacted with the WW domain of FCA. The gene encodes a protein with seven WD40 repeats and which shows significant sequence and structural similarity to a yeast protein that operates as part of a complex that cleaves and polyadenylates RNA (Simpson et al. 2003). This protein is very conserved throughout studied angiosperms, with proteins from *Sorghum bicolor* (Poales) and *Medicago truncatula* (Fabales) both showing very strong similarity. This is not surprising, as total loss of function of *FY* is lethal at the embryo stage (Henderson et al. 2005). Analysis of FY function indicated that, like FCA, it acts as a 3′-end RNA-processing factor, important at many stages of plant growth. FCA has three main functions in the induction of flowering. First, through physical interaction with FCA it interferes with *FLC* RNA accumulation, and thus allows floral meristem conversion genes to be expressed (Manzano et al. 2009). The importance of FY in this interaction is clear from transgenic experiments. Ectopic expression of *FCA* from the CaMV 35S promoter in Arabidopsis causes early flowering, but this construct has no effect in an *fy* mutant background (Simpson et al. 2003).

Clearly FY protein is essential for the FCA-induced downregulation of *FLC*. Secondly, FY is necessary for the regulation of FCA itself. The FCA/FY complex promotes premature polyadenylation within the third intron of *FCA* RNA, as opposed to the normal polyadenylation at the 3′ end of the molecule. This results in the production of a truncated RNA molecule that encodes a short inactive protein, and may also result in chromatin modification at the *FLC* locus (Quesada *et al.* 2003). This negative autoregulation of *FCA* was unexpected, and appears to be developmentally controlled, suggesting that at key times and places in the plant it may be possible to generate variable amounts of FCA protein. Certainly bypassing this autoregulation causes early flowering. The importance of FY in *FCA* autoregulation is demonstrated by the great reduction of the amount of short *FCA* transcript present in the *fy* mutant background, compared with wild type (Simpson *et al.* 2003). Thirdly, it has recently been shown that FY has a role in premature polyadenylation of the *FLC* transcript independently of any association with FCA. Feng *et al.* (2011) showed that the presence of FY caused proximal polyadenylation of *FLC* RNA and reduced total *FLC* transcript even in an *fca* mutant background.

4.2.3 FPA

FPA is another gene encoding an RNA-binding protein, and like *FCA* it has recently been shown to have a more general role in gene silencing throughout the plant life cycle (Baurle *et al.* 2007). FPA does not interact directly with FCA or FY, and appears to regulate *FLC* independently. *FPA* was isolated by positional cloning and encodes another RRM-containing protein (Schomburg *et al.* 2001). FPA contains three RRMs and a protein interaction SPOC (SPEN paralog and ortholog C-terminal) domain at the C terminus. The RRMs are all located in the N terminal region of what is a rather large protein (901 amino acids). *FPA* transcript is found throughout the plant at low levels, but is strongest in meristematic regions, particularly the inflorescence meristems (both apical and axillary). The protein is nuclear-localized. Ectopic expression of *FPA* in Arabidopsis using the CaMV 35S promoter caused

plants to flower early in short days, and generated an almost day-neutral behaviour with respect to photoperiod (Schomburg *et al.* 2001). Recent reports suggest that although FPA functions independently of FCA, the mode of action is very similar. Hornyik *et al.* (2010) investigated FPA function and concluded that the protein controls alternative RNA cleavage and polyadenylation, with consequent effects on chromatin. Sonmez *et al.* (2011) demonstrated that this activity was not just restricted to the regulation of *FLC*, but occurred more widely throughout the genome.

4.2.4 FLK

The *FLK* gene was one of the last two genes in the autonomous promotion pathway to be identified, and was isolated, using the T-DNA in an insertion mutant as a tag, in 2004. The *flk* mutant flowers extremely late, and contains no detectable *FLK* transcript but a tenfold increase in *FLC* transcript compared with wild type plants. This elevated *FLC* expression results in a downregulation of the flowering-time integrators, *FT* and *SOC1*. The gene encodes a protein with three K-homology-type RNA-binding domains, very similar to those found in animal RNA-binding proteins (Lim *et al.* 2004). The protein is predominantly nuclear localized, in keeping with its function in RNA binding. It is clear that FLK functions to downregulate *FLC* through RNA interference, just as FCA/FY and FPA do, but FLK is likely to operate through the facilitation of protein–RNA interactions, rather than through premature polyadenylation.

4.2.5 FLD

The four proteins described above all regulate FLC through RNA processing and subsequent effects on chromatin. It is therefore not surprising that other genes identified through mutant screens as members of the autonomous induction pathway have turned out to encode enzymes involved in that chromatin modification. *FLD* and *FVE* both fall into this category, and regulate *FLC* through chromatin modification. The *fld* mutant is late flowering and has elevated levels of *FLC* transcript. However, in an *flc* mutant background the *fld* mutation has no effect on flowering time, confirming that the role of FLD is to suppress *FLC*. FLD was isolated using a map-based approach, and encodes a protein containing a domain involved in chromatin remodelling. The protein is highly conserved, clearly functioning as a 'dimethylated histone H3 at lysine 4 (H3K4me2) demethylase', very similar to proteins with the same role in humans and other eukaryotes (He *et al.* 2003; Liu *et al.* 2007). *FLD* expression is not sufficient to entirely eliminate *FLC* expression, but merely to reduce it (He *et al.* 2003; Simpson 2004). The role of FLD is apparently downstream of the premature polyadenylation resulting from the activities of the RNA binding proteins. Investigation of the genetic interactions of these loci by Liu *et al.* (2010b) concluded that the premature polyadenylation appears to trigger FLD-dependent demethylation in the main body of the gene, downstream of the early polyadenylation site. This results in suppression of both sense and antisense transcription from the *FLC* locus, and many other loci throughout the genome. A recent report by Yu *et al.* (2011) indicated that FLD interacts physically with a histone deacetylase, explaining why early reports on the nature of the *fld* mutant included the observation that the *FLC* locus appeared to be hyperacetylated in the absence of *FLD*.

4.2.6 FVE

FVE also encodes a protein involved in chromatin structure and subsequent gene expression. Like *FLK*, it was isolated in 2004, using a map-based approach, and provided the last pieces of the autonomous induction pathway puzzle. The FVE protein contains six WD40 repeat domains, and shares sequence and structural elements with proteins from yeast and mammals that are present in complexes involved in chromatin modification. Like FLD, FVE has been shown to be involved in histone deacetylation of *FLC* chromatin, modifying *FLC* transcript level in the process. *FLC* chromatin was found to have excess acetylated histones in the *fve* mutant, compared with wild type (Ausin *et al.* 2004). Pazhouhandeh *et al.* (2011) explored the interactions of FVE with other proteins, and concluded that it acts as a scaffold for a CLF-polycomb repressive

complex 2 (PRC2), known to be necessary for repressive methylation of chromatin.

4.2.7 LUMINIDEPENDENS (LD)

LD was the first gene of the autonomous induction pathway to be identified at the molecular level, although we still know little about how it functions. The *LD* gene was isolated using a T-DNA insert as a tag, and encodes a homeodomain protein, a structure usually associated with DNA binding and transcriptional regulation. The simplest role that can be hypothesized for LD from this sequence is a role in transcriptional repression of *FLC* (Lee *et al.* 1994b). However, it has also been proposed that LD may function in RNA binding, rather than DNA binding. In rare cases the homeodomain motif has been shown to interact with RNA, and this mechanism would be consistent with the known importance of RNA processing in the autonomous pathway (Simpson 2004). The *LD* transcript is found throughout the plant, although it is most strongly expressed in shoot and root apical meristems and leaf primordia. An LD-GUS fusion protein was found to be nuclear localized (Aukerman *et al.* 1999). Recent reports suggest that LD may repress *FLC* transcription by inhibiting a protein that acts as a positive regulator of *FLC*. The *SUF4* (*SUPPRESSOR OF FRIGIDA 4*) locus encodes a zinc finger containing protein that interacts with the FRI-C complex and the *FLC* promoter (Kim *et al.* 2006). In the absence of the FRI-C complex (in an *fri* mutant background), the SUF4 protein is bound by LD, perhaps limiting its ability to bind the *FLC* promoter.

4.3 Other endogenous factors that influence flowering time

Although the two pathways described above are the central endogenous regulators of flowering time in Arabidopsis, many other factors can also influence the floral transition. For example, in some species the floral transition will not occur until the plant is a certain size, irrespective of the nutrients it has received (Bernier and Perilleux 2005). Developmental factors such as these only really apply to more long-lived plants, particularly perennials.

However, even short-lived annuals like Arabidopsis undergo certain life stage transitions before flowering can occur. Loss of function of genes such as *EMBRYONIC FLOWER 1* (*EMF1*) eliminates rosette formation and transforms the apical meristem into an inflorescence meristem from the point of germination (Aubert *et al.* 2001). Although *EMF1* is not thought of specifically as a flowering-time gene, because its role is clearly in controlling phase transition during development in general, it is clear from the phenotype of the *emf1* mutant that other factors besides the inhibition and autonomous induction pathways are also necessary for appropriate development and accurate timing of flowering.

CHAPTER 5

The photoperiodic pathway of floral induction

It is nearly 100 years since the formal discovery that flowering time in many plant species was regulated by day length, and yet we have only recently begun to understand how this regulation occurs. In 1920, Garner and Allard published a categorization of plants into three types: those that flowered when days were longer than a critical period, those that flowered when days where shorter than a critical length, and those whose flowering did not appear to be affected by day length. The third group, day neutral plants, flower in response to triggers considered in Chapters 4 and 6. The remaining two groups, long day plants and short day plants, experience photoperiodic induction of flowering. In this chapter we shall consider the photoperiodic regulation of flowering in Arabidopsis, which is a facultative long day plant (i.e. one whose flowering is brought forward by long days, rather than one with an absolute requirement for them). Understanding the photoperiodic induction pathway requires an analysis of how light is perceived, how time is measured, and how the two signals are integrated to activate responses. In Chapter 8 we shall contrast what we know of Arabidopsis floral induction with the systems in other species, particularly the short day plant rice (Poales).

5.1 Sensing daylight

Plants clearly have the ability to absorb light, because doing so is an essential component of photosynthesis. Leaves are adapted to function as solar panels, and thus make an ideal site for the perception of daylight. However, it is the presence of particular photoreceptors that allows plants to perceive the range of wavelengths present in incident light, and to convert that perception into signals. Extensive physiological work over many years has demonstrated that red light, far red light, and blue light are all important in photoperiodic induction, although their relative importance varies in different species and under different conditions.

5.1.1 Sensing red and far red light: phytochromes

Red light (650–670 nm) and far red light (705–740 nm) are perceived by phytochromes, which are soluble cytosol proteins that form a dimer, each member of which has a tetrapyrrole chromophore covalently bonded to the GAF subdomain in their N terminal photosensory domain (see Fig. 5.1a). Recent reports of crystallization of bacterial phytochromes suggest that the chromophore is almost buried within a pocket formed by the GAF subdomain (Wagner *et al.* 2005). This chromophore, called phytochromobilin, selectively absorbs particular wavelengths of light, allowing the whole molecule (protein plus chromophore) to act as a photoreceptor and activate downstream responses. The chromophore is an open chain of four pyrrole rings. The C terminus of phytochrome contains two repeats of a motif called a PAS domain, and the N terminus contains a single PAS domain. These domains are believed to be involved in signal relay through protein–protein interactions with other molecules, including transcription factors (Yanovsky and Kay 2003). The C terminus also contains a histidine-kinase-related domain (HKRD) with serine/threonine kinase activity. The presence of this domain has led to

Understanding Flowers and Flowering. Second Edition. Beverley Glover.
© Beverley Glover 2014. Published 2014 by Oxford University Press.

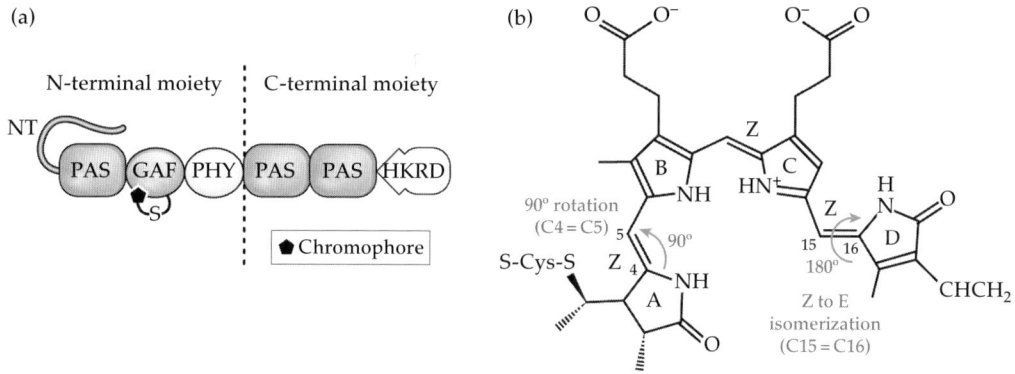

Figure 5.1 Phytochromes. (a) The chromophore is bonded to the GAF domain in the N terminal portion of the phytochrome protein. The PAS domains and the histidine kinase-related domain are towards the C terminus. Redrawn from Nagatani (2010). (b) The chromophore of phytochrome takes one of two forms, depending on the light regime. Contradictory data suggest that either the rotation or the isomerization shown occurs during the interconversion. Redrawn from Nagatani (2010). Reproduced with permission from Elsevier.

the suggestion that phytochromes activate responses by phosphorylating their target proteins (Rubio and Deng 2005). A recent report confirmed that the HKRD domain was necessary both for correct subcellular localization and for spectral sensitivity of phytochrome A (Müller *et al.* 2009).

Traditionally, phytochrome function was described physiologically and was thought to depend on interconversion of the chromophore between two states according to the available light (see Fig. 5.1b). The Pr form of the molecule absorbs red light, and is converted into the Pfr form through a conformational change. The Pfr form then absorbs far red light and is converted back to the Pr form. In the light, the transition between the two forms sets up a stable equilibrium, but in the dark Pfr declines progressively as the Pr form is more stable. The Pfr form is usually the active form, the one that causes things to happen in the plant, and so when it is absent, in the dark, the plant can no longer activate downstream light responses (Quail 2002). Although this model still holds generally true for phytochrome function, more detailed analysis in Arabidopsis has revealed the presence of five different phytochrome molecules, each of which responds to light in subtly different ways. In particular, while the Pfr form of phytochrome has been shown to be the active form for four of the five molecules, it has been proposed that a short-lived intermediate generated during the switching between Pr and Pfr may be at least

one active form of phytochrome A (Shinomura *et al.* 2000; Wang and Deng 2003). Alternatively, it may be the case that continuous far red light is able to establish and maintain a small fraction of the phytochrome A population in the Pfr form over an unusually extended period (Quail 2002).

The five phytochrome proteins in Arabidopsis are encoded by five different genes. The genes are called *PHYTOCHROME A* to *E* (*PHYA–PHYE*), and the molecules that they encode are referred to as phytochrome A, B, C, D, and E. A plant with a lesion in one of the *PHY* genes will still make the other four phytochromes and still be responsive to light. Analysis of the individual mutants has allowed dissection of the roles of the five different phytochromes, and it has become clear that the physiological model described above is oversimplified. Phytochromes B, C, D, and E all perceive red light, whereas phytochrome A distinguishes between far red light and darkness (Quail 2002). Phytochrome A is classed as a labile phytochrome, being readily degraded in the light, whereas the other four phytochromes are considered stable. Some of the phytochrome molecules also seem to be more important than others in different responses by plants to light.

The differences between the activities of the different phytochromes are attributable to differences in the protein structures and also to differences in their light lability. Phytochrome A is extremely light labile, being degraded within minutes of the light treatment needed to activate it (Nagy and Schafer 2002). It has been suggested that the phosphorylation status of the protein may influence both its lability and its activity level. This lability may serve to ensure that plants do not respond with large-scale developmental changes (such as flowering) to very brief periods of illumination. Phytochrome A is also at its most active on exposure to far red light. In contrast, the other four phytochrome molecules are all light stable and are most active on exposure to red light (Yanovsky and Kay 2003). In addition, phytochrome A protein has a daily oscillation pattern when plants are grown in short days. In short day regimes, phytochrome A protein levels peak during the night, but in long days they remain constant throughout the day (Mockler *et al.* 2003).

The importance of phytochrome A (and far red light) in the floral transition has been demonstrated using Arabidopsis plants exposed to varying light regimes. Wild type plants that are exposed to a normal short day (12 hours), followed by several hours of only far red light, flower early (at the same time as plants exposed to normal long days). However, *phyA* mutant plants flower late under this regime, as they are not able to perceive the far red light extension to the day. Exposure of wild type plants to short days followed by several hours of only red light results in late flowering (at the same time as plants exposed only to short days), indicating that far red light (and the phytochrome A which perceives it) is more important than red light (and the other phytochromes) in the flowering response. Similar experiments have shown that exposing plants to short days, but breaking up the long night with short periods of light (light breaks), results in early flowering. Although all wavelengths of light are somewhat effective as light breaks, far red light was shown to be the most effective of all (Goto *et al.* 1991; Putterill 2001). Far red light alone is insufficient to allow photosynthesis, but has been shown to be sufficient for flowering. Mockler *et al.* (2003) grew wild type and *phyA* mutant plants on nutrient media that met the plants' energy requirements in the absence of photosynthesis. Plants were kept in continuous far red light, and wild type plants flowered at the normal time, whereas *phyA* mutant plants were extremely late flowering. Clearly far red light and phytochrome A are essential for the floral

transition. However, phytochrome A is also known to play important roles in seed germination and in de-etiolation of light-deprived tissues (reviewed by Mathews 2010).

Phytochrome B is the major form of phytochrome present in light-grown tissue, and it mediates germination, de-etiolation, and shade avoidance responses. It also plays an important role in the floral transition. Phytochromes A and B are ancient protein lineages, both having arisen before the divergence of the angiosperms and the gymnosperms (reviewed by Mathews 2010). Phytochrome C has a narrower range of roles in leaf development and perception of photoperiod, and is also an ancient, although less strongly conserved, protein. Phytochrome D is found only in the Brassicaceae, following a duplication of the the *PHYB* gene in that lineage. Phytochrome E originated before the divergence of the eudicots, but is not conserved in all plant families. There is some redundancy of function in Arabidopsis, particularly between phytochromes A and E, and between phytochromes B and D (reviewed by Mathews 2010). It is also apparent that the different phytochromes have different dimerization requirements. Phytochrome A forms only homodimers, whereas phytochromes C, D, and E must heterodimerize with phytochrome B. Phytochrome B is also functional as a homodimer (Clack *et al.* 2009).

The chromophores themselves are encoded by a different set of genes, called the *HY* genes. *HY1* and *HY2* are both necessary to make phytochromobilin (the chromophore, which absorbs light), and loss of function of either of these genes results in a plant that lacks functional molecules of all the phytochromes because there are no chromophores to bond to the core proteins. These mutants are usually deficient in all phytochrome-mediated responses. Phytochromobilin is synthesized within the plastids and then released into the cytosol, where it is bonded to the main phytochrome proteins. *HY1* was isolated through positional cloning and shown to encode the enzyme heme oxygenase (Davis *et al.* 1999). Heme oxygenase cleaves the heme precursor of chromobilin into the intermediate biliverdin, in the first committed step of chromophore synthesis. There are three other genes in Arabidopsis which are predicted to encode heme oxygenases, and reports suggest that these have reduced and partially redundant functions in the generation of the chromophore (Emborg *et al.* 2006). *HY2*, isolated by map-based cloning, encodes biliverdin reductase (otherwise known as phytochromobilin synthase), which catalyses the final step of chromophore synthesis in a ferredoxin-dependent reaction (Kohchi *et al.* 2001).

The mode of action of active phytochromes is still not entirely understood, with both direct interaction with transcription factors and phosphorylation apparently being involved. However, it is clear that after the conversion of Pr to Pfr on exposure to light, all five phytochrome molecules move from the cytoplasm to the nucleus (Quail 2002). This transition is very rapid, occurring within only a few minutes of light treatment. The Pr form of PHYA is anchored within the cytoplasm by interactions with other proteins, but the Pfr form no longer interacts with the anchor and thus is free to be trafficked into the nucleus (Wang and Deng 2003; Chen and Chory 2011). In contrast, PHYB movement appears to depend upon conformational changes revealing its nuclear localization signal.

The movement of phytochrome to the nucleus is likely to be crucial to its function, since one known mode of rapid phytochrome action is through the interaction of its PAS domain with transcription factors, particularly the basic helix-loop-helix PIF (PHYTOCHROME INTERACTING) proteins. There are at least seven *PIF* genes in the Arabidopsis genome, part of a 15-member subgroup of the bHLH transcription factor family (Leivar and Quail 2011). The subfamily is named for PIF3, the first of the group identified, which was isolated from a screen for proteins to which PHYB bound. Both the active light-induced forms of PHYA and PHYB have been shown to bind PIF3 in yeast (although PHYB binds tenfold more strongly than PHYA; Ni *et al.* 1998; Zhu *et al.* 2000), and all other PIF proteins are also bound by the light-induced forms of one of the phytochrome molecules (Leivar and Quail 2011). All PIF proteins bind PHYB through a conserved N-terminal sequence called the active phytochrome B-binding (APB) motif (Khanna *et al.* 2004). PIF3 was shown to bind G-box motifs (CACGTG) in the promoters of other genes, and the other PIF proteins tested also bind the same motif

(Martinez-Garcia *et al.* 2000; Leivar and Quail 2011). The data initially suggested a positive interaction, with PIF proteins acting to recruit light-activated and nuclear-imported phytochrome proteins to the promoters of target genes, whose transcription is then directly activated. However, detailed analysis of the functional relationships of the PIF proteins and phytochrome revealed the opposite. The physical interaction of an active form of PHYB with a PIF results in rapid degradation of the PIF, within minutes. The binding results in phosphorylation of the PIF, apparently targeting it for degradation by the ubiquitin–proteasome system (reviewed by Leivar and Quail 2011). A return to the dark, and removal of the light-induced phytochrome molecules, restores PIF levels in the nucleus. The PIF proteins appear to be primarily positive transcriptional regulators, so phytochrome signalling occurs through repression of transcription of their target genes. Ectopic expression of *PIF3*, or its downregulation through antisense technology, results in perturbation of many photoresponses, including flowering time (Ni *et al.* 1998), and the PIF proteins have been shown to regulate around 1000 genes, directly or indirectly (Leivar and Quail 2011).

The influence of phytochromes on gene expression may operate through several different pathways and at several different rates. Microarray experiments have indicated that the expression profiles of 10% of all genes change in response to far red light perceived through phytochrome A. Of the genes that responded within 1 hour of treatment, 44% encoded transcription factors (Tepperman *et al.* 2001). It is unlikely that all of these genes are regulated through a PIF-mediated pathway, and although some are likely to be activated as a result of downstream processes, it is also possible that some are regulated by as yet unknown rapid response systems.

Direct interactions with transcription factors are not the only way in which phytochrome proteins function. The active form of phytochrome A has also been shown to phosphorylate other proteins besides the PIFs, including autophosphorylation of itself. The active (Pfr) form of the molecule is over twice as effective as the Pr form at causing such phosphorylation (Fankhauser *et al.* 1999), confirming that light activation is important in this response. A particular target of phytochrome phosphorylation activity appears to be the PHYTO-CHROME KINASE SUBSTRATE 1 protein (PKS1), initially isolated through its physical interaction with phytochrome A, but also shown to interact with phytochrome B. The PKS1 protein is localized to the plasma membrane (Lariguet *et al*. 2006), and its phosphorylation by active phytochromes confirms this pathway of phytochrome-mediated signalling (Fankhauser *et al*. 1999).

5.1.2 Sensing blue light: cryptochromes

Blue light is perceived through two entirely different sets of photoreceptors, the phototropins and the cryptochromes. The phototropins are not known to influence flowering time, but play roles in cellular responses such as chloroplast movement and stomatal aperture, having a significant effect on photosynthetic efficiency. However, the cryptochromes are very important for photoperiodic induction of flowering, as well as for hypocotyl growth, and have been found in animals as well as plants. Sequence analysis suggests that the plant cryptochromes are ancient, and evolved shortly after the origin of eukaryotes. This ancestral cryptochrome was then lost in the animal lineage, but shortly after the animal–plant divergence the animals acquired a new cryptochrome, by modification of an existing photolyase (Cashmore *et al*. 1999).

There are three cryptochromes present in Arabidopsis, encoded by different genes—*CRY1*, *CRY2*, and *CRY3*. Unlike the other two, CRY3 functions in the chloroplasts and mitochondria, and has very specific roles and a distinct protein structure. CRY1 and CRY2 are both nuclear-localized, with CRY2 thought to be the most active in the regulation of flower induction.

CRY1 was isolated first, from a T-DNA tagged line of the *cry1* mutant (originally known as *hy4*) (Ahmad and Cashmore 1993). The *CRY1* gene encodes a 681-amino-acid protein with an N terminal domain very similar to that of the microbial enzyme DNA photolyase. DNA photolyases are enzymes activated by blue and UVA light to bind to and repair pyrimidine dimers caused by the exposure of DNA to UVB light (Eckardt 2003). However, the plant cryptochromes do not have photolyase activity. Instead,

CRY1 associates with chromophores, primarily flavin adenine dinucleotide (FAD) and also possibly pterin (5,10-methyltetrahydrofolate), which absorb blue light (Lin 2000). Analysis of the state of FAD on purification suggests that the oxidized form is the ground state in plants in the dark. On exposure to blue light, the oxidized FAD is reduced, and the reduced form is no longer sensitive to blue light. A conformational change associated with this reduction could result in subsequent signal transduction (Gyula *et al*. 2003). However, other authors argue that a circular electron shuttle (with no oxidation or reduction) might also explain the light sensitivity of cryptochromes (for reviews of both mechanisms, see Liu *et al*. 2011a).

CRY2 (originally known as *FHA*) was isolated using the *CRY1* cDNA as a probe (Lin *et al*. 1998), and encodes a protein that is 51% identical to CRY1 in terms of amino acid sequence. However, there are clear differences between CRY1 and CRY2 function, probably as a result of their differential regulation. Both proteins are nuclear localized, but CRY2 has been shown to be the more crucial for the floral transition. CRY2 protein (but not *CRY2* mRNA) is downregulated in blue light, as a result of light-induced phosphorylation triggering protein degradation (Lin *et al*. 1998). The *cry2* mutant is late flowering in long days, while ectopic expression of *CRY2* (using the CaMV 35S promoter) results in early flowering in short days (Lin 2000).

Less is known about the signalling cascade downstream of cryptochromes than about that of phytochromes, but it is likely that there are some similarities. For example, Shalitin *et al*. (2002) showed that protein–protein interactions are important for CRY2 function, and that CRY proteins undergo blue-light-dependent phosphorylation, rather as PHYA does in far red light. Multiple serine residues of a CRY are phosphorylated in response to blue light, as a result of both autophosporylation and the activities of undescribed protein kinases. Shalitin *et al*. (2002) concluded from these data that, in the dark, CRY2 is unphosphorylated, inactive, and stable. However, when blue light activates CRY2 it autophosphorylates, triggers responses, and activates its own degradation. The importance of phosphorylation in cryptochrome-mediated photoresponses was shown by an investigation of *cry2* mutant alleles that express full length (but apparently non-functional) CRY2 proteins. In all cases the absence of blue-light-triggered responses was correlated with the absence of blue-light-induced phosphorylation (Shalitin *et al.* 2003).

There is evidence that the C terminal domains of the cryptochromes cause constitutive responses which are downregulated by the N terminal domains unless they receive blue light illumination. Yang *et al.* (2000) showed that, in the absence of the N terminal domains, the C terminal domains of CRY1 and CRY2 are constitutively active, causing photomorphogenic responses (including early flowering) even in the dark. Presumably they are maintained in an inactive state by the N terminal domains in the dark, and only released into activity on exposure to blue light.

CRY2 (but not CRY1) is ubiquitinated in response to blue light, triggering its degradation. This degradation is dependent upon the flavin chromophore—a mutant form of CRY2 that cannot bind flavin is not degraded in blue light (Liu *et al.* 2008). It has also been shown that CRY1 and CRY2 interact with COP1 (a downstream component of both phytochrome and cryptochrome signalling; see below) in the nucleus in the dark (Wang *et al.* 2001). In some *cop1* mutant backgrounds, CRY2 degradation in blue light is perturbed (reviewed by Liu *et al.* 2011a). These data have led to the suggestion that the perception of blue light by cryptochromes triggers degradation or deactivation of COP1, which eventually allows enhanced transcription of other genes (Gyula *et al.* 2003).

The interaction of the cryptochromes and the various phytochromes ensures that day and night are correctly identified in all of the different light regimes a plant may experience during its development. At least part of this interaction is physical, with phytochrome B and cryptochrome 2 physically interacting *in vitro* and co-localizing to speckles in the nucleus on exposure to light (Mas *et al.* 2000).

5.1.3 Light-induced signals

The signal transduction pathways further downstream of the photoreceptors are currently the subject of intensive characterization, mostly because of their extreme complexity. We do know a considerable amount about one downstream pathway, operating through a protein called CONSTITUTIVE PHOTOMORPHOGENESIS 1 (COP1). A large class of mutants isolated from early genetic screens in Arabidopsis show similar or identical phenotypes, including anthocyanin accumulation, chloroplast development, epidermal cell differentiation, inhibition of hypocotyl elongation, and cotyledon opening, all in the dark (Yamamoto *et al.* 1998). These responses are normally restricted to light-grown plants, but are constitutive in this class of mutants. The proteins involved are referred to as COP/DET/FUS proteins, and they act as repressors of certain responses in the dark, but are themselves deactivated in the light. A large subset of these proteins are nuclear-localized components of a ubiquitin-protease pathway that targets degradation of downstream proteins, including a key transcription factor which activates expression of many light-regulated genes (reviewed by Quail 2002).

The best studied member of this group of genes is *COP1*, which encodes a protein with a RING-finger domain and WD40 repeats, characteristic of protein–protein interactions (Deng *et al.* 1992). Many of the other *COP/DET/FUS* genes encode subunits of one of two oligomeric complexes involved in degradation of ubiquitinated proteins. The CUL4-DDB1-COP1-SPA complex and the CUL4-DDB1-DET1-COP10 complex both combine RING proteins, WD repeat proteins, and ubiquitin ligases (Chen *et al.* 2006; Chen *et al.* 2010). They both function as E3 ubiquitin–protein ligase complexes, targeting proteins for degradation by binding them to ubiquitin (Quail 2002). One of the primary targets of the COP1-containing complex is a transcription factor called HY5 (Oyama *et al.* 1997). The *HY5* gene encodes a bZIP transcriptional activator that binds to the promoters of light-responsive genes. However, HY5 protein is only present in the nucleus in the light, being virtually absent from the cell in the dark (a 15- to 20-fold difference in protein level has been measured; Osterlund *et al.* 2000). In the dark, the COP1 complex has been shown to bind HY5, targeting it for degradation. However, in the light, COP1 becomes significantly less abundant in the nucleus, allowing HY5 to function (see Fig. 5.2). A similar interaction is thought to occur between the COP1 complex and the MYB transcription factor LAF1 and the helix-loop-helix transcription factor

(a) Events in the nucleus in the dark

(b) Events in the nucleus in the light

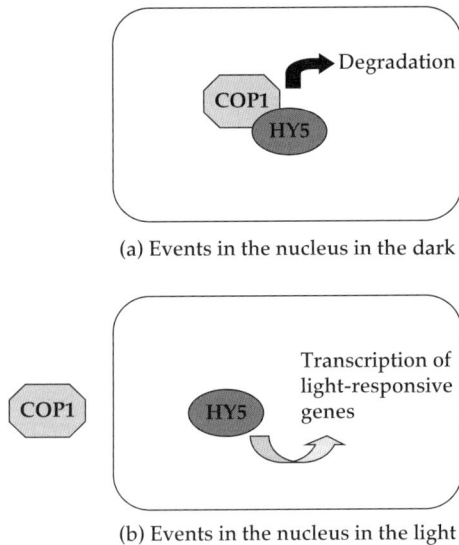

Figure 5.2 The COP1 protein mediates transcriptional responses to light. (a) In the dark, COP1 binds to the HY5 transcriptional activator in the nucleus and targets it for degradation. (b) In the light, COP1 is exported to the cytosol, allowing HY5 to activate transcription.

HFR1, both of which are also important in regulating photomorphogenesis (Seo *et al.* 2003; Duek *et al.* 2004; Yang *et al.* 2005).

The light-induced change in COP1 nuclear abundance is mediated by the activities of at least PHYA, PHYB, CRY1, and CRY2, and requires the presence of a 110-amino-acid domain in the COP1 protein called the cytoplasmic localization signal (CLS; Stacey *et al.* 1999). Analysis of protein function has shown that the CLS induces nuclear export of COP1, removing it from the nucleus in the light and preventing its activity as part of the E3 ubiquitin–protein ligase complex (Subramanian *et al.* 2004). However, some recent reports have suggested alternative ways in which the interaction of the COP1 complex with transcription factors could vary in the light and the dark, arguing that the kinetics of COP1 movement between nucleus and cytoplasm may be too slow to adequately account for the differences observed (Chen and Chory 2011).

5.2 Measuring time

As well as perceiving the quality of daylight available, photoperiodic induction of flowering depends on the ability of plants to measure time. All eukaryotes have biological clocks, although the first circadian rhythm recorded in any organism was the daily leaf movements of plants (leaves tend to be fully open in the day but fold up to some extent at night; Barak *et al.* 2000). In order for such rhythms to be truly representative of a circadian clock, and not of tissue responses to immediate light or dark conditions, they must be maintained, if imperfectly, when the organism is kept in a constant light regime. Clock-based responses are also unaffected by changing temperature, unlike most biological processes. In all organisms where the clock has been analysed it is based on a transcription/translation feedback loop, although the actual proteins involved are different in plants and animals. The basic model is one in which a clock gene autoregulates itself at a speed which takes approximately 24 hours. The clock is entrained by daily light signals to keep it running accurately. When organisms are kept in constant darkness their clock usually drifts slightly away from the 24-hour optimum. In the model, light signals occurring at the transition from dark to light every morning ensure the transcription of a clock gene (although, since the clock is basically light independent, transcription should occur even in the absence of light). This is translated into a clock protein, which activates clock-controlled genes and is responsible for all the other clock-controlled responses. However, the protein also represses the transcription of its own gene, either directly or through activation of an inhibitor of its own transcription, or through repression of an activator. When protein levels fall below a certain threshold, transcription can begin again.

It has always been thought likely that a model based on a single oscillatory gene would be too simple, and that the clock would consist of multiple interacting loops, to provide the necessary robustness. However, the basic principles of these loops are likely to follow the simple model described above.

The isolation of clock elements to test this model has involved extremely elegant genetic screens. For example, several clock genes were isolated using a screen that relied on the circadian expression patterns of certain genes. The promoter of one such gene, *CAB2*, was fused to the gene for firefly luciferase, and introduced into transgenic Arabidopsis plants.

These transgenic plants were mutagenized and then the progeny were monitored under time-lapse video cameras. Wild type plants expressed luciferase (i.e. glowed) on a 24-hour cycle, but those with mutations in clock genes glowed on arrhythmic cycles (Millar *et al.* 1995). Several criteria which a gene must meet to be considered a core part of the clock were identified. First, the gene should have an expression pattern that cycles over 24 hours; secondly, it should control its own expression in a negative feedback loop; thirdly, by clamping the gene (fixing its expression at a high or low level) it should be possible to stop the clock.

The Arabidopsis circadian clock has been shown to consist of three interlocking loops. The central oscillator is based around the activities of three main proteins (see Fig. 5.3). *CIRCADIAN CLOCK ASSOCIATED 1 (CCA1)* and *LATE ELONGATED HYPOCOTYL (LHY)* both encode transcription factors of the MYB family, although they are unusual in containing only a single copy of the MYB DNA-binding domain, whereas most plant MYB proteins contain two copies. Levels of transcript and protein of both *CCA1* and *LHY* circle on a circadian pattern, peaking at dawn. Ectopic expression of either gene from the CaMV 35S promoter downregulates the expression level of the other, and ruins the rhythmicity of the clock. The *cca1* mutant still retains rhythmic patterns of gene expression (Green and Tobin 1999), as does a loss-of-function *lhy* mutant (Mizoguchi *et al.* 2002), although in both mutants those rhythmic patterns deviate to some extent from wild type, and both flower early in short days. However, a double mutant lacking functional copies of both genes is completely arrhythmic, flowering early in short days, showing random leaf movements, and having no rhythmic patterns of gene expression (Mizoguchi *et al.* 2002). These data suggest that the LHY and CCA1 proteins are partially redundant, overlapping in function incompletely.

The *TIMING OF CAB EXPRESSION 1 (TOC1)* gene has an expression pattern that also cycles

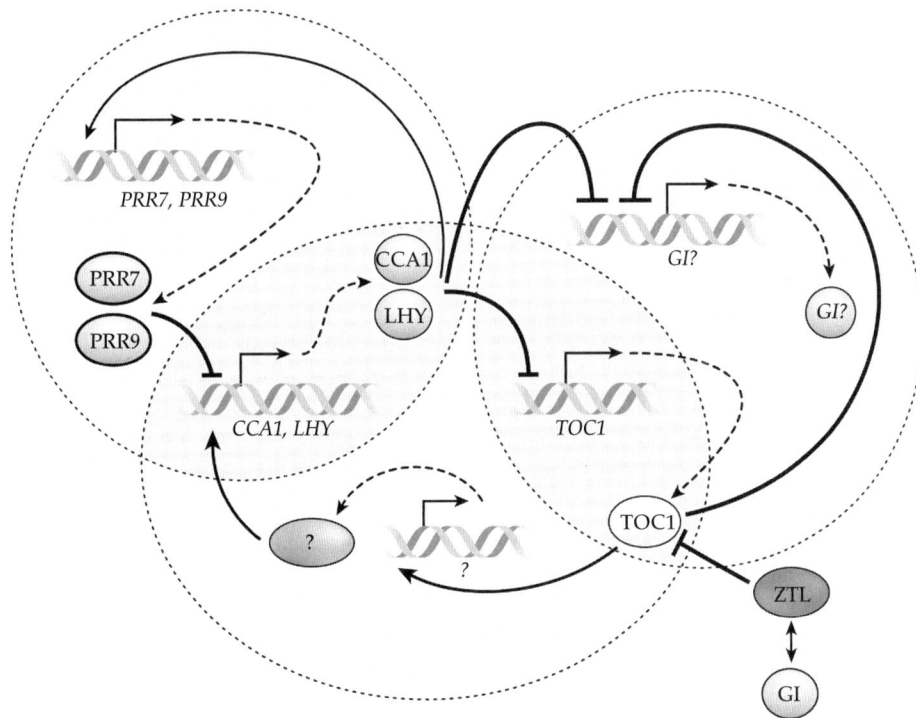

Figure 5.3 The circadian clock operates through a set of interconnecting feedback loops. TOC1, CCA1, and LHY form the central oscillator, which is made more robust by interaction with morning and evening loops. Redrawn from Harmer (2009).

over 24 hours, but with a peak at dusk. Ectopic expression of *TOC1* causes arrhythmicity, while *toc1* mutants have shorter circadian patterns (typically 21 hours in constant light) and flower late in long days (Strayer *et al.* 2000). Expression levels of both *CCA1* and *LHY* are reduced in a *toc1* mutant background, compared with wild type. *TOC1* was isolated by map-based cloning, and encodes a protein containing an N terminal motif very similar to the receiver domain of response regulators from two-component signal transduction systems (Strayer *et al.* 2000). However, the conserved aspartates that normally undergo phosphorylation in these systems are replaced by other amino acids in TOC1, so perhaps it has a different function. The TOC1 protein appears to be unable to undergo phosphorylation *in vitro*, leaving the function of the receiver-like domain unclear, and leading to the protein's current classification as a 'pseudo-response regulator' (Makino *et al.* 2000). The C terminus of TOC1 contains a domain very similar to that of the transcription factor CONSTANS (see below). Mas *et al.* (2003) showed that the gradual reduction in TOC1 protein levels after the dusk peak was caused not only by transcriptional repression by LHY and CCA1, but also by post-translational modification. The *ZEITLUPE* (*ZTL*) gene encodes a protein that contains a number of domains, including an F-box motif, and that acts as part of an E3 ubiquitin ligase complex (Somers *et al.* 2000). The *ztl* mutant has constitutive levels of TOC1 protein (although the mRNA expression pattern continues to cycle in a circadian manner). This excess of TOC1 protein is caused by the failure of TOC1 degradation in the dark, which requires a functional ZTL protein (Mas *et al.* 2003).

Putting these data together, a model has emerged in which TOC1 upregulates expression of *LHY* and *CCA1* through an unknown transcriptional regulator intermediary. The LHY and CCA1 proteins have been shown to bind to the promoter of the *TOC1* gene, where they repress its transcription. High constant levels of LHY or CCA1 (in transgenic plants) result in low constant levels of TOC1 protein. The *TOC1* promoter contains a 200-base-pair region which is essential for rhythmicity, and it is to this that LHY and CCA1 have been shown to bind (Alabadi *et al.* 2001). The entire feedback loop, from TOC1 upregulation of *LHY* and *CCA1*, through

LHY and CCA1 downregulation of *TOC1*, to the point where lack of TOC1 (assisted by ZTL-induced degradation in the dark) has resulted in sufficiently little LHY and CCA1 to allow *TOC1* transcription to begin again, takes approximately 24 hours to complete (Alabadi *et al.* 2001).

The LHY/CCA1/TOC1 central oscillator provides the basic rhythmicity to the system, but is not sufficient either to maintain a robust clock or to explain various experimental observations of plant behaviour under different light regimes. For these reasons the clock model has been expanded to include a morning loop and an evening loop, each interacting with the central oscillator (all reviewed by Harmer 2009).

Correct rhythmicity of the clock is maintained by daily entrainment. Without exposure to a light burst each morning, the clock tends to run to slightly more than 24 hours. Somers *et al.* (1998) showed that phytochrome A and phytochrome B are necessary to ensure red light signals entrain the clock, that cryptochrome 1 and phytochrome A are necessary for blue light entrainment, but that cryptochrome 2 does not seem to be involved in circadian clock input. CCA1 and LHY act as part of this morning loop. They activate transcription of several *TOC*-like genes encoding additional pseudo-response regulators that input light signals to entrain the clock. However, these PRR proteins also in turn inhibit *CCA1* and *LHY* transcription, providing a negative feedback loop in the morning (Locke *et al.* 2006).

The evening loop also provides robustness to the system, although its genetic components are not yet well understood. It is thought that TOC1, LHY, and CCA1 all negatively regulate an unknown factor (perhaps partly provided by GIGANTEA; see below) that is itself a positive regulator of *TOC1* expression. Again, this loop provides negative feedback control to regulate the amount of TOC1 protein produced in the evening (reviewed by Harmer 2009). The three interacting loops provide a circadian clock that is both sensitive and robust.

5.3 Integrating light and clock signals

An ability to perceive day length and an ability to follow a 24-hour pattern are not sufficient to explain the very precise photoperiodic induction requirements of many plant species. The two features must be integrated in such a way that the plant only flowers when the appropriate critical day length is achieved. Understanding the integration of light and clock signals has involved analysis of mutants in which photoperiodic control of flowering is perturbed, but in which other circadian or light responses are not affected—that is, those not perturbed either in the clock itself or in the photoreceptors, both of which will of course alter photoperiodic control of flower induction.

It is now believed that signals from the circadian clock and the phytochromes are integrated by the *GIGANTEA (GI)* and *CONSTANS (CO)* genes. The *co* mutant was originally identified as late flowering in long days but flowering at the normal time in short days (Koornneef *et al.* 1991). *CO* was cloned using a map-based approach (Putterill *et al.* 1995). The gene encodes a protein with two zinc finger domains, commonly found in transcription factors, and also regions indicative of protein–protein interactions. This protein appears to function as a transcriptional activator. It is expressed in both leaves and stems, at very low abundance, but not in the shoot apical meristem. *CO* expression is regulated by the endogenous circadian clock on a 24-hour cycle that peaks around 12 hours after dawn. Thus in short days *CO* expression does not overlap with the period of daylight, but in longer days there is an overlap of *CO* expression and daylight in the evening (Suarez-Lopez *et al.* 2001). The CO protein directly causes floral induction by activating expression of *FT*, one of the flowering-time integrators (see Chapter 7) and a key necessary component of the floral transition. However, CO is only able to activate *FT* expression in daylight. This means that *CO* gene *expression* is regulated by the circadian clock, but CO protein *function* is controlled by the presence of daylight. Therefore CO protein can only activate *FT* expression if it is present during a light evening—that is, in long days. In short days, *CO* is never expressed at the time which would allow it to activate *FT*. In support of this model, ectopic expression of *CO* from the CaMV 35S promoter does cause increased *FT* expression, but only during the hours of daylight. Although *CO* transcript is equally abundant as a result of this constitutive

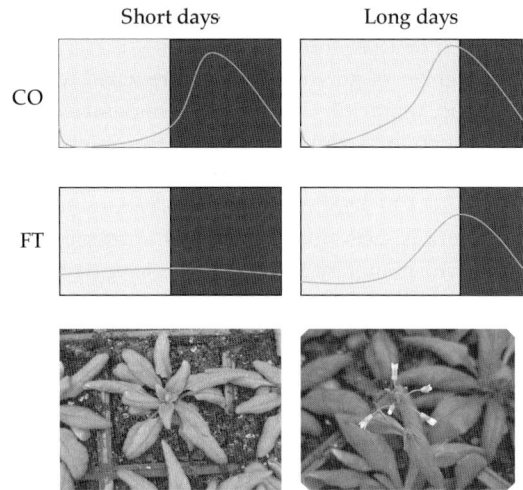

Figure 5.4 The *CONSTANS* gene (above) is expressed in a circadian pattern. It is only able to activate *FT* expression (below) when its expression coincides with daylight—that is, in long days (right-hand panel). Redrawn from Yanovsky and Kay (2003). Reproduced with permission from Nature Publishing.

promoter in both light and dark, it is only during the light period that *FT* expression is upregulated, causing the floral transition to occur (Valverde *et al.* 2004). This elegant model of integration has finally resolved the question of how photoperiod induces flowering, some 85 years after the phenomenon was formally described (see Fig. 5.4; Yanovsky and Kay 2003).

The role of light in allowing CO to activate *FT* expression was investigated by Valverde *et al.* (2004). First, they showed that blue light treatment increased *FT* transcript level in an ectopically expressing *CO* background, far red light had some effect (but less than blue light), and red light did not enhance *FT* transcription. They also showed that the *cry1*, *cry2*, and *phyA* mutants had less *FT* transcript than wild type plants, confirming that both cryptochromes and phytochrome A are necessary to allow far red and blue light to initiate CO activation of *FT* expression. The key role that light plays in CO activity is in stabilizing the protein. The CO protein produced in the evening is stable, but CO protein produced during the hours of darkness is rapidly degraded. Application of proteasome inhibitors was shown to stabilize CO protein during dark periods, suggesting that the protein is normally degraded by the proteasome unless light signals prevent this degradation (Valverde *et al.* 2004).

A number of genes have been predicted to play roles upstream of *CONSTANS*, moderating *CO* expression and thus regulating the interaction between clock-induced *CO* expression and light-controlled CO activity. The *GIGANTEA* (*GI*) gene, for example, acts upstream of *CO* (Putterill *et al.* 2004). The *gi* mutant flowers late in long days and has reduced levels of *CO* expression (Simpson 2003). *GI* was isolated by two groups, one using a T-DNA tagging approach and the other using a map-based approach. The gene encodes a large novel protein, which was initially predicted to contain multiple membrane-spanning domains (Fowler *et al.* 1999; Park *et al.* 1999). However, the GI protein was subsequently shown to be nuclear localized when fused to green fluorescent protein in both transient assays (Huq *et al.* 2000) and stable expression (Mizoguchi *et al.* 2005). The expression pattern of *GI* is circadian-regulated, peaking 8–10 hours after dawn (Fowler *et al.* 1999). This expression pattern turns out to be crucial to the function of GI in mediating the interaction between CO and light. The GI protein forms a complex with an F-box protein to degrade CDF1 (CYCLING DOF FACTOR 1), a key repressor of CO. However, since the interaction between the F-box protein and GI is activated by light (particularly blue light), and *GI* expression is controlled by the circadian clock, the complex only forms when it is light into the late afternoon/early evening—that is, in long days. This complex then acts to degrade CDF1, allowing CO activity (Sawa *et al.* 2007).

The observation that circadian rhythmicity of gene expression patterns is altered in plants either ectopically expressing *GI* or containing a *gi* mutant allele suggests that GI also feeds back into the central circadian oscillator (Mizoguchi *et al.* 2005). It is likely that other genes will also be shown to act as intermediates between *CO* expression and CO function in a variety of similar ways.

CHAPTER 6

The vernalization pathway of floral induction and the role of gibberellin

Day length is not the only environmental signal used by plants to optimize the timing of flower production. Temperature is also a factor, and particularly vernalization—that is, exposure to a prolonged period of cold. This is a sensible adaptation of plants in temperate climates to prevent the risk of autumn germination leading to flowering in winter, a time when pollinators are scarce and freezing temperatures may inhibit seed production. At the same time, the plant growth regulator gibberellin has been shown to be a powerful promoter of flowering in many species. Since gibberellins are produced by differentiating plastids, linking flowering to high gibberellin levels allows plants to coordinate reproduction with the potential for high levels of photosynthetic activity. For some time there was speculation that the vernalization-induced flower induction pathway operated through gibberellins, but we now know that, in Arabidopsis at least, vernalization and gibberellin represent two independent pathways which can both induce flowering. Chandler *et al.* (2000) showed that vernalization was not affected in mutant backgrounds which failed to make or perceive gibberellins, as well as those defective in abscisic acid and phytochrome B signalling. In this chapter we consider the two pathways separately, before looking at the evidence that they overlap. In Chapter 7 we shall integrate all of the flowering-time pathways discussed in Chapters 4, 5, and 6 to produce a holistic model of the Arabidopsis floral transition.

6.1 The vernalization promotion pathway

Photoperiodic induction is the most commonly cited and best studied example of the environmental regulation of flowering. However, flowering is likely to be accelerated by any condition that reliably indicates the passage of winter and the onset of more favourable conditions for reproduction. Although day length is one such reliable indicator, the end of a long period of cold weather is another good sign that winter has passed. Many plants flower in response to a period of between 1 and 3 months of exposure to temperatures between 1 and 10°C. The precise cold requirements vary from species to species and from plant accession to plant accession, and can be extremely specific. This strategy is adopted by many wild species, and has also been bred into many of our crop varieties, to produce lines which will overwinter in the vegetative state and flower in the spring or summer (see Fig. 6.1; Simpson and Dean 2002). Such cold-induced regulation of flowering is known as vernalization (from the Latin *vernus*, meaning 'of the spring'; Sung and Amasino 2004b).

Vernalization involves two separate processes. First, the plant must perceive the prolonged period of cold temperature. Then it must 'remember' that perception in order to induce flowering later in the spring or summer, when the cold has passed. Very few plants flower immediately on release from the cold period, so it is clear that the 'memory' of winter is retained over many weeks and months. Analysis of vernalization in Arabidopsis has been broken down into analysis of cold perception and analysis of the maintenance of the vernalized state. Although the mechanism by which a plant 'remembered' cold seemed initially the most challenging problem to solve, it has been the perception of vernalization that has proved more difficult to analyse using conventional genetic approaches.

Winter wheat

autumn	winter	spring	summer

Spring wheat

autumn	winter	spring	summer

Figure 6.1 Variation in vernalization sensitivity is used in agriculture to increase the number of harvests possible in a year. Vernalization-sensitive winter wheat (top) is sown in autumn, germinates and remains vegetative in the winter, and then flowers and sets seed in spring. Vernalization-insensitive spring wheat (bottom) is sown in spring and germinates, grows, and produces a crop in the summer, without the requirement for a period of cold.

It has long been known that prolonged periods of cold are perceived by cells in the shoot apical meristem and root apical meristem of Arabidopsis. This perception can be passed on to new daughter cells through mitotic divisions, long after the cold period has passed, but cannot be passed on through meiosis. This is a logical system which ensures that the next generation of plants starts with a fresh slate and no 'memory' of the winter their parents experienced. It is also a hallmark of many epigenetic changes, such as DNA methylation, that they are conserved through mitosis but not meiosis. Extensive experiments with *in vitro* regeneration of various vernalized tissues of *Lunaria biennis* (honesty, Brassicales) revealed that only tissues containing actively dividing cells regenerated into plants that retained a memory of vernalization (Wellensiek 1962). It seems necessary for cells to be dividing, or at least undergoing DNA replication, for vernalization to be successful (Sung and Amasino 2004b). Since the apical meristems are the source of the most active cell division within the plant, it is hardly surprising that vernalization seems to operate through these tissues.

Initial analysis of cold-induced flowering in Arabidopsis established that a requirement for vernalization segregated as a monogenic trait, which mapped to the *FRIGIDA* (*FRI*) locus (Simpson and Dean 2002). *FRI* was introduced in Chapter 4. The main function of the protein is to enhance expression of *FLC*, the main repressor of flowering in Arabidopsis. FLC represses flowering by downregulating the flowering-time integrator genes, such as *FT*. Having established that many Arabidopsis plants with a summer annual habit (i.e. those that do not need a cold period to induce flowering) contain mutations in *FRI* or *FLC*, attention focused on the mechanism of vernalization interaction with these genes (Simpson and Dean 2002). The first intron of *FLC* is necessary for maintenance of *FLC* repression after cold treatment, but other regions of the gene are needed for the initial downregulation, confirming the hypothesis that cold repression and maintenance of that repression are two different processes (Sheldon *et al.* 2002). Vernalization results in a reduction in *FLC* transcript levels in the plant, and this reduced expression is maintained after plants are transferred to warmer environments. Increasing periods of cold exposure result in increasing downregulation of *FLC*, a quantitative effect which appears to be related to the extent of the initial downregulation of *FLC*, rather than the security of maintenance of the repressed state (Sheldon *et al.* 2006). The maintenance of the *FLC* repressed state through mitosis, and the temporal separation between the timing of cold treatment and the onset of flowering, led to the hypothesis that vernalization operated through epigenetic control of *FLC* expression.

In support of the suggestion of an epigenetic method of response to cold treatment, prolonged growth in cold temperatures was shown to reduce methylation level across the genome. It was further shown that genome-wide demethylation, whether induced by chemical means, through the use of transgenes, or in mutant plants, promoted earlier flowering in Arabidopsis (Finnegan *et al.* 1998; Sheldon *et al.* 2000a). These data seemed to contradict the idea that vernalization induced an epigenetic repression of *FLC* transcription, until Sheldon *et al.* (1999) showed that plants expressing an antisense *METHYLTRANSFERASE 1* gene, and with only 15% of the wild type level of genomic methylation, flowered early and had reduced *FLC* transcript levels. Since demethylation usually allows transcription, rather than represses it, the authors concluded that methylation inhibits a repressor of *FLC* expression, and that such methylation is reduced by vernalization. Finnegan *et al.* (2005) conducted a detailed analysis of the effects of demethylation and vernalization on the *FLC* locus. They showed that both processes resulted in reduced acetylation of histones H3 and H4 in the promoter of *FLC* and in neighbouring genes. However, DNA demethylation also resulted in repression of other MADS box genes with roles in flowering, lending support to the idea that methylation inhibits a repressor of MADS box genes in general, and may therefore affect flower induction in a rather indirect way. More recent work has demonstrated that vernalization results in a number of repressive marks being placed on the *FLC* chromatin, particularly methylation of histone 3 lysine 9 (H3K9) and histone 3 lysine 27 (H3K27), and that these modifications are only maintained in mitotically active cells (Finnegan and Dennis 2007).

Vernalization is conceptually and mechanistically different from cold acclimation, where short periods of cold temperatures induce responses that prepare plants for freezing temperatures. It is essential that these two processes are discrete, to ensure that plants are not tricked into flowering by short cold spells in autumn. However, both cold acclimation and vernalization must occur at temperatures above freezing, as both require metabolic activity which would be severely inhibited by subzero temperatures (Sung and Amasino 2004b). Given the rapid temperature changes experienced by plants in many habitats, cold acclimation can be induced by only a short period of chilling, with one day sufficient for Arabidopsis. In contrast, vernalization of Arabidopsis is optimal after 30–40 days of continuous cold, a period of time sufficient to overcome the FRI-induced enhancement of *FLC* expression (Lee and Amasino 1995). Cold acclimation may operate through changes in membrane fluidity at different temperatures, altered calcium ion fluxes, and protein phosphorylation, although none of these processes are sufficiently stable to induce the memory of winter that is part of vernalization (Sung and Amasino 2004b). Further evidence for the mechanistic differences between cold acclimation and vernalization

comes from the perfectly normal expression, and response to cold, of acclimation-induced genes in mutants compromised in vernalization (*vrn1* and *vrn2*; see below). Liu *et al.* (2002) also showed that ectopic expression of *CBF1*, the first transcription factor gene to be upregulated upon cold acclimation, has no effect on *FLC* transcript levels or on flowering time. It is possible that vernalization and cold acclimation do share some elements of a common pathway, even if the majority of their actions seem to be separate, although the data are currently ambiguous. Some reports suggest that the *HIGH EXPRESSION OF OSMOTICALLY RESPONSIVE GENES 1* (*HOS1*) gene is involved in both processes, with *hos1* mutants showing reduced cold acclimation and thus reduced freezing tolerance, reduced *CBF* expression, early flowering, and reduced *FLC* transcript levels (Lee *et al.* 2001). The *HOS1* gene encodes a RING-finger protein that acts as an E3 ligase required for ubiquitin-mediated protein degradation (Dong *et al.* 2006). This protein is translocated from the cytoplasm to the nucleus in response to exposure to cold (Lee *et al.* 2001). However, more recent analyses suggest that the link between *HOS1* and vernalization is an artefact of the role that *HOS1* plays in general stress responses. Bond *et al.* (2011) demonstrated that *VIN3*, the first induced gene in the vernalization response (see below), behaves normally in a *hos1* mutant background. These authors also found little difference in flowering time and *FLC* transcript levels between wild type and *hos1* mutants, leaving the link between cold acclimation and vernalization in doubt.

How plants remember exposure to a period of cold was solved before we understood how they recognize cold, primarily due to analysis of two mutants, the *vernalization* (*vrn*) mutants, in which flowering is not induced in response to a prolonged period of cold. The *vrn1* mutant experiences a reduction in *FLC* expression during its exposure to cold, but on return to warmer conditions the *FLC* transcript level recovers and flowering is not induced (Levy *et al.* 2002). This suggests that the mutant can perceive cold but then fails to retain a memory of it. *VRN1* was isolated by map-based cloning and encodes a protein containing two B3 DNA-binding domains and a nuclear localization signal—that is, a protein with several hallmarks of a transcription factor. GFP fusion experiments confirmed that VRN1 is nuclear localized, supporting the idea that it is involved in transcriptional regulation. Levy *et al.* (2002) showed that the expression levels of *VRN1* mRNA did not change on exposure to cold temperatures, confirming that the gene plays a role in maintenance of the vernalized state, rather than induction of it. Ectopic expression of *VRN1* from the CaMV 35S promoter induced early flowering without vernalization, and even earlier flowering with a cold treatment. Finally, Levy *et al.* (2002) showed that VRN1 protein binds DNA, including *FLC*, *in vitro*, and proposed that VRN1 binds to the *FLC* locus *in vivo*, targeting complexes (possibly including VRN2; see below) to it which cause methylation and subsequent inactivation of the gene.

The *vrn2* mutant, which is vernalization-insensitive, also experiences cold-dependent reduction in *FLC* transcript levels, but return of *FLC* expression on increasing temperature (Gendall *et al.* 2001). *VRN2* was isolated by map-based cloning and encodes a protein with a nuclear localization signal and several features indicative of a member of the polycomb group of proteins. Polycomb proteins regulate gene transcription by modifying chromatin structure, and are active in both plant and animal development (Gendall *et al.* 2001; Chanvivattana *et al.* 2004). Just like *VRN1*, *VRN2* expression pattern is not affected by vernalization, and the protein is nuclear localized at all times. VRN2 has been shown to interact physically with other polycomb group proteins, and is therefore thought to function as part of a polycomb complex to maintain the chromatin of the *FLC* gene (to which it may be targeted by VRN1) in a state incompatible with transcription through histone modifications (Chanvivattana *et al.* 2004).

The observation that *VRN1* and *VRN2* are expressed even in the absence of vernalization indicated that they could not be responsible for the cold-induced repression of *FLC* expression, but only for its maintenance following the action of some other protein. In 2004, Sung and Amasino reported the identification of VERNALIZATION INSENSITIVE 3 (VIN3) as the solution to this mystery. In the *vin3* mutant, repression of *FLC* expression does not occur in response to vernalization, and there is thus no cold effect for the plant to remember. *VIN3* expression is induced quantitatively on exposure to a long period of cold temperature, and as *VIN3* transcript levels increase in the plant, *FLC* transcript levels decrease. *VIN3* expression ceases after the plant is returned to warmer temperatures, becoming undetectable within 3 days of warmth, at which point the VRN2 complex takes over the maintenance of the induced state. *VIN3* was isolated by map-based cloning and encodes a plant homeodomain-finger-containing protein (PHD protein), a structure usually involved in chromatin remodelling or DNA binding (Sung and Amasino 2004a). The VIN3 protein has been shown to bind to the 5′ region of the *FLC* locus as part of the VRN2 complex (Wood *et al.* 2006). It appears that VIN3 is necessary to target the VRN2 complex to *FLC*, but that once targeted the complex remains in place even after *VIN3* transcription stops due to a return to warmth. During vernalization, acetylation levels of *FLC* chromatin decrease in response to the binding of the VRN2 complex, and this deacetylation is followed by methylation. In the *vin3* mutant neither deacetylation nor methylation of *FLC* occurs, whereas the *vrn1* and *vrn2* mutant lines experience deacetylation of *FLC* but reduced methylation (Sung and Amasino 2004a). In the *vrn1* mutant, methylation of lysine 9 on histone H3 is lost, while in the *vrn2* background no histone H3 methylation occurs (Bastow *et al.* 2004).

Like VRN1 and VRN2, LIKE HETEROCHROMATIN PROTEIN 1 (LHP1) is also necessary to maintain repression of *FLC* after exposure to cold, although it may in addition have wider roles in other aspects of plant development. LHP1 is a chromodomain protein, very similar to animal and yeast proteins that maintain chromatin in particular states. LHP1 protein is enriched in the chromatin of *FLC* after exposure to cold, binding to methylated histone, and remains in position after the plant is returned to warm temperatures. The involvement of LHP1 as well as polycomb group proteins in the flower induction process confirms that the epigenetic control of *FLC* operates through mechanisms very similar to those present in animals and fungi (Mylne *et al.* 2006; Sung *et al.* 2006).

These experiments have allowed us to build up a complete picture of the vernalization induction of flowering in Arabidopsis. During a prolonged

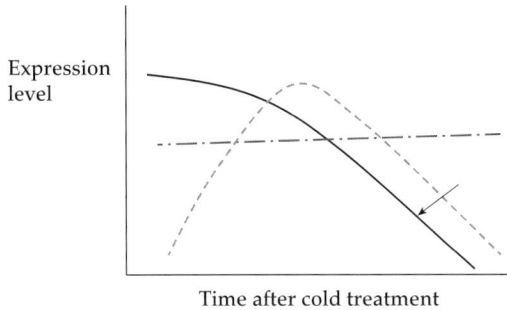

Figure 6.2 Graph showing expression levels of various genes following vernalization. *FLC* transcript levels (solid line) start high, repressing flowering. *VIN3* (dashed line) is induced by cold treatment, and represses *FLC* expression, before expression levels decline again. *VRN1* and *VRN2* (dots and dashes) expression is unaffected by vernalization. However, the VRN1 and VRN2 proteins (arrow) act to maintain the repressed state of the *FLC* locus.

cold treatment, *VIN3* expression is initiated, and the VIN3 protein causes deacetylation of *FLC* chromatin. VIN3 protein is associated with VRN2 in a multiprotein complex which may target *FLC* chromatin for repression through histone methylation (Wood *et al.* 2006). VRN1 and VRN2, along with LHP1, are responsible for methylation of the modified *FLC* gene, preventing its transcription, and this methylation is stable on return to warmer temperatures, providing the 'memory' of winter (see gene expression patterns in Fig. 6.2). In the absence of *FLC* expression the flowering-time integrators can be expressed and flowering can be induced.

It is worth noting that a second, poorly characterized, vernalization induction pathway operates in Arabidopsis. Loss of *FLC* activity in a null mutant line does not totally remove the vernalization response. Polycomb proteins have also been shown to repress the activity of different MADS box genes—*AGL19*, *AGL24*, *MAF1*, and *MAF2*. AGL19 induces expression of floral meristem identity genes (see Chapter 9), bypassing both FLC and the flowering-time integrators that it represses. Although this pathway has been little studied, recent data suggest that VIN3 may play a role in maintaining the repressed state of *AGL19* (Schoenrock *et al.* 2006; reviewed by Alexandre and Hennig 2008).

As was mentioned in Chapter 3, the natural variation between different ecotypes of Arabidopsis living in different habitats has recently been the focus of much attention. Studies of variation in flowering time have very frequently isolated *FRI* and *FLC* as key loci determining ecotypic differences, confirming the importance of the vernalization pathway in fitting plants to their local environments (e.g. Strange *et al.* 2011).

6.2 The gibberellin promotion pathway

Almost all of the classic plant growth regulators have been linked with the floral transition. There is evidence from a range of species that, under appropriate conditions, gibberellin, auxin, cytokinin, and brassinosteroids can each influence flowering time. This is hardly surprising, as levels of each of the major plant growth regulators (PGRs) are affected by environmental conditions, including nutrient and water availability and light quality. General changes to plant development as a result of these environmental stimuli may include a direct effect on flower induction, or an indirect one operating through changes in source/sink relationships of various tissues. However, within Arabidopsis at least there is evidence to suggest that gibberellins directly promote flower induction, and that this promotion operates through a pathway which is genetically distinct from the autonomous, photoperiodic, and vernalization induction pathways discussed previously.

Gibberellins are tetracyclic diterpenoids, and over 130 different forms of gibberellin have been described from plants, fungi, and bacteria (Gomi and Matsuoka 2003). They are synthesized in differentiating chloroplasts, and the biosynthetic pathway leading to their production has been characterized by biochemical and molecular genetic methods. The primary signal that high gibberellin levels send to the plant is one of good conditions. In high-quality light, and with sufficient water and nutrients available, young leaves develop rapidly and more chloroplasts differentiate, leading to increased gibberellin levels. Gibberellins are perceived by soluble receptors within the cell (AtGID1a, AtGID1b, and AtGID1c in Arabidopsis). When bound by gibberellins, the gibberellin receptors also interact with DELLA proteins (of which five are encoded by the Arabidopsis genome) and target them for degradation by the 26S proteasome (Nakajima *et al.* 2006). DELLA proteins are named for a group of amino acids which make up a conserved motif within their sequence. These proteins are nuclear-localized negative regulators of transcriptional responses—their degradation in the presence of gibberellin allows gibberellin-induced genes to be expressed.

The first piece of evidence in support of the presence of a gibberellin induction pathway is the observation that treating Arabidopsis plants with exogenous gibberellins promotes earlier flowering. In fact, application of gibberellin seems to promote the floral transition in all rosette plants investigated, but has a different effect in non-rosette plants. Plants with elongated stems at the vegetative stage seem to experience further stem elongation when treated with gibberellins, but do not flower early (Corbesier and Coupland 2005). The molecular mechanism of action of exogenous gibberellin is as yet unclear, but there is evidence that gibberellin application may induce expression of *LEAFY* and *SOC1* (both flowering-time integrators, while *LFY* is also classed as a floral meristem identity gene; see Chapters 7 and 9). Blazquez *et al.* (1998) showed that growth on 1% sucrose enhanced expression of a GUS reporter gene fused to the promoter of *LEAFY*. Although treatment with gibberellin alone had no obvious effect on reporter gene activity, the combination of gibberellin and 1% sucrose resulted in much stronger (more than double) reporter gene expression than was caused by sucrose alone. Eriksson *et al.* (2006) further showed that GA_4 is the active gibberellin which induces upregulation of *LEAFY* and therefore flowering, and that its concentration at the shoot apical meristem increases shortly before the floral transition occurs. More recent reports also suggest that gibberellins can enhance expression of *FT,* the gene encoding the mobile promoter of flowering that travels to the meristem after long days are perceived in the leaves (see Chapter 5; Hisamatsu and King 2008), and the *SPL* genes activated in the meristem at the floral transition (Jung *et al.* 2012).

Gibberellin promotes flowering in short days through a GAMYB-dependent activation of *LEAFY.* The gibberellin-responsive MYB transcription factor, AtMYB33, is necessary to allow gibberellin to activate *LEAFY* expression, and is presumably responsible for direct transcription of *LEAFY* in response to gibberellin signals. Appropriate *LFY* expression is dependent upon a MYB recognition motif within the *LFY* promoter, and this motif is bound by AtMYB33 (Blazquez *et al.* 1998; Blazquez and Weigel 2000; Gocal *et al.* 2001). Since DELLA proteins act as the molecular switches through which gibberellin signalling occurs, they are also important in the flowering response. Achard *et al.* (2006) showed that DELLA proteins delay the onset of flower production under high salt conditions, supporting a role for gibberellin in the general induction of flowering through the degradation of these repressors.

Further evidence for the role of gibberellins in flower induction, and particularly of their independence from the other floral induction pathways, comes from mutant studies. The *spindly* (*spy*) mutant has constitutively active gibberellin signalling, and flowers early (Jacobsen and Olszewski 1993). The mutant is also resistant to inhibitors of gibberellin synthesis, and the gene encodes an enzyme involved in post-translational protein modification (an O-linked N-acetylglucosamine transferase), which acts as a negative regulator of gibberellin signalling (Jacobsen *et al.* 1996; Thornton *et al.* 1999). Loss of this negative regulation of gibberellin signalling results in an early transition to flowering. Further evidence comes from the work of Chandler *et al.* (2000), who analysed double mutant plants containing lesions in the autonomous pathway promoter *FCA* and lesions in gibberellin synthesis or perception. Plants that are *fca* mutants have a strong vernalization requirement, as the autonomous promotion pathway is perturbed and *FLC* expression remains high. Double mutants between *fca* and *ga1*, and between *fca* and *gai*, were produced and analysed. The *GA1* gene encodes ent kaurane synthase, the enzyme required for the first committed step of gibberellin synthesis, and so *ga1* mutants produce no gibberellin. The *ga1* mutant is late flowering in long days, and rarely flowers at all in short days. However, the double mutants were still responsive to vernalization and photoperiod, confirming that gibberellin induces flowering through a separate pathway. The *GAI* gene is required for the perception of gibberellin, and *gai* mutants can grow on toxic levels of gibberellin. Again, *gai/fca* double mutants were responsive to vernalization and photoperiod, indicating that gibberellin promotes flowering through an independent pathway. Although much remains to be discovered about the way in which gibberellin promotes the floral transition in Arabidopsis, these genetic studies make it clear that the gibberellin induction pathway must be thought of as distinct from the autonomous, photoperiodic, and vernalization pathways.

Whatever the details of gibberellin signalling, it is clear that gibberellin does not promote flowering through repression of *FLC*, unlike the autonomous and vernalization induction pathways (see Figure 6.3). *FLC* transcript levels are unchanged in response to gibberellin application and subsequent early flowering (Sheldon *et al.* 2000b), indicating that gibberellin operates downstream of FLC. In conjunction with the observed role of gibberellin

Vernalization

FLC

GA ⟹ Flowering integrators

GA ⟹ Floral meristem identity genes

Figure 6.3 Despite early hypotheses that gibberellin and the vernalization promotion pathway were parts of the same process, it is now clear that they operate at different points in the pathway that represses flowering. Vernalization represses the activity of the floral repressor FLC, whereas gibberellin bypasses FLC and directly activates expression of the flowering integrators and floral meristem identity genes.

in activating *LEAFY* expression, this has led to the proposition that gibberellin is able to bypass the autonomous inhibition pathway and induce flowering by direct activation of flowering-time integrators or even floral meristem identity genes. Such a direct response might be seen as a sensible response of a short-lived plant to the information that high levels of gibberellin provide—namely that the young leaves are vigorous and growing, and that there is plenty of light for photosynthesis. Since good conditions are by no means guaranteed in the life of a short-lived annual, it may be beneficial to be able to respond rapidly to such conditions when they arise.

6.3 Does gibberellin act in the vernalization promotion pathway as well as independently?

The proposal that vernalization may promote flowering through gibberellin is a long-standing one in the literature, and should be considered distinct from analysis of the gibberellin promotion pathway. Evidence in favour of a role for gibberellin in the vernalization pathway was provided by Hazebroek *et al.* (1993), who observed an increase in the activity of an enzyme of gibberellin synthesis following vernalization. Using shoot apical meristems of *Thlaspi arvense* (pennycress, Brassicales), they measured an increase in the activity of ent-kaurenoic acid hydroxylase after a prolonged cold treatment. This increase resulted in increased synthesis of a gibberellin, GA_9, albeit a less biologically active one than many. A model was proposed in which vernalization resulted in methylation changes to genes encoding enzymes of gibberellin biosynthesis (including ent-kaurenoic acid hydroxylase), allowing enhanced gibberellin production and thus early flowering (Sheldon *et al.* 2000a). Such a model is consistent with what we know of how vernalization works, independently of gibberellin. However, there is little other evidence to support a role for gibberellin in the vernalization promotion pathway, and the genetic data in Arabidopsis indicate quite clearly that the pathways usually function independently of one another.

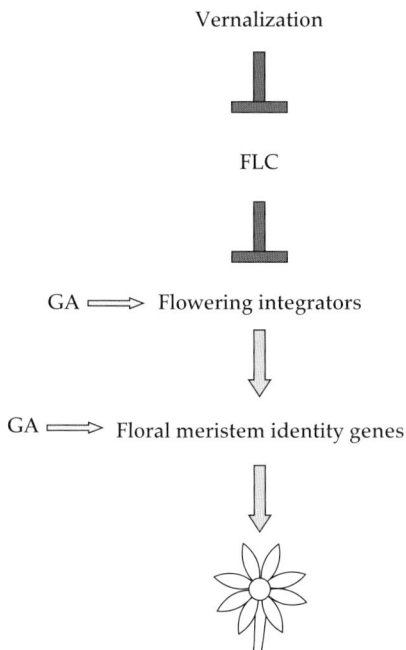

CHAPTER 7

Integrating the *Arabidopsis thaliana* flower induction pathways

In the previous three chapters we have discussed the various pathways that regulate the floral transition in Arabidopsis. Chapter 4 considered the floral inhibition pathway, which prevents the plant from flowering until it has grown vegetatively and acquired sufficient photosynthate to fuel seed production. It also considered the autonomous induction pathway which ensures that the plant will always flower eventually, even if it does not experience inductive environmental conditions. Chapter 5 discussed the photoperiodic promotion pathway, which encourages the floral transition in long days, indicative of summer and good growth conditions. Chapter 6 considered the influence of vernalization on flowering, a pathway which operates in ecotypes inhabiting more northerly latitudes, where floral induction occurs after winter has passed. It also discussed the ability of gibberellin, the plant growth regulator produced by young leaves in response to good conditions, to bring forward the floral transition. An average Arabidopsis plant will experience inhibitory and promotive signals from each of these pathways to a greater or lesser extent at any one point in its life. When the promotive signals (from any combination of the induction pathways) become stronger than the inhibitory signals the transition to flowering will occur. In this chapter we shall investigate how the pathways are brought together to result in a single response, through the activities of the flowering-time integrator genes.

7.1 Integrating the flowering-time pathways

The various flowering-time pathways can be thought of as converging on a set of genes, known as the flowering-time integrators. It is tempting to think of the pathways converging on the activity of *FLC*, the floral repressor, but in fact FLC is situated immediately upstream of the point of convergence. As Fig. 7.1 shows, although several of the pathways do moderate *FLC* activity, others bypass it, and so the true point of convergence cannot be at FLC itself. Instead, integration of the different pathways is achieved by the genes that are downregulated by FLC. The identification of these genes, a set of loci each of whose function or expression is regulated by more than one of the input pathways, was described by Simpson and Dean (2002) as 'a key step forward in our understanding of flowering time control.'

Before moving on to discuss the details of the flowering pathway integrators themselves, it is useful to summarize how the pathways converge on these genes. The inhibition pathway culminates in strong FLC expression and therefore strong repression of the flowering-time integrators. The autonomous promotion pathway results in downregulation of *FLC*, through a combination of mRNA degradation and chromatin remodelling, and thus allows activation of the flowering-time integrators. The vernalization promotion pathway also operates through epigenetic downregulation of *FLC* by VIN3, and its maintenance in a non-transcribed state by the VRN2 complex. This also allows induction of the flowering-time integrator genes. To this point these three pathways (inhibition, autonomous induction, and vernalization) have all directly targeted *FLC*. However, the remaining two pathways (photoperiodic promotion and gibberellins) both act on the genes downstream of FLC. The photoperiodic promotion pathway bypasses FLC, and in response to

Vernalization

FLC

Autonomous
induction

GA ⟹ Flowering integrators ⟸ Photoperiodic
induction

GA ⟹ Floral meristem identity genes

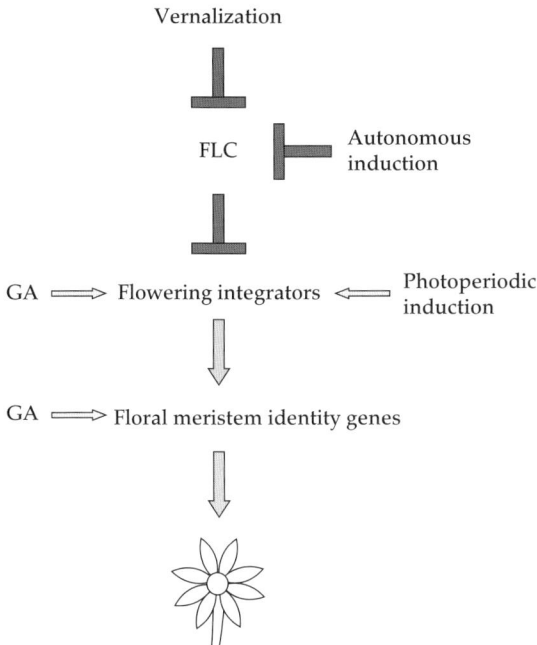

Figure 7.1 The four flowering induction pathways converge on the flowering-time integrators, either directly or through repression of FLC.

circadian clock-regulated *CONSTANS* (CO) expression resulting in stable CO protein translation in the presence of the daylight-activated forms of phytochromes and cryptochromes, the flowering-time integrator genes are directly activated. Gibberellin is also able to directly induce expression of flowering-time integrators, bypassing the FLC block. It can be seen, then, that the pathways all converge on the activity of a small subset of flowering-time integrator genes, immediately downstream of FLC, either directly or through modification of FLC activity. At any one time point these genes will be experiencing some degree of inhibition as a result of FLC action (varying in its strength according to the relative strength of signal from the inhibition, autonomous promotion, and the vernalization pathways), and some degree of activation as a result of gibberellin-induced signals and/or CO activity. When the combined activation signals are stronger than the remaining FLC-based inhibition, the flowering-time integrators will be expressed and the floral transition will occur.

7.2 Function of flowering-time integrators

It can be seen that the flowering-time integrators are critical to the Arabidopsis floral transition. There are three main genes described within this category, and we shall discuss the function of each in turn. However, it is useful to note first that all of them act, either directly or in combination with other proteins, as transcriptional regulators, and they promote flowering by activating the expression of genes which operate at the meristem to commit it to the reproductive state. The activity of each of the flowering-time integrators is based around the expression of a subset of these 'floral meristem identity genes' (FMI genes). The FMI genes will be discussed in detail in Chapter 9, when we consider the physiological and molecular changes that occur at the meristem as the plant becomes committed to the reproductive state.

Not only is the regulatory activity of each flowering-time integrator slightly different, but also the output pathways that lead into them vary slightly (Jack 2004). Fig. 7.2 shows a schematic of how each of the five flowering-time pathways interacts with the flowering integrators. Of the three integrator genes, *SOC1* is the most crucial hub, being positively regulated by the photoperiodic induction pathway and by gibberellins, and negatively regulated by FLC (and thus integrating the vernalization, autonomous promotion, and inhibition pathways). *FT* is positively regulated by the photoperiodic induction pathway, but negatively regulated by FLC, and thus also integrates the vernalization, autonomous promotion, and inhibition pathways. Only gibberellins have no known direct effect on *FT*. *LFY* is positively regulated by gibberellins, and also by SOC1, and thus also integrates everything upstream of SOC1 with gibberellin-induced promotion.

7.2.1 FLOWERING LOCUS T (FT)

The *ft* mutant was one of the original late flowering lines identified by Koornneef *et al.* (1991). The mutant has normal expression of *LEAFY* (*LFY*), but a double mutant between *lfy* and *ft* has substantially reduced expression of *APETALA1* (*AP1*). Both *LFY* and *AP1* function as floral meristem identity genes,

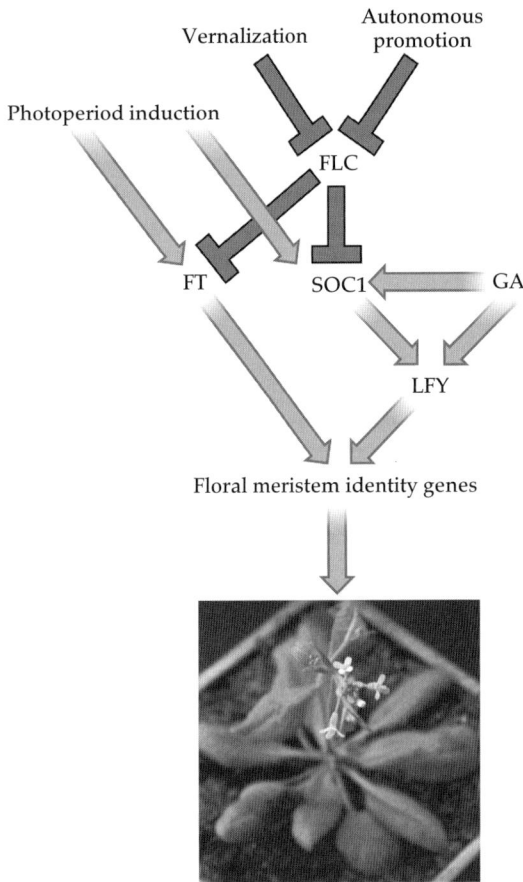

Figure 7.2 The different floral induction pathways are integrated slightly differently by SOC1, FT, and LFY. Only SOC1 integrates all four pathways. LFY responds directly to gibberellin (GA), as well as indirectly to everything that regulates SOC1. FT integrates all pathways except GA, to which it is not responsive (redrawn from Jack 2004).

necessary for the conversion of the meristem to the reproductive state, and are discussed in more detail in Chapter 9. *LFY* also acts as a flowering pathway integrator (see Section 7.2.3 below). However, in the context of the *ft* mutant, both *AP1* and *LFY* can be thought of as markers for the floral transition, so reduced *AP1* expression in an *ft/lfy* mutant background confirms the importance of *FT* in regulating flowering.

The *FT* locus was isolated by two independent research groups in 1999. Kardailsky *et al.* (1999) generated activation-tagged lines of plants, with an

enhancer of transcription transformed randomly into the genome. One of the early flowering lines they selected contained an insertion that mapped to the same place as the *FT* locus, and they were able to use the insertion to isolate the gene. Kobayashi *et al.* (1999) also isolated the *FT* locus, this time using a T-DNA insertion line. The *FT* locus encodes a small globular protein related to membrane-associated mammalian proteins that can bind hydrophobic ligands. Accordingly it was predicted to play a role in signal transduction, through binding to other molecules, and has since been shown to act as a transcriptional regulator in a complex with more conventional DNA-binding proteins. The protein is over 50% identical at the amino acid level to the protein encoded by the floral meristem identity gene *TERMINAL FLOWER 1* (discussed in more detail in Chapter 9). Indeed, Hanzawa *et al.* (2005) showed that only a single amino acid controlled the difference between the protein functioning as FT or as TFL1, providing evidence for extreme functional lability of this class of protein. The single amino acid is part of a region which may act as a molecular surface for interactions with other molecules, including the promoters of target genes (Ahn *et al.* 2006).

Kardailsky *et al.* (1999) conducted an analysis of the effects of ectopic expression of *FT* from the CaMV 35S promoter, and concluded that *FT* expression was sufficient to induce flowering. The transgenic plants showed early and excessive expression of *AP1*, although only in the floral primordia, suggesting that *AP1* activation requires the presence of another protein not present in vegetative tissues. The transgenic plants also contained higher levels of *LFY* transcript than wild type, a surprising observation in view of the normal *LFY* expression seen in the *ft* mutant. Ectopic expression of *FT* was sufficient to rescue a *constans* mutant, confirming the hypothesis that FT acts downstream of CO, to integrate the photoperiodic induction pathways with the other floral induction pathways. Strikingly, the combination of ectopic expression of *FT* and ectopic expression of *LFY* resulted in plants that entirely bypassed the vegetative stage of development. These double transgenic lines produced a single terminal flower on germination, and produced no leaves at all, only the cotyledons formed during embryogenesis (Kardailsky *et al.* 1999).

In the endogenous situation, *FT* expression itself is induced by the photoperiodic induction pathway, and repressed by FLC (and therefore indirectly induced by the vernalization and autonomous promotion pathways). The gibberellin promotion pathway does not seem to operate through FT. Evidence for the photoperiodic induction of *FT* was provided by Kobayashi *et al*. (1999), who fused a rat glucocorticoid receptor to CO and ectopically expressed the construct in Arabidopsis. This experimental approach localizes the transgenic protein in the cytoplasm, until an artificially applied steroid binds the receptor, allowing the protein to move into the nucleus. Since CO functions as a transcription factor, it can only be active when inside the nucleus. Induction of CO activity in these plants by treatment with the steroid caused rapid induction of *FT* expression within 12 hours. As discussed in Chapter 5, CO protein is only stably produced and therefore able to induce *FT* expression in daylight, providing the photoperiodic control to the system (Valverde *et al*. 2004). Recent reports suggest that *FT* expression is restricted, until CO protein levels reach a threshold, by the activity of the negative regulators TEMPRANILLO (TEM) 1 and 2. *TEM1* and *TEM2* encode transcription factors with AP2 domains which bind directly to the *FT* promoter. Ectopic expression reduces *FT* transcription and delays flowering, while loss of function increases *FT* transcript levels and promotes flowering. The competition between TEM and CO is probably won by CO in long days when sufficient stable protein is present to activate *FT* transcription (Castillejo and Pelaz 2008).

However, *FT* is not solely responsive to the photoperiodic induction pathway. At the same time the *FT* locus receives inhibitory signals from FLC (and is therefore influenced by the vernalization, autonomous induction, and inhibition pathways). This was most simply demonstrated by Hepworth *et al*. (2002), who showed that ectopic expression of *FLC* resulted in a total absence of *FT* transcript in transgenic plants. More recently, Helliwell *et al*. (2006) showed that FLC protein binds to the first intron of *FT* as part of a multiprotein complex. This complex is assumed to function to inhibit *FT* transcription.

Recent reports have suggested an additional role for FT in regulating the floral transition. Although not classically regarded as one of the pathways to flower induction, warm temperature has been shown to induce early flowering in Arabidopsis (Balasubramanian *et al*. 2006). Mutant or transgenic lines with reduced or increased *FT* activity are often perturbed in this response, suggesting a role for FT in the integration of temperature signalling into the decision of when to flower (Blazquez *et al*. 2003; Lee *et al*. 2007).

One of the most satisfying features of the *FT* gene is the demonstration that its protein can move from induced leaves to the shoot apical meristem, where it evokes the flowering response. This movement has been shown by a number of authors using a number of different techniques. Corbesier *et al*. (2007) made transgenic Arabidopsis plants expressing an FT:GFP fusion protein from a promoter expressed only in phloem companion cells, and observed that whereas mRNA was detected only in the companion cells, GFP was detectable in the vasculature, moving towards the shoot apical meristem. In a reverse experiment, Mathieu *et al*. (2007) fused FT to three copies of yellow fluorescent protein (YFP), making a protein too large to be trafficked between cells. Expression of this construct from the CaMV 35S promoter induced early flowering, as did expression of it from a shoot apical meristem-specific promoter, but expression from a companion cell-specific promoter did not induce early flowering, suggesting that immobilization of the protein in the companion cells inhibited its normal function. Jaeger and Wigge (2007) further showed the role of FT protein movement in flower induction by attaching a nuclear localization signal to the protein, restricting it to the nucleus of the cell in which it was synthesized. Expression of this construct from the CaMV 35S promoter was sufficient to induce flowering, but expression from a companion cell-specific promoter was not. Previous experiments by Huang *et al*. (2005) and others had shown that FT activates its own expression in the shoot apical meristem in a positive feedback loop. Furthermore, once activated, expression of endogenous *FT* is stable and does not require any additional factors to maintain it. These data may finally shed some light on the identity of the mysterious 'florigen' beloved of the early plant physiologists. *FT* is transcribed and translated in leaves in response to appropriate signals from the various flowering-time pathways.

The protein that is produced is mobile, travelling through the phloem to reach the shoot apical meristem. It should therefore also be graft-transmissible. Once the FT protein reaches the meristem it then induces expression of *FT* in those cells, and this expression is stable. These results show us how the evocation of flowering in the meristem can be caused by a mobile signal, and how such evocation can be stable once it occurs.

One further protein is required to allow FT to activate expression of the floral meristem identity genes. Recent reports have identified a basic leucine zipper (bZIP) transcription factor which interacts with the FT protein and is necessary for the floral transition to occur. *FD* was initially identified by Koornneef *et al.* (1991) as necessary for the floral transition, but has been little studied since. In 2005, two groups reported that *FD* is expressed at the shoot apical meristem prior to the floral transition, and encodes a bZIP transcription factor that interacts physically with FT protein. The presence of the FD protein is necessary for the induction of *AP1* expression (i.e. for the induction of floral meristem identity) by FT (Abe *et al.* 2005; Wigge *et al.* 2005). *FD* expression is upregulated once FT protein arrives at the meristem, an upregulation previously prevented by FLC (Searle *et al.* 2006). These data about *FD* function provide further support for the idea of a meristem 'waiting' for the arrival of *FT* mRNA and ready to be stably converted to the flowering state.

7.2.2 SUPPRESSION OF OVEREXPRESSION OF CONSTANS 1 (SOC1)/AGAMOUS-LIKE 20 (AGL20)

Like *FT*, *SOC1* was described by several different groups in the same year. It was given two different names, each reflecting the method by which it was isolated. It will be referred to as *SOC1* throughout this work, as that is the name used most frequently in the literature, although *AGL20* (*AGAMOUS-LIKE 20*) provides more information about the molecular identity of the encoded protein. *SOC1* was analysed by Borner *et al.* (2000) because it encodes a MADS box transcription factor with an expression pattern indicative of a role in the floral transition. In particular, *SOC1* expression is induced in the shoot apical meristem as the floral transition occurs. However, *SOC1* expression is reduced in mutant lines with lesions in genes active in the photoperiodic and autonomous induction pathways. Borner *et al.* (2000) screened for and isolated an insertion mutant with an insertion in the *SOC1* locus, and described the phenotypic consequences of mutation of the gene. Loss of function mutants are late flowering, whereas ectopic expression of *SOC1* results in an early flowering phenotype. Ectopic expression of *SOC1* is sufficient to overcome mutations in the long day and autonomous flowering pathways, allowing plants to flower at the normal time (Mouradov *et al.* 2002). Expression of *SOC1* was also found to be upregulated by gibberellin application.

Also in 2000, Lee *et al.* identified *SOC1* from a screen for suppressors of *FRI* activity. Using a transgenic approach involving the random insertion of transcriptional enhancers into the genome, they established that ectopic expression of *SOC1*, without other genetic changes, was sufficient to induce early flowering. These authors showed that *SOC1* expression was positively regulated by the vernalization, autonomous, and photoperiodic flower induction pathways.

In the same year, Samach *et al.* (2000) also isolated *SOC1* as a gene with a rapid increase in expression level following induction of CO activity using a steroid-activated form of CO in transgenic plants. *SOC1* transcript increased in abundance threefold within 4 hours of steroid treatment, and the authors then showed that *SOC1* was expressed at 10 times higher levels in plants ectopically expressing *CO* than in wild type plants. These data indicate a role for SOC1 in integrating the photoperiodic induction pathway.

SOC1 expression is induced by the photoperiodic pathway and by gibberellin, and is repressed by FLC (and therefore indirectly induced by the autonomous and vernalization promotion pathways, and repressed by the inhibition pathway). Unlike *FT*, then, *SOC1* is a true integrator of all of the flowering-time pathways.

Gibberellin has been shown to directly enhance *SOC1* expression (Borner *et al.* 2000), particularly in short days, and SOC1 then activates *LFY* expression. SOC1 binds directly to the promoters of three genes encoding SQUAMOSA PROMOTER BINDING PROTEIN-LIKE transcription factors—*SPL3*, *SPL4*, and *SPL5*—and activates their transcription (Jung *et al*. 2012). These three genes promote the floral transition when expressed. In short day conditions, gibberellin signalling induces expression of the *SPL3*, *SPL4*, and *SPL5* genes in a SOC1-dependent manner, providing a mechanism through which SOC1 integrates gibberellin induction into the floral transition (Jung *et al*. 2012). Not surprisingly, inhibition of gibberellin synthesis results in reduced expression of the *SPL* genes.

The photoperiodic pathway activates *SOC1* through the actions of CO (Samach *et al*. 2000), and analysis of the *SOC1* promoter has indicated that overlapping promoter elements respond to FLC and CO signals, allowing repression and activation to occur at the same time, creating the necessary conflict at the flowering-time integrator (Hepworth *et al*. 2002). Within the *SOC1* promoter a MADS box binding motif was identified, and FLC was shown to bind to this site *in vitro*. Mutation of this site *in planta* prevented FLC repression of *SOC1* expression. This particular region of the *SOC1* promoter was not bound by CO, but a nearby region responded to CO activity (Hepworth *et al*. 2002). FLC represses *SOC1* expression by binding the *SOC1* promoter with SHORT VEGETATIVE PHASE (SVP), another MADS box protein that seems to act together with FLC to produce an inhibitory complex (Li *et al*. 2008).

It has also been shown that removal of FLC-based repression of *SOC1* is insufficient to result in strong expression of the gene. Moon *et al*. (2003) showed that vernalization (and thus reduction in FLC activity) was not sufficient to induce *SOC1* expression in a gibberellin-deficient mutant line (*ga1*, deficient in the first enzyme of gibberellin biosynthesis) grown in short days. Therefore, without a photoperiodic induction signal, gibberellin is necessary to induce *SOC1* expression.

Taken together these data describe a gene whose expression is tightly controlled by multiple inputs converging on its promoter, and whose protein product can also induce transcription of other genes in a manner dependent on the presence of other signals and proteins.

7.2.3 LEAFY (LFY)

The third flowering-time integrator, *LFY*, is more usually described as a meristem identity gene (see Chapter 9), primarily because it has been shown to act downstream of SOC1, which activates its transcription. However, it can also respond directly to gibberellin signalling, and so is also thought of as the third flowering-time integrator (Jack 2004). Through SOC1, *LFY* responds to signals from the photoperiodic, vernalization, and autonomous promotion pathways, but these are all integrated directly by SOC1. However, *LFY* expression is also directly induced by the application of gibberellin. Blazquez *et al*. (1998) showed that treatment with gibberellin alone had no effect on the activity of a reporter gene fused to the *LFY* promoter, but that the combination of gibberellin and 1% sucrose resulted in much stronger (more than double) reporter gene expression than was caused by sucrose alone. Eriksson *et al*. (2006) demonstrated that GA_4 is the active gibberellin that induces upregulation of *LEAFY*. It has also been shown that gibberellin promotes flowering in short days through a GAMYB-dependent activation of *LFY*. The gibberellin-responsive MYB transcription factor, AtMYB33, is necessary to allow gibberellin to activate *LFY* expression, and is presumably responsible for direct transcription of *LEAFY* in response to gibberellin signals. Appropriate *LFY* expression is dependent upon an MYB recognition motif within the *LFY* promoter, and this motif is bound by AtMYB33 (Blazquez *et al*. 1998; Blazquez and Weigel 2000; Gocal *et al*. 2001). A separate promoter element is responsible for *LFY* induction in response to long days (i.e. the photoperiodic promotion pathway operating through SOC1; Blazquez and Weigel 2000).

The activity of *LFY* as a floral meristem identity gene will be discussed in more detail in Chapter 9.

These three flowering-time integrators may not represent the complete picture, and others may remain to be discovered. Evidence for the presence of other floral pathway integrators comes from genetic analysis. Moon *et al*. (2005) showed that a triple mutant between *soc1*, *ft*, and *lfy* was still able to flower under long day conditions. Although the triple mutant flowered extremely late, and the flowers that it produced were developmentally abnormal, its ability to flower at all suggests that there is some other route through which floral induction signals can be directed to the floral meristem identity genes. Further genetic screens for enhancers of the flowering-time integrator mutant phenotypes may identify additional integrators. However, in the mean time the flowering-time integrators provide a satisfying mechanism by which the signals from multiple different flowering-time pathways can be integrated and translated into the developmental response of meristem transition and flower production.

CHAPTER 8

Flower induction beyond *Arabidopsis thaliana*

In the previous four chapters we have discussed the various pathways that regulate the floral transition in Arabidopsis, and their integration to generate a developmental response. Arabidopsis is an excellent model for analysis of the floral transition, because it is responsive to so many variables. Although not all plants respond to all of photoperiod, vernalization, gibberellins, and endogenous signals, all plants do respond to at least one of these. Our current depth of understanding of all of these pathways in Arabidopsis allows us to explore whether the processes are similar in a range of species, or whether novel signalling modules are at work. In this chapter we look at the evidence that the same or similar pathways operate in other species. We shall look in most detail at the short day plant rice, and then also consider a number of perennial species, with a final look at Gregor Mendel's famous genetic model, the garden pea (see Fig. 8.1).

8.1 The *Arabidopsis* flower induction model in other species

It is all too easy to forget that model species are only intended to provide the beginning of the answers to questions, rather than the full story. Although we now have a good picture of how the floral transition is controlled in Arabidopsis, this does not mean that we understand the control of flowering in all plants. The focus for the next few decades of flowering-time research will be to investigate the extent to which the Arabidopsis flowering-time model applies to other species, and to explore what innovations other species have evolved to fit their floral transitions to their particular environments. One

crucial question that must be addressed is the extent to which the long day photoperiodic induction pathway seen in Arabidopsis operates to control flowering in short day plants (i.e. those that flower when the days are shorter than a certain maximum length). It will also be important to assess the extent to which each of the other Arabidopsis pathways functions to induce flowering in day neutral plants. The best studied model apart from Arabidopsis is currently rice, which has been analysed both because of its economic importance and because it is a short day plant. However, work is now progressing on a range of species that show different flowering responses, including perennial species for which the floral transition is an annual problem requiring a reversion to the vegetative state afterwards.

8.2 Flower induction in rice: a model short day plant

Rice (*Oryza sativa*, Poales; see Fig. 8.2a) is one of the world's most important food crops, providing staple carbohydrate and protein for half the world's human population. It is also a facultative short day plant, which synchronizes its flowering and reproduction with the rainy season in its usual habitats (Putterill *et al.* 2004). Although it may seem strange for a wind-pollinated plant to flower under wet conditions, seed set is enhanced when sufficient water is available, and damp conditions are important for germination of rice seed. The practice of flooding rice fields has primarily evolved to reduce weeds, pests, and pathogens, as rice itself can tolerate flooded conditions but has no specific need for them. Rice has no vernalization requirement, as it

Figure 8.1 Species in which the floral transition has been studied. (a) *Pisum sativum* (Fabales). Photo by Rasbak. (b) *Triticum* species (Poales). Photo by Optograph. (c). *Populus* species (Malpighiales). Photo by Matt Lavin. (d) *Arabis alpina* (Brassicales). Photo by Franz Xaver. (e) *Oryza sativa* (Poales). Photo by C.T. Johansson. (f) *Beta vulgaris* (Caryophyllales). Photo by Forest and Kim Starr. All images from Wikimedia Commons. See also Plate 3.

is adapted to life in warm environments. It is therefore of immense economic importance to establish how the short-day-induced flowering of rice occurs, since this is the only significant environmentally controlled promotive pathway that it contains.

The floral transition in rice makes an interesting comparison to that of Arabidopsis not only because it is a short day plant, but also because it is a monocot. The monocots diverged from the eudicot lineage very early in the history of the angiosperms (see Chapter 1), so a comparison of the control of flowering in the two lineages provides useful information about what is conserved and what is unique to each system. This in turn provides insight into the floral transition in early angiosperms, and into the novel pathways that have evolved since.

Analysis of the rice floral transition has primarily been conducted through mapping of quantitative trait loci (QTLs), and, to a lesser extent, by using a candidate gene approach to investigate the function of orthologues of Arabidopsis genes. These approaches have been significantly enhanced in recent years by the publication of the rice genome sequence in 2002, allowing ready identification of sequences similar to those regulating flowering in Arabidopsis, and also facilitating identification of QTLs following coarse mapping.

The first gene involved in the rice floral transition to be studied at the molecular level was isolated in 2000. The *PHOTOPERIODIC SENSITIVITY 5 (SE5)* gene encodes a protein related to HY1 of Arabidopsis, and is needed for the synthesis of the chromophores on all phytochrome molecules (Izawa *et al.* 2000). The *se5* mutant flowers early in long days and in short days. These data confirmed that phytochrome-induced signals in long days normally delay flowering in rice, whereas in Arabidopsis they activate it. However, analysis of the rice genome sequence suggests that the light receptors themselves differ slightly from those in Arabidopsis.

(a)

(b)

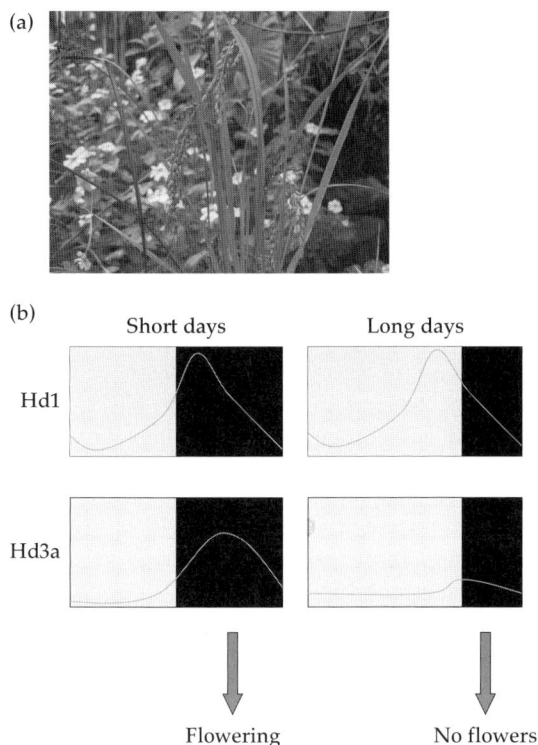

Figure 8.2 (a) A rice plant in flower (foreground). (b) The *Hd1* gene (above) is expressed in a circadian pattern. It is only able to activate *Hd3a* expression (below) when its expression coincides with darkness—that is, in short days (left panel). Compare with Figure 5.4.

Specifically, while orthologues of *PHYA, PHYB,* and *PHYC* are apparent in the rice genome, there is no sign of *PHYD* or *PHYE* orthologues. As noted in Chapter 5, *PHYD* is thought to be a Brassicaceae-specific gene, whereas *PHYE* shows limited conservation in different plant families (Mathews 2010). *PHYA, PHYB,* and *PHYC* are the three ancient members of the phytochrome family. In contrast, three genes encoding cryptochromes appear to be present, compared with the two in Arabidopsis (Izawa *et al.* 2003). *OsCRY1a* and *OsCRY1b* share significant sequence similarity with Arabidopsis *CRY1,* while *OsCRY2* is apparently the orthologue of Arabidopsis *CRY2* (Zhang *et al.* 2006). This suggests that detailed analysis of day length measurement in rice will reveal small differences in the ways in which different wavelengths of light are perceived in Arabidopsis.

Analysis of the rice genome has revealed that genes with very similar sequences to *CCA1* and *TOC1* are present, suggesting that the rice circadian clock functions in an essentially similar way to the Arabidopsis one. The rice *CCA1* orthologue was shown to have a circadian-regulated expression pattern similar to that of Arabidopsis *CCA1* (Izawa *et al.* 2002). Further studies have indicated that expression profiles of rice orthologues of Arabidopsis clock-associated genes generally overlap with those of their Arabidopsis counterparts, leading to a general conclusion that the circadian clock in rice (and indeed in other angiosperms) is the same as that in Arabidopsis (reviewed by Song *et al.* 2010).

Many of the other components of the Arabidopsis photoperiodic induction pathway also seem to be present in rice, although they function in slightly different ways. A rice orthologue of *GIGANTEA, OsGI,* was identified by analysis of genes differentially regulated in wild type and *se5* mutant plants (Hayama *et al.* 2002). The protein encoded by *OsGI* is very similar to that encoded by *GI,* and it appears that the proteins fulfil similar roles, although the downstream outcomes of their activities are divergent. Ectopic expression of *OsGI,* using the CaMV 35S promoter, delays flowering, whereas ectopic *GI* expression in Arabidopsis induces flowering. In the transgenic rice plants, expression levels of *Hd1* (the *CO* orthologue; see below) were increased, suggesting that *OsGI* functions to upregulate *Hd1,* just as *GI* functions to upregulate *CO.* However, the *FT* orthologue, *Hd3a,* was expressed at lower levels in these transgenic plants, whereas ectopic expression of *GI* and enhanced *CO* expression increases *FT* transcript levels in Arabidopsis. These data led to the conclusion that the same pathway components are present in rice as in Arabidopsis, but that the central role of *CO* in inducing *FT* expression seems to be reversed in rice, with enhanced *Hd1* expression resulting in decreased *Hd3a* activity. *OsGI* itself behaves much as Arabidopsis *GI* does, peaking in expression in the evening and activating expression of the *CO* orthologue, *Hd1* (Hayama *et al.* 2003; Simpson 2003). However, it may also play an additional role not shown by Arabidopsis GI. Recent reports indicate that OsGI also activates expression of *Early heading date 1* (*Ehd1*) at dawn, through a blue-light-mediated response (Itoh *et al.* 2010). *Ehd1* encodes a B type response regulator protein with DNA binding activity, and it has no clear orthologue in the Arabidopsis genome (Doi *et al.* 2004). Its expression is also necessary for *Hd3a* expression (*FT*; see below), suggesting that *Ehd1* has evolved as a second signal through which GI-transduced circadian signals can result in the expression of a florigen-like signal.

Loci involved in flower induction in rice are traditionally named *Heading date* (*Hd*) loci, and two of the *Hd* loci encode the rice orthologues of *CO* and *FT*. Yano *et al.* (2000) isolated *Hd1* by map-based cloning and found that it encodes the rice *CO* orthologue, a protein very similar in sequence to CO itself. *Hd1* expression is controlled by the circadian clock in the same way that *CO* expression is, with maximum transcript occurring at night in short days, but in the later afternoon/early evening in long days. *Hd3a* encodes a protein with strong sequence similarity to FT, the crucial flowering-time integrator in Arabidopsis. Although expression of *Hd3a* first appears in the leaf in short days, the protein has been shown to move to the apex, just as FT does in Arabidopsis. As was discussed in Chapter 5, in Arabidopsis *CO* expression in the light causes activation of *FT* expression, but *CO* expression in the dark (i.e. in short days) does not induce *FT* activity. However, in rice the situation appears to be entirely the opposite (compare Fig. 8.2b with Fig. 5.4). In the dark, *Hd1* expression induces *Hd3a* expression, which then induces flowering. However, when *Hd1* expression occurs in the light (i.e. in long days) it does not induce *Hd3a* expression and flowering does not occur (Izawa *et al.* 2002; Kojima *et al.* 2002). Indeed, exposure to a single 10-minute burst of light during the night is sufficient to suppress *Hd3a* expression and delay flowering (Ishikawa *et al.* 2005). It is thought that phytochrome B signalling is important in determining whether *Hd1* expression results in *Hd3a* expression or repression (Ishikawa *et al.* 2009).

No orthologue of *FD* has been identified in the rice genome, raising questions about the mode of action of *Hd3a* (reviewed by Tsuji *et al.* 2011). Since FT-based induction of flowering requires an association between FT and a DNA-binding protein, it is likely that some other DNA-binding protein fulfils the same role in rice.

Since rice is a facultative short day plant, rather than an obligate one, it must have a floral promotion pathway that operates in long days, just as the Arabidopsis endogenous promotion pathway does. This long day promotion pathway has only recently been described (Komiya *et al.* 2009), and is entirely distinct from the Arabidopsis endogenous pathway. In long days, rice flowers an average of 30 days later than in short day conditions. This is partly due to the repression of *Hd3a* expression by Hd1 in the light. The fact that this is an active repression, not just absence of activation, is demonstrated by the flowering time of the *hd1* mutant—late in short days, but earlier than wild type in long days. Phytochrome signalling is thought to be necessary to determine whether Hd1 acts as a repressor or activator of *Hd3a*. *Hd3a* expression levels are also kept low in long days by the activity of Ghd7. This small protein with a CCT domain has no orthologue in Arabidopsis, and the expression of the gene encoding it is strongly upregulated in long days. The protein functions as a repressor of transcription of *Ehd1*, which, as noted earlier, is itself an activator of *Hd3a* expression (Doi *et al.* 2004; Itoh *et al.* 2010). The combination of repression by Hd1 and inhibition through loss of the inductive Ehd1 signal means that a rice plant in long days has no Hd3a activity. Instead, flowering is induced in long days by *RICE FLOWERING LOCUS T 1* (*RFT1*), the closest paralogue of *Hd3a* in the rice genome, encoding a protein that is 91% identical to Hd3a. Downregulation of *RFT1* results in delayed flowering in long days. In the wild type plant, *RFT1* expression increases in the leaves in long days, and the protein then moves through the phloem to the apical meristem. Thus it seems that rice has acquired a second florigen, using one FT-like protein (Hd3a) to promote flowering in short days and another (RFT1) to promote flowering in non-inductive long day conditions (Komiya *et al.* 2009).

The rice orthologue of the second major Arabidopsis flowering-time integrator plays a rather different role from its Arabidopsis counterpart. The *OsSOC1* gene (also known as *OsMADS50*) was isolated by a screen for MADS box genes with similarity to Arabidopsis *FLC* (Tadege *et al.* 2003), and was found to map to the known flowering-time locus *Hd9*. In fact the OsSOC1 protein has greater sequence similarity to SOC1 than to FLC, and a number of experiments have indicated that its function is more similar to that of SOC1 than to that of FLC. Expression of *OsSOC1* increases at the floral transition, as would be expected for a flowering-time integrator, while ectopic expression of *OsSOC1* in Arabidopsis complements the *soc1* mutant. Ectopic expression of *OsSOC1* in a wild type Arabidopsis background induces early flowering. In contrast, suppression of *OsSOC1* by RNA interference (RNAi) methods causes late flowering in rice (Lee *et al.* 2004). Since Arabidopsis *SOC1* expression is downregulated by FLC, Tadege *et al.* (2003) investigated potential FLC/OsSOC1 interactions in rice. They were able to identify a CarG box in the *OsSOC1* promoter, which appeared to be suitable for FLC binding, but they could not identify an *FLC* orthologue by library screening, searching through EST collections, or analysis of the rice genome sequence. However, ectopic expression of Arabidopsis *FLC* in rice delays flowering by 2–4 weeks, and also delays upregulation of *OsSOC1*. These data suggest that another MADS box protein, recognizing similar binding motifs to FLC, may be involved in *OsSOC1* downregulation (Andersen *et al.* 2004). The differences between Arabidopsis and rice SOC1 become apparent when their modes of action are considered. Whereas SOC1 in Arabidopsis integrates the flower induction pathways, OsSOC1 apparently operates further upstream. It is necessary for the expression of *Ehd1*, the response regulator protein that transduces clock signals such as CO, and also for the expression of *RFT1*, the long day florigen (Komiya *et al.* 2009). Expression of both genes is prevented in an *OsSOC1* mutant background, placing *OsSOC1* firmly within the photoperiodic induction pathway, where it is mainly active in leaves, not in the apical meristem.

The *LEAFY* orthologue in rice also appears to differ from its Arabidopsis counterpart. *RICE FLORICAULA/LEAFY* (*RFL)* promotes flowering, and when its expression is downregulated in an RNAi line the flowering response is strongly suppressed (Rao *et al.* 2008). However, it is clearly upstream of the general integration of flowering signals, as its expression increases that of both *OsSOC1* and *RFT1*, and seems to be most effective in the leaves.

Overall, the photoperiodic induction pathway in rice seems remarkably similar to that of Arabidopsis, with a few minor additions, suggesting that the evolution of long and short day plants has involved tinkering with the basic mechanisms of photoperiodic induction, rather than the development of two distinct systems. However, since rice is not sensitive to vernalization it is perhaps not surprising that no clear orthologues of *FLC*, *FRI*, *VRN1*, or *VRN2* have been identified in the rice genome, suggesting that no equivalent vernalization-promotion pathway is present (Izawa *et al.* 2003). Perhaps most surprising is the discovery that the flowering of rice in non-inductive long days is induced not by a pathway similar to the Arabidopsis endogenous floral induction pathway, but by a second *FT* gene induced over a longer time frame in non-inductive conditions. In fact the absence of the *FLC* module for floral inhibition means that the Arabidopsis endogenous pathway could not operate in rice, and this second *FT* pathway is an evolutionarily parsimonious solution.

8.3 Flower induction in wheat and barley: a novel vernalization pathway

The photoperiodic control of flower induction in other cereals so far appears to be similar to the pathways identified in rice and Arabidopsis. However, many cereals have an additional vernalization requirement, which rice does not share. For example, both wheat and barley (both Poales) need around 60 days of cold to induce flowering. In both species, spring varieties, which are planted in the spring and flower immediately without the need for vernalization, have been bred. The difference between these spring varieties and traditional vernalization-requiring winter varieties has been mapped to three major QTLs, named *VERNALIZATION1*, *VERNALIZATION2*, and *VERNALIZATION3* (*VRN1*, *VRN2*, and *VRN3*). These loci should not be confused with the Arabidopsis *VRN1* and *VRN2* loci, to which they are not orthologous (Andersen *et al.* 2004).

The *VRN1* loci have been identified in both wheat and barley, and renamed *TmAP1* (or *WAP1* or *TaVRT-1*) and *BM5*, respectively (Danyluk *et al.* 2003; Trevaskis *et al.* 2003; Yan *et al.* 2003). Both genes encode orthologues of the Arabidopsis floral meristem identity gene *APETALA1* (*AP1*). Transcripts of these genes are upregulated by exposure to cold in winter varieties of wheat and barley, but are constitutively expressed at high levels in spring varieties. These findings were unexpected, as they appear to suggest that a key component of the Arabidopsis floral meristem identity system is involved in a much earlier stage of floral induction in cereals, acting to induce the floral transition in response to vernalization. The identification of a CArG box in the promoter of wheat *TmAP1*, and the observation that spring varieties usually have a deletion of part of the promoter in this region, suggests that another MADS box protein acts upstream of this AP1-like factor (Yan *et al.* 2003).

Wheat *VRN2* was isolated in 2004, using a map-based approach (Yan *et al.* 2004). The *VRN2* locus contains two genes, both encoding proteins with a zinc finger domain and a CCT domain. There are no clear orthologues of *VRN2* in Arabidopsis. In contrast to the behaviour of *VRN1*, the *VRN2* locus is a dominant repressor of flowering, which is downregulated by vernalization. No *VRN2* transcript is detectable after vernalization has occurred. Loss of function of *VRN2* results in the production of spring lines of wheat with no requirement for cold induction. Again, it was surprising that a protein acting as a dominant repressor of flowering, and inactivated by cold, was not the expected direct orthologue of the Arabidopsis protein (i.e. FLC).

The *VRN3* gene was isolated through a map-based approach, and the only annotated gene found in the appropriate genomic regions of both wheat and barley was an *FT*-like sequence (Yan *et al.* 2006). Analysis of genetic interactions showed that VRN3 activates expression of *VRN1*, in a module very like the FT-induced expression of floral meristem identity genes in Arabidopsis (see Chapter 7). In wheat, VRN3 interacts with an FD-like protein, FDL2, and the *VRN1* promoter, to activate *VRN1* transcription (Li and Dubcovsky 2008). Both genetic and transgenic experiments further suggest that VRN2 represses *VRN3* expression, blocking the floral transition until its own levels decline after exposure to cold (Yan *et al.* 2004, 2006). Experimental data show that reduced *VRN2* expression is associated with increased *VRN3* expression, and that increased

VRN2 expression is associated with reduced *VRN3* expression.

Perhaps surprisingly, this signalling pathway apparently functions in a loop. Loukoianov *et al.* (2005) demonstrated that *VRN1* expression results (either directly or indirectly) in reduced *VRN2* expression. Distelfield *et al.* (2009) proposed that *VRN1* is the primary point at which vernalization acts. By enhancing *VRN1* expression, a period of cold removes the VRN2 block to flowering, activates the VRN3 induction of flowering, and at the same time allows VRN1 to begin activating the changes at the apical meristem necessary to cause flower development. This signalling cascade provides a mechanism to induce flowering after cold, and confirms the position of FT (VRN3 in wheat or Hd3a in rice) as a key integrator of floral induction pathways in the cereals.

The data emerging for wheat and barley suggest that, unlike photoperiodic induction, vernalization has evolved multiple times and that each evolutionary event has involved the recruitment of similar but distinct genes from the flowering-time pathways. In the cereals the FT and AP1 proteins play recognizable roles, but an absence of FLC/FRI means that vernalization operates through the FT pathway with independently recruited controls.

8.4 Flower induction in perennials

Most crop plants are annuals, like Arabidopsis. However, for many plant species a perennial life cycle is the norm. In these species, flowering is followed by a return to vegetative growth and the cycle is repeated, often many times over many years. Perennials are often polycarpous, with some meristems committing to the flowering state each year while others remain vegetative. Polycarpy requires an extra level of control, allowing some meristems to undergo the floral transition but preventing others from doing so. This is believed to be achieved through a range of competence or juvenile/adult mechanisms, with only competent (adult) meristems undergoing the floral transition (Battey and Tooke 2002). The number of meristems undergoing the floral transition may also be determined by environmental factors. For example, in the wallflower (*Cheiranthus cheiri*, Brassicales), variations in the temperature perceived by the apical meristem result in the production of varying numbers of axillary meristems competent to adopt the floral fate, as opposed to those constrained to the vegetative fate (Battey and Tooke 2002). In this way the total numbers of flowers and total reproductive output can be adapted to the varying conditions of different years, a sensible system for a species which has many opportunities to flower and should optimize maximum flowering to coincide with optimal environmental conditions.

8.4.1 Flower induction in *Arabis alpina*

Arabis alpina is a perennial relative of Arabidopsis, in the Brassicaceae, and has therefore been used as a model system in which to examine perenniality. In order to flower, *Arabis alpina* requires vernalization, with at least 12 weeks of cold necessary to induce a full flowering response. However, there is no apparent photoperiodic induction pathway in this species, implying that the passage of winter is the only requirement to induce flowering each year. Wang *et al.* (2009) identified a mutant lacking the vernalization requirement and able to flower continuously for 12 months. This mutant, *perpetual flowering 1* (*pep1*), was found to contain a lesion at the *AaFLC* locus. Analysis of *AaFLC* revealed a very different response to vernalization compared with that of Arabidopsis *FLC*. Although *AaFLC* transcript levels were high before vernalization, and reduced during the cold treatment, on return to warm temperatures transcript levels returned to normal. Similarly, Wang *et al.* (2009) showed that histone methylation of *AaFLC* in response to vernalization mirrored that of Arabidopsis *FLC*, but whereas these repressive chromatin marks persist in Arabidopsis after a return to warm temperatures, at the *AaFLC* locus they were lost after the end of the cold period. The authors concluded that the basic mechanism of vernalization in *Arabis alpina* matched that of Arabidopsis, but that the loss of *AaFLC* repression on the return to warmth allowed AaFLC to repress all subsequent meristems that had not been induced to flower during the first winter, ensuring they grew vegetatively until the following winter cold allowed the next repression of *AaFLC* and the next flush of flowering.

8.4.2 Flower induction in poplar

Trees are the extreme examples of the perennial form, with some species making the transition between vegetative growth and flowering many hundreds of times within an individual life cycle. The publication of the full genome sequence of *Populus trichocarpa* (poplar, Malpighiales) (Tuskan *et al.* 2006) provided the opportunity to explore this repeated cycle using candidate gene approaches (as trees are not generally amenable to genetic approaches). Like rice, poplar has no clear orthologue of *FLC* (Leseberg *et al.* 2006). Instead, attention has been focused on the roles of a pair of recently duplicated *FT*-like genes, *FT1* and *FT2*.

Using a series of transgenic and environmental manipulations of the transcription profiles of these two genes, Hsu *et al.* (2011) demonstrated that FT1 induces the flowering response in response to vernalization (much as occurs in wheat and barley), and that FT2 induces vegetative growth in the summer and the transition to dormancy in autumn through photoperiodic control (see Fig. 8.3). In a naturally growing poplar tree, *FT1* transcript is abundant only in winter. If trees are transplanted in November into a 25 degrees Centigrade regime the levels of *FT1* trancript do not rise, irrespective of day length, confirming that this is a cold-induced response rather than a short-day response. Induction of *FT1* expression from a heat shock promoter resulted in flowering within 30 days, demonstrating that this cold-induced gene promotes the flowering response.

In contrast, *FT2* transcript is found in the leaves and buds of wild-grown poplar trees in the spring and summer growing season, when days are long and the weather is warm. Transplanting trees in November into a growth room at 25 degrees Centigrade did not induce *FT2* expression unless the day length was also increased to 16 hours. This indicates that *FT2* expression is controlled by a photoperiodic pathway effective in long days, much like the Arabidopsis one. However, FT2 does not induce flowering. Ectopic expression of *FT2* using a heat shock promoter enhanced vegetative growth, but only resulted in a few small indications of flowering if expression levels were increased significantly above those found in the wild. Hsu *et al.* (2011)

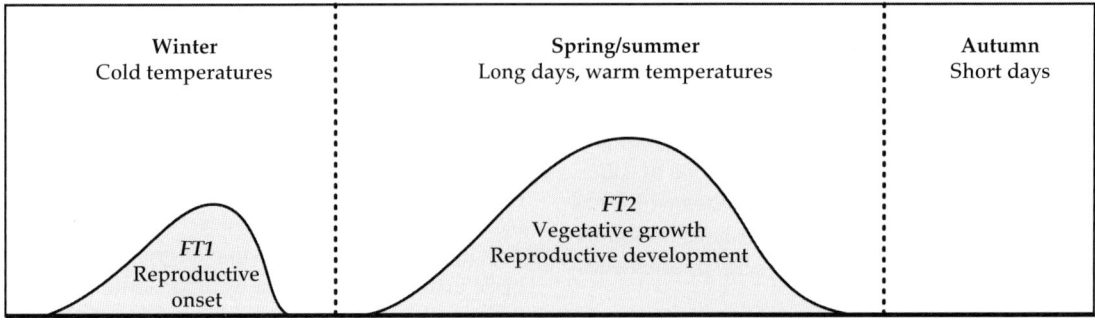

Figure 8.3 Expression patterns of poplar *FT1* and *FT2*. *FT1* transcript is cold-induced, and activates the floral transition. *FT2* transcript is long day-induced, and activates vegetative growth.

concluded that FT2 was a regulator of vegetative growth in long days. However, its loss of expression in autumn when the days are getting shorter is also associated with the transition to dormancy. If expression levels of *FT2* in the transgenic plants were maintained at 'summer-like' levels during the short days of winter, vegetative growth did not cease, identifying a second role for *FT2* in allowing the onset of dormancy when its photoperiod-controlled transcription ceases.

More detailed analyses are needed to determine how cold induces *FT1* expression and how long days induce *FT2* expression, and to explore how both proteins induce the appropriate developmental response. A previous study demonstrated the presence of a *CO* orthologue in poplar, with a circadian-regulated expression pattern much like that seen in Arabidopsis (Böhlenius *et al.* 2006), so it is likely that at least some of the upstream control of expression of the *FT* genes is conserved with Arabidopsis. Hsu *et al.* (2011) did note that FT1 and FT2 differ in amino acid sequence in a loop of the protein which is known to be crucial to its function in Arabidopsis (see Chapter 9), suggesting that minor modifications of duplicated proteins might be sufficient to generate novel developmental control.

8.4.3 Flower induction in biennial sugar beet

The biennial life form is quite common for herbaceous plants, and usually involves both a vernalization response (to ensure that flowering occurs only in the second year) and a photoperiodic response (to ensure that flowering occurs in the summer of the second year). *Beta vulgaris* has provided a useful model for the study of the floral transition in biennial species, as its importance as a crop has resulted in a good investment in genetic and genomic resources. It is also interestingly placed phylogenetically, being a member of the Caryophyllales, an order towards the base of the Asterids. It appears that beet uses two paralogues of *FT* to regulate its two flowering responses. Recent work by Pin *et al.* (2010) showed that both *BvFT1* and *BvFT2* are expressed in the leaves of beet, but at different times. *BvFT1* is expressed during the juvenile phase of development, and represses the bolting (stem elongation) and flowering response. It does this very strongly—ectopic expression of *BvFT1* in either beet or Arabidopsis repressed flowering, indicating that the protein function is clearly distinct from that of Arabidopsis FT. Its expression levels fall after exposure to a period of cold, suggesting that it is acting as a negative regulator of flowering which is itself repressed by vernalization. *BvFT2*, in contrast, is expressed in mature leaves during the growing season, and is required for the floral transition. Ectopic expression of *BvFT2* induced flowering in Arabidopsis and beet, and downregulation of *BvFT2* in beet using an RNAi approach prevented flowering even in long days. This gene seems to be acting in a more Arabidopsis-like fashion, and the authors proposed that BvFT1 prevents flowering by repressing *BvFT2* transcription until vernalization releases the block. One surprising observation was that *BvFT2* RNAi plants (with reduced *BvFT2* expression) were still able to flower in response to vernalization—suggesting that an alternative pathway can cause the floral transition in the absence of BvFT2 activity. Interestingly, Pin *et al.* (2010) noted that the two *FT* paralogues in beet differ in sequence in the same loop of the protein as that also shown to be significant for determining protein function in Arabidopsis and poplar (see above; also see Chapter 9).

8.5 Flower induction in legumes

Flowering time has been studied in legumes (Fabales) since Gregor Mendel developed the garden pea as a model system for the analysis of inheritance. The rationale for studying flowering time in pea is based on its unique suitability for both genetic and physiological studies. Although it is a long day plant, like Arabidopsis, it has the added benefit of being easy to graft, thus allowing researchers to investigate whether something missing from a mutant line is mobile. Mobile signals will travel from wild type root stocks into mutant shoots grafted to them, and rescue the mutant. Immobile signals will not do so. Extensive genetic stocks are also available in pea.

A number of flowering-time mutants have been described in pea, and these have been reviewed by Weller *et al.* (1997) and Weller *et al.* (2009). Analysis of these mutants has led to the identification of a number of flowering-time pathways, some of which are very similar to those of Arabidopsis. For example, the *gigas* (*gi*) mutant flowers late in long days, but later still in short days, indicating that pea contains an autonomous floral induction pathway, operating in both long and short days. Grafting a *gi* shoot on to a wild type root revealed that the signal produced by the *GI* gene is mobile, as the mutant shoot flowered at the normal time following this graft (Beveridge and Murfet 1996). Recent work has shown that *GI* encodes an FT-like protein, PsFTa1 (Hecht *et al.* 2011). Thus pea contains an autonomous floral induction pathway that operates through a mobile, FT-like, signalling molecule.

Similarly, the *STERILE NODES* (*SNE*), *DIE NEUTRALIS* (*DNE*), and *PHOTOPERIOD* (*PPD*) genes

are all required for inhibition of flowering in short days. Mutation in any of these genes causes early flowering in short days, but does not affect flowering time in long days. Grafting experiments with wild type root stocks and shoots of these mutants revealed that all three genes induce transduction of a mobile, graft-transmissible signal. Pea therefore also contains a floral repression pathway, operating only in short days, which functions through the transmission of a signal from vegetative tissue to the apex (Weller *et al.* 1997).

Analysis of the pea genome, alongside that of other legumes, has shown that the *FT* gene family is apparently extended in these plants. Pea contains five *FT*-like loci (Weller *et al.* 2009; Hecht *et al.* 2011). One of these has been shown to correspond to the *GIGAS* locus, and to induce the flowering response in both long and short days, but the functions of the other four are unknown. Given the data in species such as beet and poplar which show that FT signalling can induce different developmental transitions, it will be important to establish the function of each of these paralogous genes in order to fully understand the floral transition in pea.

The floral transition has recently been analysed in other legume species, with the models *Medicago truncatula* and *Lotus japonicus* being particularly amenable to genetic and genomic approaches. For example, Laurie *et al.* (2011) analysed the functions of the five *FT* loci in *Medicago*, demonstrating that three of these genes could complement the Arabidopsis *ft* mutant. They concluded that only *MtFTa1* expression was upregulated by both long days and vernalization, and that the phenotypes both of plants ectopically expressing *MtFTa1* and of plants with a transposon-induced mutation in the gene indicated that MtFTa1 is the sole inducer of flowering in response to vernalization, but that it works partially redundantly with other signals to induce flowering in response to long days.

8.6 Flower induction in other species

The breadth and depth of understanding of the Arabidopsis floral transition has facilitated a candidate gene approach to flowering in many species.

This approach has been further enhanced by the enormous amounts of genomic data that have emerged from expressed sequence tag (EST), transcriptome, and genome projects over the last decade. Accordingly, reports have indicated that many of the major players in the Arabidopsis story, particularly FT, are also involved in flowering in species as diverse as sunflower (Asterales), maize (Poales), and various legumes (Fabales). Although the details of control of expression, mode of action, and interaction with other proteins vary from species to species, it is becoming apparent that many of the molecular switches in the floral transition are conserved across the angiosperms.

Development of Flowers

CHAPTER 9

Changes at the shoot apical meristem in response to floral induction

Section IIA of this book was concerned with the multiple environmental and developmental stimuli that combine to induce the transition from the vegetative state to the floral state. Section IIB turns to look at the consequences of that transition. In this chapter we consider the changes that occur at the shoot apical meristem once the decision to flower has been taken by the plant. These changes involve the expression of a succession of floral meristem identity genes, which convert the meristem to the reproductive state. Floral meristem identity genes can be defined as those genes that specify the floral fate of lateral meristems arising from a reproductive shoot apical meristem. At the same time, it is important that apical meristem indeterminacy is maintained, to allow multiple flowers to be produced. Later chapters in this section will analyse the development of the floral organs from a floral meristem, and the development of the all-important gametes within the reproductive organs themselves.

9.1 Physiological changes at the shoot apical meristem

Once floral induction has occurred, and the plant is committed to flowering, many physiological changes occur at the shoot apical meristem (SAM) to prepare it for the developmental transition it must undergo. Without these physiological changes, the floral transition will not usually occur successfully. An increase in the rate of RNA synthesis is one of the earliest events that occur in the SAM, in order to support the transcription of novel genes. The rate of protein synthesis also increases (Jacqmard *et al.*

1972). Both changes are predictable, as approximately 50% of the transcripts in a typical floral organ are not found in leaves of the same species, highlighting the need for the transcription and translation of flower-specific genes. It has been shown that application of chemical inhibitors of RNA or protein synthesis to the SAM of *Sinapis alba* (white mustard, Brassicales) is sufficient to prevent the flowering response, even when the plant has been exposed to the usual inductive conditions. Recent high-throughput approaches have started to pin down the precise transcriptional changes that occur at the SAM after floral evocation. For example, Torti *et al.* (2012) used laser microdissection along with deep sequencing of cDNA populations to identify 202 genes induced in the Arabidopsis SAM at the point of floral transition.

RNA and protein synthesis are not the only metabolic processes that occur as the meristem undergoes its transition to the reproductive state. Rises in sucrose and ATP levels have also been recorded in *S. alba*, presumably to support the increased rate of RNA and protein turnover (Bodson 1985; Bodson and Outlaw 1985). Increases in invertase activity and mitochondrion number have also been correlated with the change in meristem state.

The cell cycle is another important process affected by the change in meristem state. The rate of cell division within the meristem has been shown to increase following floral induction in several species. For example, in *S. alba* exposure to one long day is sufficient to induce the photoperiodic floral promotion pathway, and leads to the plant flowering. The single long day also results in a mitotic wave across the SAM, caused by a shortening of the G2 phase of cycling cells and a return to cycling by non-cycling G2 cells. G2 is the cell cycle phase between S (DNA synthesis) and M (mitosis), and is where many meristematic cells 'rest' between cell cycles. This shortening of the G2 phase is followed by a shortening of G1 and S phases, which together speed up cell cycling and also synchronize it so that all cells in the SAM are cycling together. Meristematic cell cycle length is reduced from approximately 86 hours to around 32 hours as a result of these changes, although it is not clear why the more rapid division of cells is necessary for a successful floral transition (Gonthier *et al.* 1987).

9.2 Shoot apical meristem anatomy

One of the key changes that occur at the SAM in response to the floral transition is an alteration in the expression patterns of certain genes. Many such expression pattern changes were discovered by comparing mutant and wild type plants, and others have been described more recently using transcriptomic approaches (Torti *et al.* 2012).

The Arabidopsis SAM is a typical, if small, example (see Fig. 9.1). It consists of around 100 cells, and that number remains constant despite continuous cell division and cell differentiation throughout the life of the plant. The SAM is composed of three cell layers. The outermost layer is called the L1, and is a single cell thick. These cells divide anticlinally to generate the epidermis of the plant. Below the L1 layer is the L2, usually 2–3 cells thick, and again composed of cells which predominantly divide anticlinally in the meristem, but also divide periclinally in the leaf primordia to generate the mesophyll layers. The L3 layer may be many cell layers thick. Cells in the L3 divide periclinally and anticlinally. Since these cell layers are maintained throughout the life of the SAM and the plant that it produces, all cells derived from a single layer are clonally

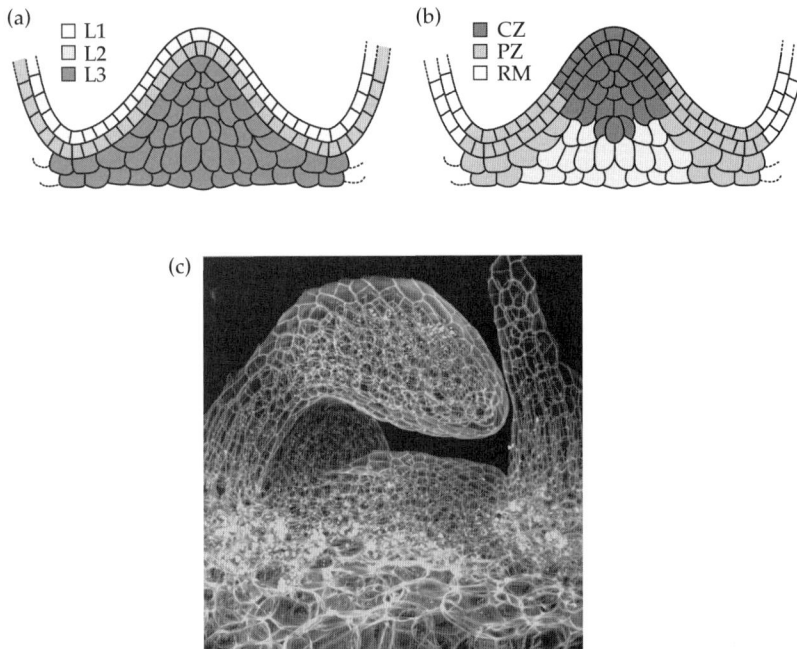

Figure 9.1 Anatomy of the shoot apical meristem (SAM). (a) The SAM is made up of three cell layers—L1, L2, and L3. (b) It can also be divided into a central zone (CZ), peripheral zones (PZ), and rib meristem (RM), with lateral meristems arising at the sides. (c) Three-dimensional reconstruction of an Arabidopsis SAM, with developing leaf primorida. The tissue was fixed, stained using a propidium iodide pseudo-Schiff reaction to label cell walls, mounted in a chloral hydrate-based medium, and imaged using a Leica SP scanning laser confocal microscope. Image kindly supplied by John Runions and Jim Haseloff (University of Cambridge).

related to one another. This can be seen in chimaeric mutants, where, for example, the L1 cells may be lacking a gene required for pigment synthesis but the L2 and L3 cell layers are wild type (reviewed by Tilney-Bassett 1986).

Besides its layers, the SAM can also be divided into three functional zones. In the very centre of the SAM is the central zone, or stem cell niche, characterized by cells with a low rate of division relative to the rest of the SAM. These cells are thought of as a back-up population, which can enter into more rapid mitosis should cell proliferation in the rest of the SAM not meet the needs of various developmental programmes. The peripheral zone encircles the central zone like a doughnut. The peripheral zone is where the majority of cell division occurs, and on the edges of the peripheral zone cells are recruited to various developmental programmes and become part of leaves or other plant organs. At the base of the meristem is the region known as the rib meristem. Periclinal cell divisions within the rib meristem generate the stem of the plant, and ensure that the SAM stays at the top of the developing structure.

Extensive analysis of SAM function and maintenance in Arabidopsis has been conducted in recent years, and this has established a well-supported model for its functioning and regulation (reviewed by Stahl and Simon 2010). The *WUSCHEL* (*WUS*) gene is expressed in the organizing centre, a group of cells below the central zone, and encodes a homeodomain transcription factor necessary for the maintenance of cell division in the central and peripheral zones. The central zone cells secrete CLAVATA3 (CLV3), a small glycopeptide that acts as a ligand for the CLAVATA1 (CLV1) leucine-rich repeat receptor kinase, present on the outer membranes of cells in the organizing centre and peripheral zones. Interaction of CLV3 and CLV1 triggers cell differentiation in response to excess cell division, and also reduces expression of *WUS*. Together these genes act as a feedback loop that serves to keep meristem size constant under all environmental conditions.

9.3 Gene expression patterns in the shoot apical meristem

The expression patterns of genes thought to play a role in conversion of the SAM can be analysed using a number of techniques. One option is to fuse the promoter of a gene of interest to a reporter gene, such as GUS (β-glucoronidase, which stains blue) or GFP (green fluorescent protein, which fluoresces under appropriate illumination), and transfer the construct into a transgenic plant (Jefferson *et al.* 1987). GUS staining kills the tissue, but GFP allows the monitoring of expression patterns and their changes in a living plant (Haseloff and Amos 1995). For both reporters there can be difficulties associated with the length of time for which the signal lasts, making accurate analysis of temporal changes in expression pattern difficult. These techniques are also only possible in a system where it is easy to isolate promoter sequence and plant transformation is routine. They are often used in Arabidopsis, but less frequently in other species. A good alternative is to use *in situ* hybridization to investigate expression pattern. In this technique, plant tissue is fixed, dehydrated, and embedded, as for normal microscopy, and then sectioned. Thin sections of tissue are placed on microscope slides, and are then probed with the antisense strand of a transcript of interest, usually fused to a coloured marker. Coloured stain can then be observed through a microscope in cells expressing the gene of interest. An advantage of *in situ* hybridization is that it is possible to probe serial sections with different genes, giving an accurate impression of the expression patterns of multiple genes at a particular time point (Jackson 1992).

Genes may have one of three possible expression patterns with regard to the floral transition at the meristem. Some genes have fixed patterns of expression, which do not change as the meristem undergoes the transition. These genes are likely to be very important in general meristem function, and may play roles in maintaining meristem activity. Genes such as the *CLV* genes mentioned earlier may fall into this category, and do not play specific roles in the floral transition.

Other genes may show modified patterns of expression. These genes are expressed in the SAM at both the vegetative and reproductive stages, but the layers or zones in which they are expressed vary as the SAM undergoes the floral transition. Such genes are likely to be involved in processes that require modification when the meristem changes state, such as the control of phyllotaxy (the positioning of lateral organs and lateral meristems). One example of an Arabidopsis gene with a modified expression pattern is *CDC2*, which encodes a protein kinase that controls both the G1 to S phase cell cycle transition and entry into mitosis, and so is essential for cell division and growth. The gene is expressed throughout the meristem in the vegetative state, but transcript is concentrated in the rib meristem region of floral meristems, arising either on the flanks of the SAM or when a final terminal flower is produced (Martinez *et al.* 1992). This pattern corresponds to the shift in maximum mitotic activity to the rib region in the floral meristem, to build the carpels and receptacle of the mature flower.

The final category of genes contains those with stage-specific expression in the apical meristem. Such genes are only expressed either following the floral transition, or before the floral transition. These are the key genes whose expression is necessary to convert the meristem to the floral state, and it is these genes that we call the floral meristem identity genes.

9.4 Floral meristem identity genes act downstream of the flowering-time integrators

In Chapter 7, three key Arabidopsis flowering-time integrator genes were introduced. The *FT* gene encodes a membrane-associated protein which integrates signals from the photoperiodic induction pathway (through CO), and the autonomous promotion, inhibition, and vernalization pathways (all through its repression by FLC). The FT protein moves from the leaves to the SAM. The *SOC1* gene encodes a MADS box transcription factor that is repressed by FLC (and thus integrates the autonomous promotion, inhibition, and vernalization pathways) but activated by long days (through CO) and by gibberellin. The third flowering-time integrator, *LEAFY*, primarily acts as a floral meristem identity gene whose expression is induced by SOC1. However, it can also be directly activated by gibberellin, causing it to be classed as a flowering-time integrator as well. The three flowering-time integrators activate the expression of the floral meristem identity genes, which are essential for conversion of

the meristem to the reproductive state. These genes are not usually expressed prior to the floral transition, and are therefore genetically and developmentally downstream of the flowering-time integrators.

9.5 Floral meristem identity genes

In this section we shall consider the function of each of the major floral meristem identity genes in turn, looking at both the Arabidopsis genes and their orthologues in *Antirrhinum majus* (Lamiales). Antirrhinum is a good genetic model, with active transposons generating many mutants, and has large flowers in which morphological abnormalities are easily spotted. Meristem conversion has been studied almost as extensively in Antirrhinum as in Arabidopsis, and the combination of data from both species has allowed important advances in our understanding of the subject. A key characteristic of floral meristem identity is that many of the genes are involved in mutual regulatory interactions. These interactions make any discussion of floral meristem identity more complex, but they serve in nature to provide buffering against environmental noise and to ensure that meristem conversion occurs correctly and does not revert once it has occurred (Blazquez *et al.* 2006). Although *LEAFY*, *APETALA1*, and *CAULIFLOWER* are considered to be the three most crucial floral meristem identity genes, we shall also discuss the roles of *TERMINAL FLOWER 1*, *APETALA 2*, *UNUSUAL FLORAL ORGANS*, and *AGAMOUS*, all of which are necessary for the development of normal, determinate floral meristems on the flanks of a normal, indeterminate inflorescence meristem.

9.5.1 LEAFY (LFY)

The central role of *LFY* in the conversion of the meristem is a major unifying principle of angiosperm development, with *LFY* orthologues identified in many species from diverse parts of the plant kingdom (Jack 2004). Interestingly, *LFY* appears to be a plant-specific gene, with no clear homologues in any non-plant genome. It has recently been shown that its role in the moss *Physcomitrella patens* is to ensure cell division within the developing embryonic sporophyte, which arrests at the first cell division following fertilization in a mutant lacking both of the moss *LFY* orthologues (Maizel *et al.* 2005; Tanahashi *et al.* 2005). In gymnosperms there are also two orthologues of *LFY,* and expression patterns suggest that they play a role in meristem function and conversion (Mouradov *et al.* 1998; Vazquez-Lobo *et al.* 2007; reviewed by Moyroud *et al.* 2010). In angiosperms the LFY protein is both necessary and sufficient to specify a floral meristem fate in many (but not all) developmental contexts, and the protein has a dual role in integrating signals from the floral inductive pathways (see Chapter 7) and acting as a floral meristem identity gene by activating downstream genes controlling floral organ development.

LFY is the best characterized gene involved in conversion of the meristem to the floral state. Plants that are *lfy* mutants produce inflorescence shoots where their flowers should be, subtended by bracts. Some flowers do develop in the mutant lines, but they are abnormal and have characteristics of secondary inflorescences, including indeterminacy. Thus the basic conversion of the meristem to the reproductive state has occurred in these lines, but the flowers themselves cannot be properly produced without *LFY* activity. *LFY* was isolated using the Antirrhinum orthologue (*FLORICAULA, FLO*) as a probe in the genomic region to which *LFY* had been mapped (Weigel *et al.* 1992). The Antirrhinum *FLO* gene had been cloned first from a transposon tagged mutant line (Coen *et al.* 1990). The Arabidopsis and Antirrhinum proteins are 70% identical at the amino acid sequence level, and the *flo* mutant has a phenotype similar to that of the *lfy* mutant, with bracts subtending ectopic inflorescence shoots (see Fig. 9.2a). However, the *flo* mutant is more extreme in phenotype than even a null *lfy* mutant, as the conversion of flowers to inflorescence meristems is complete in Antirrhinum. Neither protein has real sequence similarity to previously described proteins with known biochemical functions. Since their initial isolation they have been shown to function as transcription factors, binding DNA directly through a helix-turn-helix motif hidden in an unusual protein fold (Hames *et al.* 2008). This function is consistent with their role in the activation of downstream genes involved in the development of floral organs, and the LFY protein has been shown to be

Figure 9.2 Early floral meristem identity mutants. (a) The Antirrhinum *flo* mutant (left) has inflorescence shoots produced in place of the flowers found in the axils of wild type bracts (right). Image kindly supplied by Enrico Coen (John Innes Centre). (b) The Arabidopsis *ap1* mutant is slightly better converted to the floral form, with indeterminate floral structures arising from the meristem. See also Plate 4.

localized to the nucleus, as expected for a transcription factor.

LFY is expressed strongly in young floral organ primordia and floral meristems, but only in inflorescence meristems at the initial point of transition, consistent with its role in specifying floral identity (Weigel *et al.* 1992). Strangely, although *FLO* is also expressed in floral meristems and bract primordia, it is only expressed in the organ primordia that will become sepals, petals, and carpels, but is absent from the developing stamen primordia (Coen *et al.* 1990). Both genes are initially expressed in the inflorescence meristem at the floral transition, but rapidly resolve outwards to a region where the lateral meristems and bracts are specified.

Experiments to investigate the putative function of *LFY* in vegetative meristems demonstrated that the LFY protein is a developmental switch sufficient to convert all lateral meristems into solitary flowers, and even to convert the SAM into a single flower, although this latter conversion takes some developmental time to complete (Weigel and Nilsson 1995). Ectopic expression of *LFY* in transgenic Arabidopsis plants using the CaMV 35S promoter results in the conversion of all lateral or axillary meristems into flowers. Thus the axillary meristems in the axils of the rosette leaves form precocious flowers, instead of remaining dormant and forming branches later in development. Similarly, the lateral meristems in the axils of the cauline leaves also form single flowers. Eventually, the SAM is also converted to a terminal flower. Strikingly, the same construct was shown to have the same dramatic effects on meristem identity in other species, including tobacco (*Nicotiana tabacum*, Solanales). Ectopic expression of *LFY* in aspen (*Populus tremula*, Malpighiales) resulted in the production of flowers in the axils of leaves, and the early conversion of the SAM to a terminal flower. Whereas aspen trees ordinarily take around 20 years to flower, these conversions all occurred within a few months (Weigel and Nilsson 1995). Similarly, ectopic expression of *LFY* in citrus trees (Sapindales) has also been shown to reduce their time to flowering, from over 6 years to within 1 year (Pena *et al*. 2001).

The immediate cellular consequences of LFY activity are not yet well described, although we do know that LFY is directly involved in the activation of expression of *AP1* and *CAL* (see below), as well as *AGAMOUS* and *APETALA3*. Recent reports suggest that it also activates the expression of a number of other genes, including *LATE MERISTEM IDENTITY 1* (*LMI1*), which encodes a homeodomain leucine-zipper transcription factor. LMI1 is required for activation of *CAL* expression, and in its absence the meristem defects of weak *lfy* mutant alleles are substantially worsened (Saddic *et al*. 2006). Progress in understanding LFY function has been much enhanced by the development of a model for the protein's DNA binding specificity (Moyroud *et al*. 2011). Application of this model to the Arabidopsis genome identified 2677 genes adjacent to LFY binding sites, including key floral regulators such as *AP1* and *TFL1* (see below). It seems likely that LFY has such a strong effect on meristem identity because it activates transcription of a suite of

other genes, each of which is crucial for the proper development of a floral meristem.

Recent studies of LEAFY function in a wider range of angiosperm species have revealed additional roles for the protein. Specification of inflorescence architecture is related to *LFY* expression pattern in a range of species. Both Arabidopsis and Antirrhinum produce racemes (in which the SAM remains indeterminate and the flowers develop from axillary meristems). However, many other plants, including petunia and tobacco (both Solanales), produce cymes (in which the primary meristem forms a flower and the axillary meristem becomes the new inflorescence meristem, before itself becoming a flower when a new axillary meristem is ready to take over, and so on indefinitely). In species with cymes, LFY interacts with the UFO homologue (see Section 9.5.6) to specify floral identity, and the *UFO* homologue is only expressed in the apical meristem, and then in the subsequently dominant meristems, after the axillary meristems which will take over indeterminate growth have been formed (Lee *et al*. 1997; Souer *et al*. 1998; reviewed by Moyroud *et al*. 2010). Similarly, studies in maize (Poales), pea (Fabales) and Arabidopsis have suggested that LFY also has a role in specifying meristem growth, more generally than just at the floral stage. Moyroud *et al*. (2010) speculated that the functions of LFY in stimulating meristematic growth and specifying floral meristem identity might be separable, with different degrees of importance in different species. Finally, a recent report by Yamaguchi *et al*. (2012) implicated LFY in adaxial identity of cells in the pedicel, influencing floral positioning.

9.5.2 APETALA 1 (AP1)

Plants with mutations in *AP1* have phenotypes very similar to that of the *lfy* mutant, suggesting that *AP1* is also necessary for the floral identity of the meristem. The phenotype is, if anything, slightly less dramatic, with a fuller conversion of the inflorescence meristem to the floral form (see Fig. 9.2b). Sepals do form in *ap1* mutants, but they are leaf-like, and secondary flowers form within the axils of these sepals. In turn, the first whorl organs of these secondary flowers may also form tertiary flowers in their axils (Mandel *et al*. 1992). *AP1* encodes a transcription factor which is a member of the MADS box family, and was isolated by a combination of mapping and a candidate gene approach (Mandel *et al.* 1992). MADS box transcription factors, such as SOC1 and FLC (see Chapters 4 and 7), are characterized by a region of 50 amino acids containing a DNA-binding and dimerization domain. The MADS box family is very large, with over 40 members apparent in the Arabidopsis genome. *AP1* is expressed in the floral meristem, from the first time point at which they are visible on the flanks of the inflorescence meristem, and remains expressed in petal and sepal primordia during flower development (Mandel *et al.* 1992).

The Antirrhinum orthologue of *AP1* is *SQUA-MOSA (SQUA)*, which encodes a MADS box transcription factor with 68% sequence identity to AP1 (Huijser *et al.* 1992). The *squa* mutant is more extreme than the *ap1* mutant, but slightly less dramatic than *flo*. Most floral meristems are converted to inflorescence meristems, in a reiterating pattern, but the plant does occasionally produce anomalous flowers. Like *AP1*, *SQUA* is also expressed in the wild type floral meristem, but not the inflorescence meristem, and it remains expressed in all developing floral organs except the stamens.

In both Arabidopsis and Antirrhinum, double mutant phenotypes between *lfy/ap1* and *flo/squa* are more extreme phenotypically than either single mutant. The *lfy/ap1* double mutant is very similar phenotypically to the *flo* single mutant, in that the conversion of floral meristems to inflorescence meristems is essentially complete.

Genetic analyses have indicated that *AP1* functions downstream of *LFY* in the conversion of the meristem to the floral state. Ectopic expression of *LFY* does not confer any phenotypic change in an *ap1* mutant background, implying that LFY works through *AP1* (Weigel and Nilsson 1995). In contrast, ectopic expression of *AP1* from the same CaMV 35S promoter does induce early flowering and conversion of meristems to the floral state, even in a *lfy* mutant background. The flowers produced by these plants are abnormal, with perturbations in organ identity and development. These data confirm that *AP1* functions downstream of *LFY*, but that the pathway is not entirely linear, with LFY being necessary for the activation of certain floral organ

identity genes not regulated by AP1 (Mandel and Yanofsky 1995). Nonetheless, ectopic expression of *AP1* in citrus was able to induce early flowering, just as ectopic expression of *LFY* was (Pena *et al.* 2001). Further evidence for *LFY* acting upstream of *AP1* comes from work by Wagner *et al.* (1999), who showed that inducible activation of *LFY* resulted in an almost immediate upregulation of *AP1* expression. Furthermore, the *AP1* promoter contains a region bound by LFY *in vitro*, suggesting that the regulation of *AP1* expression by LFY is direct (Parcy *et al.* 1998). More recently, Pastore *et al.* (2011) showed that the activation of *AP1* expression by LFY involved a physical interaction of LFY with the LATE MERISTEM IDENTITY 2 MYB transcription factor (also known as AtMYB17). LFY is not the only regulator of *AP1* expression, since (as was described in Chapter 7) the FT-FD protein complex also directly activates *AP1* expression.

Matters are slightly less clear-cut in Antirrhinum, where the *squa* mutant contains normal amounts of *flo* transcript and vice versa (Huijser *et al.* 1992), and the single mutants are so much more extreme than in Arabidopsis that the double mutant suggests much less synergy between the genes. It has been suggested that in Antirrhinum the FLO and SQUA proteins act together in the induction of downstream genes, rather than in a more linear pathway.

AP1 exerts its affect on meristem identity through its role as a transcriptional activator of other genes. The most recent and exhaustive study (Kaufmann *et al.* 2010) identified a total of 249 genes which were both significantly differentially expressed in response to *AP1* expression and had promoter regions that the AP1 protein bound *in vitro*. These genes included a disproportionately high number of sequences encoding further transcription factors, and also included *LFY*, *FD*, and *TFL1* (see below), suggesting that AP1 is involved in a transcriptional feedback loop with its own regulators.

The function of AP1 in floral meristem identity is apparently ancient, at least within the eudicots. Pabon-Mora *et al.* (2012) used virus-induced gene silencing to explore the role of the *AP1*-like genes in two species of the Ranunculales (opium poppy and Californian poppy), and found that the genes were necessary for correct transition of the meristem.

9.5.3 CAULIFLOWER (CAL)

The *CAL* gene is probably the least well understood of the floral meristem identity genes. Because it represents a Brassicaceae-specific duplication of *AP1*, around 60 million years ago, no comparative analysis with systems such as Antirrhinum has been possible (Lowman and Purugganan 1999). *CAL* was isolated in a search for genes with sequence similarity to *AP1*, and encodes a MADS box transcription factor that is 76% identical in amino acid sequence to AP1 (Kempin *et al*. 1995). The *cal* mutant has no discernible phenotypic differences from a wild type plant, but the *ap1/cal* double mutant plant has a very striking phenotype. Where flowers should be, the plant continues to proliferate inflorescence meristem, from which no organs develop. This results in the production of heads of meristematic tissue several millimetres across and resembling miniature vegetable cauliflowers. CAL is therefore clearly necessary for the correct transition of the inflorescence meristem to the floral meristem and the production of flowers. However, it is also apparent that *AP1* can compensate for the loss of *CAL*, since the *cal* single mutant is phenotypically wild type. The partial redundancy between these genes has made determining the function of *CAL* more difficult. A recent analysis of the four main functional domains of AP1 and CAL, generating chimaeric proteins containing all possible combinations of the domains from the two proteins, concluded that the K domain (a region of MADS box proteins involved in interactions with other proteins; see Chapter 11) was particularly significant in differentiating the functions of AP1 and CAL (Alvarez-Buylla *et al*. 2006).

CAL mRNA accumulates in young floral meristems, but does not persist in any floral organ during development, unlike *AP1*. This difference suggests that *CAL* cannot compensate for the loss of *AP1* in its later activities, such as the activation of particular floral organ identity genes (Kempin *et al*. 1995). One clear function of *CAL* is to enhance *AP1* activation of *LFY* expression. Although genetic and molecular data agree that *AP1* acts downstream of *LFY*, in the *ap1/cal* double mutant *LFY* expression levels are significantly reduced, especially in the early stages of its expression (Bowman *et al*. 1993). This has led to the conclusion that the AP1/CAL duo enhances *LFY* expression, perhaps operating as a positive feedback loop to ensure the full conversion of the meristem to the floral state.

A *CAL* orthologue has been isolated from *Brassica oleracea*, and it has been shown that the variety grown for vegetable cauliflowers (*B. oleracea* var. *botrytis*) contains a mutation within the *CAL* coding sequence, resulting in the production of a truncated 150-amino-acid protein rather than the wild type 255-amino-acid version (Kempin *et al*. 1995).

9.5.4 TERMINAL FLOWER 1 (TFL1)

The *TFL1* gene is required for a less obvious aspect of meristem conversion, namely the maintenance of the apical meristem in the inflorescence state, and therefore indeterminate. When *LFY* is first expressed, its transcript can be found right up into the apex of the plant, but it is shortly afterwards restricted to the flanks of the apical meristem, where the lateral floral meristems arise. Without this restriction, LFY would convert the apical meristem to the floral state and the plant's growth would stop. The TFL1 protein is involved in repressing the floral nature of the apical meristem, and maintaining it in the inflorescence state and indeterminate. In *tfl1* mutants, up to five ordinary flowers are made and then the apical meristem itself becomes floral, producing a single terminal flower. In the *tfl1* mutant the *LFY* and *AP1* genes are expressed in apical regions of the meristem as well as in the usual peripheral regions, indicating that one role of TFL1 is to repress *LFY* and *AP1* expression in the apex. This appears to work through a feedback loop. Before the vegetative to reproductive transition, *TFL1* is expressed in the apical meristem but at low levels and not in the apex itself. It is upregulated in the apex itself following the floral transition, probably by LFY, and then represses *LFY* and *AP1* expression in the apical meristem (Bradley *et al*. 1997). Conversely, in the *lfy* mutant, *TFL1* is expressed in the ectopic shoots that are formed where flowers should be, and produced from the axillary meristems on the flanks of the apical meristem (Ratcliffe *et al*. 1999). So, in the absence of the LFY-induced enhancement of *TFL1* expression in the SAM itself, and exclusion from lateral meristems, the gene is expressed in lateral meristems which remain indeterminate. Further evidence for the feedback loop comes from transgenic experiments. Ectopic expression of *LFY* or *AP1* from the CaMV 35S promoter results in a reduction in *TFL1* transcript levels in the apical meristem (Liljegren *et al.* 1999), whereas ectopic expression of *TFL1* from the same promoter results in reduced levels of *AP1* and *LFY* transcript (Ratcliffe *et al.* 1998). This feedback loop is clearly essential to maintain the apical meristem in the indeterminate state while allowing determinate floral meristems to form on its flanks.

The Antirrhinum orthologue of *TFL1* is *CEN-TRORADIALIS* (*CEN*). *In situ* hybridization experiments have revealed that *CEN* is expressed 2 to 3 days after *FLO*, and not at all in *flo* mutants. Thus in Antirrhinum it also appears that *FLO* expression induces *CEN* expression, and the CEN protein then excludes *FLO* expression from the apical region of the meristem, keeping it in the inflorescence form.

The *CEN* gene was cloned before *TFL1*, using a transposon tagged mutant line of Antirrhinum (see Fig. 9.3; Bradley *et al.* 1996). Using this sequence as a

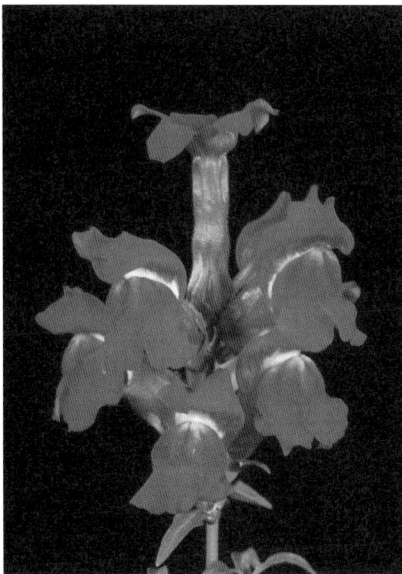

Figure 9.3 Maintaining indeterminacy. The *cen* mutant in Antirrhinum experiences conversion of the indeterminate inflorescence meristem to a determinate floral meristem. Image kindly supplied by Enrico Coen (John Innes Centre).

probe, *TFL1* was isolated the following year (Bradley *et al.* 1997), and was concurrently isolated from a T-DNA insertion mutant by another group (Ohshima *et al.* 1997). The two proteins are 70% identical, and there are also very similar proteins present in the rice genome. The biochemical nature of TFL1 and CEN is not yet clear, but they have most similarity to animal phosphatidylethanolamine-binding proteins, which associate with membrane protein complexes. This suggests that they are involved in signal transduction at the cell membrane. However, TFL1 is also 55% identical to the flowering-time integrator FT, which clearly functions in a transcriptional complex with FD (Kardailsky *et al.* 1999). Recent reports confirm that TFL1 also colocalizes with FD in the nucleus, and appears to function as part of a transcriptional repressor complex, even though it has no DNA binding ability itself (Hanano and Goto 2011). These data suggest that FT and TFL1 operate by mediating transcriptional regulation, but they play opposite roles in the conversion of the meristem (Kardailsky *et al.* 1999). As was discussed in Chapter 7, Hanzawa *et al.* (2005) showed that only a single amino acid controlled the difference between the protein functioning as FT or as TFL1. The single amino acid is part of a region which may act as a surface for interactions with other molecules (Ahn *et al.* 2006).

Indeterminate inflorescences and determinate inflorescences are widely scattered throughout the plant phylogenetic tree, suggesting either that indeterminacy evolved multiple times, or that it was lost multiple times. The presence of the same mechanism to maintain an indeterminate meristem, operating through the same proteins, in two species as phylogenetically distant as Antirrhinum and Arabidopsis, suggests that indeterminate inflorescences may be the ancestral state, at least within the eudicots, with indeterminacy lost in those species with determinate inflorescence meristems (Bradley *et al.* 1997). Iwata *et al.* (2012) recently demonstrated that the continuous flowering (extremely indeterminate) varieties of rose (Rosales), thought to have arisen in China some time between AD 1000 and AD 1800, result from a transposon insertion in the *TFL1* orthologue. The wild type gene confers a once-flowering phenotype, where the meristem is indeterminate in spring only.

9.5.5 APETALA 2 (AP2)

The *AP2* gene has two main roles in flower development. It is required for the maintenance of the floral
identity of the meristem, and so is classed as a floral
meristem identity gene, but it is also involved later
in sepal and petal development, so is also classed
as an organ identity gene (see Chapter 10). In this
section we shall focus only on the role of *AP2* in
specifying floral meristem identity. In long days,
ap2 mutant plants are phenotypically normal with
regard to meristem identity, but in short days they
resemble *ap1* mutants, with sepals converted to
leaves and indeterminate flowers developing inside
these leaves, often in a reiterating pattern (Bowman
et al. 1993). Thus without signals from the photoperiodic induction pathway, *ap2* mutant plants fail
to convert their inflorescence meristems into floral
meristems. This suggests that the AP2 protein is
required for the conversion of inflorescence meristems into floral meristems, but only when the plant
is flowering in response to the autonomous or gibberellin promotion pathway, not the photoperiodic
promotion pathway. It may be that high levels of *FT*
expression can bypass the need for *AP2* activity (see
Chapter 7).

AP2 encodes a protein with very little similarity
to anything else that had been functionally characterized when it was cloned. However, it has a
domain that includes a nuclear localization signal,
suggesting that it might be a transcription factor.
It also contains a domain initially thought to be involved in protein–protein interactions. This domain
is 68 amino acids long, and is repeated directly, giving a central 136-amino-acid domain, which has
come to be called the AP2 domain (Jofuku *et al.*
1994). A variety of evidence suggests that the AP2
domain functions as a DNA-binding domain, supporting the idea that AP2 is a transcriptional regulator. The AP2 domain contains a region predicted to
form an amphipathic alpha helix, and tobacco AP2
domains have been shown to bind DNA (Ohme-
Takagi and Shinshi 1995). Dinh *et al.* (2012) showed
in particular that the second of the two repeats in
the AP2 domain binds an AT-rich target sequence.
Recent reports indicate that AP2 can function both
as a transcriptional activator and as a transcriptional repressor, and interacts with an unusually large
number of target genes (Yant *et al.* 2010).

Genes encoding over 150 proteins with AP2-like
domains have now been identified in the Arabidopsis genome, and the functions of most of them are
unknown (Wessler 2005). However, there is evidence
that AP2 itself may regulate the activity of some of
these genes, as *ap2* mutant plants have altered expression patterns of some *AP2*-like genes (Okamuro
et al. 1997). It is also becoming clear that AP2 has
roles in some parts of vegetative plant development, perhaps through the activation of *AP2*-like
genes. These roles will be considered more fully in
Chapter 10. The AP2 domain was initially thought
to be plant-specific, but recent evidence points to
its existence in cyanobacteria, bacteriophages, and
ciliates (Wessler 2005). The cyanobacterium AP2-like
protein has been shown to bind DNA *in vitro* (Magnani *et al.* 2004). It has been suggested that the plant
AP2-like proteins were derived by lateral gene transfer from a prokaryote, as they do not appear to be
present in any other eukaryote (apart from a single
ciliate) and they contain few or no introns, a sign of
a recent prokaryotic ancestor (Magnani *et al.* 2004).

AP2 is expressed continuously in both inflorescence and floral meristems, and in all four types
of floral organs, as well as in the stem and leaves
(Jofuku *et al.* 1994). However, this expression pattern
does not reflect the true activity of AP2, as recent
analysis has shown that it is post-transcriptionally
repressed in certain tissues by a microRNA (Chen
2004). This repression only really affects the role of
AP2 in floral organ identity, so will be discussed in
more detail in Chapter 10.

Antirrhinum orthologues of *AP2* were not identified through mutant screens, but by isolation of
AP2-like sequences and characterization of their
function. Two *AP2*-like genes have been identified,
and are called *LIPLESS 1* (*LIP1*) and *LIP2*. Plants
with mutations in either gene are phenotypically
wild type. A double mutant between both genes has
defects in floral organ identity but none in meristem conversion, suggesting that *AP2* and its Antirrhinum orthologues have divergent roles in floral
meristem identity (Keck *et al.* 2003).

9.5.6 UNUSUAL FLORAL ORGANS (UFO)

The *UFO* gene has an effect on meristem identity in
both long and short days. When *ufo* mutant plants

are grown in long days, the flowers have increased numbers of organs in all whorls, suggesting that the determinacy of the floral meristem is disrupted. When *ufo* mutants are grown in short days, the flowers are replaced by shoots which are intermediate in appearance between flowers and inflorescences, again suggesting that the determinacy of the meristem, and specifically its conversion from inflorescence to floral, is affected by this gene. *UFO* seems to be necessary, particularly in short days, to ensure the full floral identity of the lateral meristems, although its role is not as important as that of earlier floral meristem identity genes such as *LFY*, without which no conversion occurs at all (Levin and Meyerowitz 1995; Wilkinson and Haughn 1995).

The Antirrhinum orthologue of *UFO* is *FIM-BRIATA* (*FIM*), and the *fim* mutant is very similar to the *ufo* mutant, although the phenotype is less extreme. The *fim* mutant shows defects in floral organ development and reduced determinacy of the floral meristems, but does not have reversion of floral meristems to an inflorescence fate in the way that the *ufo* mutant has. *FIM* was isolated from a transposon tagged mutant line of Antirrhinum and found to encode a protein with very little similarity to proteins then in the available databases (Simon *et al.* 1994). *UFO* was isolated using *FIM* as a probe, and the two proteins were found to be 60% identical (Ingram *et al.* 1995). However, later analyses suggested that the FIM and UFO proteins contain an F-box motif, a domain present in cell cycle-associated cyclin proteins and some transcription factors, and which is known to provide substrate specificity to certain types of ubiquitin ligases. FIM/UFO may therefore act to target proteins for degradation. FIM has been shown to physically bind three members of a small family of proteins called the FIM-associated proteins (FAPs), which are related to human and yeast proteins involved in the targeting of proteins for degradation, particularly during cell cycle progression (Ingram *et al.* 1997). It has also been shown that the phenotypic consequences of ectopic expression of *UFO* from the CaMV 35S promoter are suppressed by mutations in the COP9 signalosome, a complex that selects proteins for degradation. One role of UFO and FIM therefore seems to be in the targeted degradation of particular proteins (Wang *et al.* 2003). If this targeted degradation is focused

on transcription factors or their cofactors then the functional consequence of UFO activity could be in transcriptional regulation. Chae *et al.* (2008) showed that UFO interacts physically with the C terminus of LFY, and that this complex binds the promoter of *APETALA 3* (a floral organ identity gene; see Chapter 10). For *AP3* transcription to occur, Chae *et al.* (2008) showed that proteasome activity was necessary, strongly suggesting that UFO functions by targeting inhibitory factors at the *AP3* promoter for degradation and allowing transcriptional activation by LFY to occur.

Both *UFO* and *FIM* are expressed in the floral meristems of the two species, consistent with their role as specifying the determinate floral nature of the meristem. *UFO* and *FIM* are expressed later in the development of the floral meristem than some other meristem identity genes, including *LFY*, but clearly earlier than the organ identity genes. Expression of both *UFO* and *FIM* initiates in the centre of the floral meristem and then resolves into a ring, leaving a clear area at the centre of the apex (Simon *et al.* 1994; Ingram *et al.* 1995). Neither of the genes is expressed in the apical meristem itself, but only in the lateral floral meristems. Later in development expression is restricted further to a few cells at the base of the sepal and petal primordia. The expression pattern of the *FAP* genes overlaps that of *FIM* itself, adding weight to the idea that they interact *in vivo* (Ingram *et al.* 1997).

FIM acts downstream of FLO and SQUA, with a complete absence of *FIM* transcript in the meristems of *flo* mutants, and a great delay in its timing of expression in *squa* mutants (Simon *et al.* 1994). UFO has also been shown to act with LFY. Even before the biochemical analyses of Chae *et al.* (2008) described earlier, it had been observed that in an *lfy* mutant background there is no phenotypic difference between *ufo* mutants, plants that are wild type at the *UFO* locus, and plants ectopically expressing *UFO* (Lee *et al.* 1997). These data led to the suggestion that LFY and UFO work together to activate downstream genes involved in floral organ development.

In contrast, UFO is also known to act upstream of AP1. Hepworth *et al.* (2006) showed that the floral meristems arising from a *ufo* mutant inflorescence meristem were initially bract-like as a result of the limitation of *LFY* and *AP1* transcript to a very small number of cells. Expression of *UFO* from the *AP1* promoter was unable to complement this phenotype, suggesting that UFO is required before AP1 is active.

The petunia orthologue of *UFO* is called *DOUBLE TOP* (*DOT*), and has a stronger role in meristem conversion than is seen for *UFO* or *FIM*. The *dot* mutant produces no floral meristems, resembling an Arabidopsis *lfy* mutant, whereas ectopic expression of *DOT* converts the SAM into a single flower (Souer *et al.* 2008). In this system, with an inflorescence architecture of cymes rather than racemes, it is *DOT* activity that is crucial for floral meristem identity, with the *LFY* orthologue (*ABERRANT LEAF AND FLOWER, ALF*) dependent on DOT for its activity and function. Together ALF and DOT activate expression of the floral organ identity genes in petunia (Souer *et al.* 2008).

9.5.7 AGAMOUS (AG)

The final floral meristem identity gene, *AG*, is very similar to *AP2* in that it is better known for its role in floral organ identity, but does have a limited role in ensuring the complete determinacy of the floral meristem. Flowers that are *ag* mutant are always indeterminate. They produce a whorl of sepals, two whorls of petals, and then start again, producing indefinite numbers of whorls of sepals and petals. Under short day conditions the floral meristem of an *ag* mutant may revert to an inflorescence meristem, producing an inflorescence shoot from within the flower. AG therefore plays a role not only in the determinacy of the floral meristem but also in the maintenance of the floral state of the meristem, after the switch has occurred as a result of the activities of the other proteins (Yanofsky *et al.* 1990).

The *AG* gene was isolated from a T-DNA tagged line, and encodes a MADS box transcription factor. The gene is expressed slightly later than *LFY* and *AP1*, and appears in the central region of the floral meristem, and then later in the developing stamen and carpel primordia in whorls 3 and 4 (Yanofsky *et al.* 1990).

Unusually, although the 5′ promoter region of *AG* is important for the regulation of its expression, so are some intragenic regions. Sieburth and Meyerowitz (1997) used a GUS reporter gene fused to

various elements of the *AG* genomic sequence to demonstrate that the large (3 kilobases) second intron is necessary for the correct spatial expression of *AG*. A reporter gene expressed solely from the *AG* promoter was expressed in vegetative tissues as well as in the floral meristem, indicating that the second intron contains sequence which represses vegetative expression of *AG*.

AG expression is known to be activated by LFY (Parcy *et al.* 1998), and the LFY protein has been shown to bind *in vitro* to the second intron of *AG* (Busch *et al.* 1999). The subsequent restriction of *AG* transcript to the third and fourth whorls of the developing flowers is the result of the combination of LFY and WUSCHEL (WUS) activity. *WUS* encodes a homeodomain transcription factor necessary to specify cellular proliferation in meristems, and without which shoot apical meristems terminate prematurely (Laux *et al.* 1996). The WUS protein has also been shown to bind to the second intron of *AG*, but only in the presence of LFY, and acts in a negative feedback loop (Lohmann *et al.* 2001). WUS activity induces *AG* expression in the correct primordia, but AG then downregulates *WUS* expression in those cells, causing them to stop proliferating and causing the meristem to assume a determinate state. Recent work has shown that this downregulation occurs through AG binding to the *WUS* locus and recruiting polycomb group genes that cause histone methylation and transcriptional silencing of *WUS* (Liu *et al.* 2011b). In the *ag* mutant the lack of this *WUS* downregulation results in the maintenance of the indeterminate state of the floral meristem (Lenhard *et al.* 2001). This feedback loop does not occur in a vegetative meristem, due to the lack of LFY activity.

Ectopic expression of *AG* in Arabidopsis alters not only flower morphology but also the numbers of organs in whorls and the numbers of whorls, confirming the role of AG in inducing the determinate meristem state.

The Antirrhinum functional orthologue of *AG* is *PLENA* (*PLE*), which was isolated from a transposon insertion line and encodes a protein that is 64% identical to AG (Bradley *et al.* 1993). The *ple* mutant phenotype is less extreme than *ag*, showing altered floral organ development and some indeterminacy, but not the reversion of floral meristems to the inflorescence state. The role of PLE seems to be more focused on floral organ identity and meristem determinacy than on maintenance of the floral meristem state. However, the regulation of the two genes is very similar, with a transposon insertion in the second intron of *PLE* causing ectopic expression of the gene in vegetative tissues, and recent reports confirming the presence of a LFY binding site in the intron (Causier *et al.* 2009). Differences in protein function may be attributable to the observation that PLE and AG represent members of different ancient lineages, both derived originally from the same MADS box transcription factor, but having experienced extensive divergence in the intervening time

(Kramer *et al.* 2004; Causier *et al.* 2005; for a more detailed discussion see Chapter 10).

To summarize, the conversion of the lateral meristems from inflorescence to floral following the transition of the apical meristem from vegetative to reproductive requires the activity of a suite of floral meristem identity genes. The timing of the expression of these genes suggests that they function sequentially. The flowering-time integrators, acting in response to the signals from the various floral promotion and inhibition pathways, activate the floral meristem identity genes, with *LFY*, then *AP1*, and then *AG* gradually committing the meristem further and further towards the floral state.

Development of the floral organs

In Chapter 9 we discussed the floral meristem identity genes, which act sequentially to commit the lateral meristems on the flanks of an inflorescence meristem to the floral state. This involves inhibiting the indeterminacy of the meristems, so that they do not continue to proliferate, by down-regulating WUS, the homeodomain transcription factor that maintains cell division within the meristem. It also involves activating the expression of the genes required for the specification of particular floral organs—that is, the floral organ identity genes. The pathway to flower development is a cascade, with flowering-time integrators activating floral meristem identity genes which in turn activate floral organ identity genes. The floral organ identity genes themselves encode transcription factors, and are predicted to activate the many structural genes required to form the new organs, as well as further regulatory genes controlling processes such as pigment synthesis. The expression patterns of the floral organ identity genes determine the positions of developing organs, and those expression patterns are themselves determined in part by cadastral genes (genes whose function is to specify the domain of expression of other genes). Floral organ identity genes act together as transcriptional complexes to ensure expression of the correct structural genes in each whorl of the developing flower.

10.1 The original ABC model of flower development

The ABC model of flower development was formally proposed in 1991, and was based on concurrent genetic studies in Arabidopsis and Antirrhinum (Coen and Meyerowitz 1991). The model has been integral to our modern understanding of flower development, and has been shown to apply, at least in basic form, to all angiosperm species tested (see Chapter 11 for a discussion of floral diversity and divergence from the model). The model rests on Goethe's foliar theory of the flower (see Chapter 2) in assuming that all floral organs are modified leaves, and that without the expression of key floral organ identity genes, the organs in the flower would develop as leaves.

Angiosperm flowers usually have four whorls. A whorl is the domain in which a single type of floral organ is produced, and may consist of one or more concentric circles (if one or more rings of each organ are produced). In some species the whorl boundaries are blurred, and in dioecious species one whorl of reproductive organs is absent from the adult stage (although it may be identifiable early in development). Some basal angiosperms produce flowers with a spiral arrangement of their floral organs, but even in these cases it is usually possible to distinguish between the spirals which make up each of the four sets of organs. According to the original ABC model, in the outer whorl (called whorl 1), only A function genes are expressed on top of the usual leaf developmental programme genes. In the presence of A function genes, the primordia undergo modifications to the basic leaf developmental programme which result in their development into sepals. In the second whorl in (whorl 2), both A and B function genes are expressed on top of the standard leaf developmental genes. In the presence of both A and B function genes, the primordia undergo more dramatic modifications that result in them developing into petals. In the third whorl (whorl 3) the A function genes are no longer expressed. The B function genes remain active, and C function genes are now also expressed. In the presence of both B and C function genes the primordia are modified even more dramatically and develop into stamens. Finally, in the innermost whorl (whorl 4) the B function genes are downregulated and only C function genes are expressed. In the presence of C function genes alone the primordia are modified and develop into carpels. This model is shown diagramatically in Fig. 10.1a.

There are two key points about the model that must be grasped in order to understand its simplicity. First, and most importantly, the model states that organ identity is a function of the ABC genes expressed in developing primordia, not a function of the position of the whorl of primordia within a flower. The organs do not develop directly as a result of the position of their primordia, but only as a result of the expression of the genes. For example, the model predicts that loss of B function gene expression in whorl 2 would result in whorl 2 organs developing as sepals. This is because the gene expression pattern in that whorl would be A alone, and A alone specifies sepal identity, even though the position of the primordia (in whorl 2) would lead one to predict the development of petals. This point underpins all of the subsequent discussion of

how the genes within the model work. The second key point about the model is that it assumes that the boundaries of A and C function gene activity are set by mutual antagonism, rather than by reference to external factors. A and C function do not overlap because each downregulates the other, and so they maintain a strict boundary. However, in the absence of either function (in a mutant, for instance) the expression of the other function would extend all the way across the meristem. Thus, in a mutant with a deletion of the C function genes, A function gene expression would be expected to invade whorls 3 and 4, with consequent effects on the identity of the organs in those whorls.

The model was derived from analysis of floral homeotic mutants with the wrong organs developing in particular whorls. The genetic screens which identified these mutants were conducted independently in Antirrhinum and Arabidopsis, by Enrico Coen's group in the John Innes Centre in Norwich, England, and Elliot Meyerowitz's group in CalTech, California. Meyerowitz's Arabidopsis work was based on a variety of mutant collections, generated by EMS and X-rays, while Coen's Antirrhinum work focused on transposon mutagenized plants. However, both groups identified mutants that fell into the same three classes, which they named A, B, and C, and which are described in Fig. 10.1b.

Mutants in class A had carpels instead of sepals in whorl 1, stamens instead of petals in whorl 2, normal stamens in whorl 3, and normal carpels in whorl 4. Mutants in class B had normal sepals in whorl 1, sepals instead of petals in whorl 2, carpels instead of stamens in whorl 3, and normal carpels in whorl 4. Class C mutants had normal sepals in whorl 1, normal petals in whorl 2, petals instead of stamens in whorl 3, and sepals instead of carpels in whorl 4. Although various mutants that showed abnormal development of particular organs were found, these were the only three classes of homeotic mutants observed—that is, only these three categories of mutants experienced complete conversion of one type of organ into another.

The ABC model was produced from these mutants by assuming that each mutant class carried a mutation in a particular gene (named A, B, or C). Each gene was postulated to be normally expressed (in a wild type plant) in the whorls that were

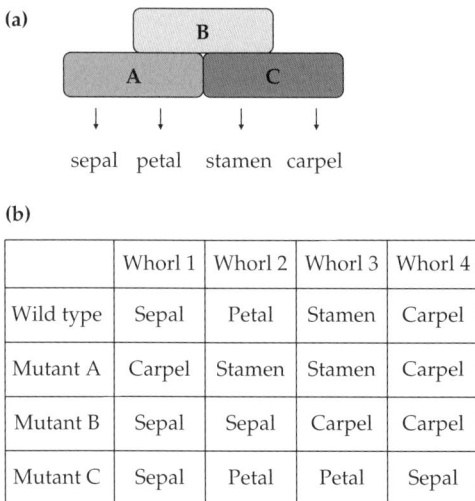

(a)

sepal petal stamen carpel

(b)

	Whorl 1	Whorl 2	Whorl 3	Whorl 4
Wild type	Sepal	Petal	Stamen	Carpel
Mutant A	Carpel	Stamen	Stamen	Carpel
Mutant B	Sepal	Sepal	Carpel	Carpel
Mutant C	Sepal	Petal	Petal	Sepal

Figure 10.1 The original ABC model. (a) Diagram of the expression patterns of three gene functions in four whorls, generating four wild type organs. (b) Table describing the organs found in the four different whorls of plants with mutations in A, B, and C function genes.

phenotypically disturbed in the mutant. So genes in class A were predicted to be active in whorls 1 and 2, genes in class B are active in whorls 2 and 3, and genes in class C are active in whorls 3 and 4. These expression patterns produce the model described above, and are sufficient to explain the phenotype of the B function mutant (loss of B function leads to an expression pattern of A, A, C, C, which the model predicts would generate sepal, sepal, carpel, carpel—precisely the phenotype observed in a B function mutant). To explain the phenotypes of the A and C function mutants it is necessary to postulate the mutual repression of the A and C function genes, so that loss of one causes the other to spread across the meristem. A C function mutant would then have the expression pattern A, AB, AB, A, which would lead to a phenotype of sepal, petal, petal, sepal. The reverse applies to the A function mutant. Finally, the model was tested using combinations of mutants. Coen and Meyerowitz (1991) used the model to predict the phenotypes of double mutant combinations in Arabidopsis, and then generated those mutants through genetic crosses. All double mutants carried the predicted phenotype, confirming the validity of the model. A triple mutant containing neither A, B, nor C gene activity produced four whorls of leaves where its floral organs should be, lending support to Goethe's foliar theory.

10.1.1 A function genes

The concept of A function is the part of the ABC model that has stood up least well to detailed molecular analysis. Extension of the model to a wide range of species has also revealed difficulties with conventional A function. Here we shall focus on A function in Arabidopsis, and comment on the situation in Antirrhinum. Chapter 11 discusses a revised concept of A function in the light of studies in a broad range of species.

In Arabidopsis there are two A function genes, although both play other roles in plant development. The *APETALA 2 (AP2)* gene (introduced in Chapter 9) acts as an A function gene, as well as playing its role in floral meristem identity. The *ap2* mutant has carpels/stamens/stamens/carpels in its four whorls, the expected phenotype of an A

function mutant if C function has spread across the meristem (Bowman *et al.* 1991). Indeed, *in situ* hybridization revealed that the C function gene, *AGAMOUS (AG)*, is expressed in whorls 1 and 2 in an *ap2* mutant background, confirming this aspect of the model (Drews *et al.* 1991). The *AP2* gene encodes the only non-MADS box protein in the ABC model, a protein containing the core AP2 domain that has since been identified in many other proteins (Jofuku *et al.* 1994). Recent reports have demonstrated that AP2 functions as a transcription factor, with the second repeat of the AP2 domain binding AT-rich DNA sequences present in the intron of the C function gene *AGAMOUS* (Dinh *et al.* 2012; see Section 10.1.3). *AP2* is expressed throughout the plant and is required for activation of other genes involved in diverse aspects of plant development. *AP2* is also directly involved in other aspects of plant development. For example, correct AP2 function is necessary for maintenance of the stem cell population in the vegetative shoot apical meristem of Arabidopsis (Wuerschum *et al.* 2006). The expression of *AP2* throughout the plant, and particularly in all four whorls of the flower, initially seemed to contradict the ABC model, which predicted a whorl 1- and 2-specific expression pattern (Jofuku *et al.* 1994). However, it has since been shown that *AP2* is translationally repressed by a microRNA (miR172) that is most active in whorls 3 and 4, leaving a functional AP2 protein present only in whorls 1 and 2 (Chen 2004). Careful *in situ* hybridization analysis further revealed that *AP2* expression is much stronger in the outer floral whorls than in the inner ones, irrespective of the activity of miR172, and that the function of miR172 appears to be to mop up any *AP2* transcript that has leaked into whorls 3 and 4 (Wollmann *et al.* 2010). Therefore, within the flower, *AP2* meets all the criteria of an A function gene, being active only in whorls 1 and 2 (even if it is transcribed more widely), repressing C function expression, and being necessary for correct sepal and petal development. It is only when the other roles of *AP2* are considered, including the specification of floral meristem identity and diverse roles in vegetative development, that it becomes clear that classic A function is an oversimplification of the role of *AP2*.

The second A function gene in Arabidopsis is *APETALA 1 (AP1)* (which was also introduced in

Figure 10.2 The original ABC mutants. (a) *apetala 1* (A function in Arabidopsis). (b) *ovulata* (A function in Antirrhinum) (right) compared with wild type (left). (c) *apetala 3* (B function in Arabidopsis). (d) *deficiens* (B function in Antirrhinum) (right) compared with wild type (left). (e) *agamous* (C function in Arabidopsis). (f) *plena* (C function in Antirrhinum) (right) compared with wild type (left). Antirrhinum images kindly supplied by Enrico Coen (John Innes Centre). Image (e) provided by Ian Furner (University of Cambridge). See also Plate 5.

Chapter 9). The *ap1* mutant flower has the classic carpel/stamen/stamen/carpel phenotype (shown in Fig. 10.2a.), although the more important role of the gene in floral meristem identity may mask the floral homeotic phenotype except in long days (Mandel *et al.* 1992). *AP1* encodes a MADS box transcription factor, and is the first of the ABC genes to be expressed in the floral meristem, because of its essential role in establishing the floral identity of that meristem (Mandel *et al.* 1992). The initial expression of *AP1* is in all four floral whorls, but it is repressed by the activity of the C function gene, and confined to whorls 1 and 2 later in flower development (Gustafson-Brown *et al.* 1994). It has recently been shown that this restriction of *AP1* transcript is also a direct result of the activity of the B function genes (*APETALA 3* and *PISTILLATA*; see below), whose protein products bind to the *AP1* promoter

(Sundstrom *et al.* 2006). Unlike AP2, AP1 has little role in the repression of C function genes. In *ap1* mutant flowers the C function gene *AGAMOUS* is not expressed ectopically in whorls 1 and 2, indicating that AP1 is not involved in the repression of *AG* activity (Weigel and Meyerowitz 1993). *AP1* therefore meets some of the criteria of an A function gene, being restricted in its expression pattern to whorls 1 and 2 by C function gene activity (and B function activity) and being necessary to specify correct sepal and petal development, but it does not itself repress C function activity, indicating that the different aspects of A function activity can be separated.

The situation in Antirrhinum is even more complex. The original Antirrhinum A function mutant, *ovulata*, has the same phenotype as *ap1* and *ap2* mutant flowers (see Fig. 10.2b; Carpenter and Coen 1990), but molecular analysis revealed that this phenotype results from ectopic expression of the Antirrhinum C function gene *PLENA (PLE)*, as a result of a transposon insertion within an intron of *PLE* (Bradley *et al.* 1993). No other A function mutants have been identified from Antirrhinum mutagenesis programmes, although multiple examples of the B and C function mutants have been obtained.

To investigate the possibility that genetic redundancy between several A function genes was responsible for the absence of a mutant, Keck *et al.* (2003) isolated two *AP2*-like genes from Antirrhinum and investigated their function. The genes were named *LIPLESS 1* and *LIPLESS 2* (*LIP1* and *LIP2*), and encode proteins that are 72% identical to one another, with each being 65% identical to AP2. A reverse genetic approach was used to identify transposon insertions in each gene, but both mutants were found to be phenotypically wild type. However, the double mutant created by crossing them did show phenotypic abnormalities. In the *lip1/lip2* double mutant, sepals were converted to leaf-like organs and petals were much reduced, with no lips and underdeveloped lobes. Stamens were also shorter than wild type, and the ovules were carpelloid in structure, resulting in female sterility. Both genes were found to be expressed in sepal primordia early in development, and then later in developing petals and carpels. The C function gene *PLENA* was expressed normally in whorls 3 and 4 in this double mutant, and its expression did not extend into whorls 1 and 2. These data suggest that LIP1 and LIP2 act together to specify some aspects of A function in Antirrhinum. They are clearly necessary for correct sepal and petal development, and are expressed in whorls 1 and 2. However, unlike AP2 they play no role in repressing the activity of C function genes, and so loss of their function does not result in homeotic conversion of organs.

The AP1 orthologue in Antirrhinum is *SQUA-MOSA (SQUA)* (which was introduced in Chapter 9). However, the *squa* mutant does not have an A function mutant phenotype, instead displaying an extreme conversion of floral meristems to inflorescence meristems. This suggests that SQUA is involved primarily in the establishment of floral meristem identity, and does not play much of a role in organ morphogenesis, unlike AP1.

10.1.2 B function genes

B function has been the simplest of the ABC functions to understand. Molecular analysis has revealed that there are two B function genes, both of which encode MADS box transcription factors, and that these proteins act as a heterodimer. The system is conserved between Arabidopsis and Antirrhinum, and indeed genes from one species can act in place of the equivalent gene from the other species. The B function genes do not seem to play any other role in plant development, and their expression profile is restricted to the second and third whorl of the developing flower. The only area of B function activity that still requires significant analysis is the regulation of their expression within such tightly defined boundaries, as this regulation does not appear to occur as a result of their interactions with other ABC genes.

The first B function gene in Arabidopsis is *APETALA 3 (AP3)*. The *ap3* mutant has the classic B function phenotype of sepals/sepals/carpels/ carpels (see Fig. 10.2c). AP3 was cloned using the Antirrhinum B function gene *DEFICIENS (DEF)* as a probe (Jack *et al*. 1992). DEF had previously been isolated by subtractive hybridization using *def* mutant and wild type tissue (Sommer *et al*. 1990).

The *def* mutant also has the classic B function phenotype (see Fig. 10.2d). The proteins encoded

by *AP3* and *DEF* are 61% identical, and both of them encode MADS box transcription factors. *In situ* hybridization revealed that both *AP3* and *DEF* are expressed in the second and third whorl organ primordia from a very early stage, right through to development of the mature petals and stamens (Jack *et al*. 1992; Schwarz-Sommer *et al*. 1992). In fact, analysis of a temperature-sensitive *def* mutant line, in which *DEF* activity can be repressed at various stages of development by transferring plants to different temperature regimes, revealed that DEF activity is necessary throughout petal development. Loss of *DEF* expression in later stages of petal development results in a reversion to more sepaloid characters, such as green streaks at the edges of the tissue (Zachgo *et al*. 1995). Although transgenic experiments are rarely conducted in Antirrhinum, which is very difficult to transform, ectopic expression of *AP3* in Arabidopsis using the CaMV 35S promoter resulted in the conversion of the carpels to stamens, confirming that B and C activity is sufficient to induce stamen formation. In the fourth whorl of these transgenic plants, expression of the other Arabidopsis B function gene, *PISTILLATA (PI)*, was activated, suggesting that *AP3* and *PI* regulate each other's expression.

The *PI* gene also encodes a MADS box transcription factor, and was isolated through its similarity to the second B function gene of Antirrhinum, *GLOBOSA (GLO)* (Goto and Meyerowitz 1994). *GLO* had been previously isolated in a search for novel MADS box transcription factors in Antirrhinum (Trobner *et al*. 1992). The two proteins are 58% identical at the amino acid level, and mutation of either gene results in plants that are phenotypically identical to *def* and *ap3* mutants. Both genes are also expressed throughout the petals and stamens from early development to maturity (Trobner *et al*. 1992; Goto and Meyerowitz 1994). The apparent identity of the *ap3* and *pi* (or *def* and *glo*) mutant phenotypes was surprising in view of the differences in the gene sequences. Outside the MADS box domain, the proteins encoded by the two B function genes in each species are quite dissimilar. However, molecular analysis established that DEF and GLO function as a heterodimer, and only activate transcription of downstream genes once they have heterodimerized. Since the heterodimer is absent from both mutants, they are phenotypically identical (Trobner *et al*. 1992). The same has proved to be true of AP3 and PI, which also function as a heterodimer. It has also been shown that the two B function genes in each species are responsible for the maintenance of each other's transcription. Although the transcription of *DEF* and *GLO* is initially regulated by other factors, once they are both active they take over the regulation of one another, and the same is true of *AP3* and *PI* (Schwarz-Sommer *et al*. 1992; Goto and Meyerowitz 1994).

The activation of B function gene expression is primarily achieved through the actions of floral meristem identity genes. In Arabidopsis, LEAFY has been shown to induce *AP3* expression. The *lfy* mutant has reduced levels of *AP3* expression and a reduced domain of *AP3* activity (Weigel and Meyerowitz 1993). The UFO protein is also necessary for *AP3* expression, as the *ufo* mutant contains reduced levels of *AP3* transcript and fewer petals and stamens (Levin and Meyerowitz 1995). Transgenic experiments have shown that LFY and UFO together are sufficient to activate *AP3* expression. Although ectopic expression of *LFY* and *UFO* from the CaMV 35S promoter results in plants that die before flowering, it was possible to show that the transgenic seedlings expressed *AP3* by adding an *AP3* promoter fused to a reporter gene to the experiment. In the transgenic seedlings, reporter gene activity was observed, indicating that *AP3* expression had occurred (Parcy *et al*. 1998). It is also apparent that B function gene expression is repressed during vegetative growth by the activity of DELLA proteins. Gibberellin directly inactivates DELLA proteins, and thus causes flower development by allowing expression of *AP3* and *PI* (Yu *et al*. 2004). This direct activation of B function genes (and also *AGAMOUS*, the C function gene) is a secondary route through which gibberellins induce flowering, alongside their role in activating floral meristem identity genes (discussed in Chapter 6).

One current area of fruitful research into B function is the analysis of downstream genes. Until recently, subtractive and differential methods had been used to identify a small number of genes acting downstream of the B function genes. However, high-throughput techniques have recently been applied to this problem, and have revealed that a striking number of genes appear to be direct or indirect targets of B function activity. In Arabidopsis, microarray analysis has indicated that 200–1100 genes depend on the activity of *AP3* and *PI* for their expression, the number varying with developmental stage and experimental design (Zik and Irish 2003; Wellmer *et al*. 2004; Wuest *et al*. 2012). Mara *et al*. (2010) found that three genes negatively regulated by the AP3-PI heterodimer were required for photomorphogenesis, suggesting that their downregulation by B function might allow petals to develop without reference to the light regime (which is significant in regulating leaf size and shape). Wuest *et al*. (2012) used conditional expression of an artificial microRNA to downregulate *AP3* and *PI* at different developmental stages, and discovered that early expression of the B function genes was necessary for maximal transcriptional effects on other genes, but that later expression was particularly important for petal (as opposed to stamen) development. In Antirrhinum, macroarrays have been used to explore genes differentially expressed between sepal and petal development, and at different stages of petal development, using the temperature-sensitive *def-101* allele (Bey *et al*. 2004). Over 500 genes were found to be differentially regulated during petal development, and of those that were upregulated, the most frequent groups were of genes involved in cell cycle, cell growth, or secondary metabolism, all essential features of petal development.

The striking similarity of the sequences and mode of action of the B function genes in Arabidopsis and Antirrhinum led Irish and Yamamoto (1995) to investigate whether the proteins they encode could cross species boundaries. They created a construct containing the *DEF* cDNA fused to the *AP3* promoter, and transferred it into Arabidopsis plants mutant for *AP3*. The construct was able to rescue a weak mutant phenotype completely, and a strong mutant phenotype partially, confirming that B function has been conserved to a very high degree throughout eudicot evolution.

10.1.3 C function genes

C function is relatively simple, in that it is controlled by a single gene in each species analysed. That gene is also involved in floral meristem identity, maintaining the determinate nature of the meristem, and so, like A function genes, C function genes are classified as both floral meristem identity genes and floral organ identity genes. The C function mutants have a sepal/petal/petal/sepal phenotype (see Figs. 10.2e and f), although the indeterminacy of the floral meristem in the mutants makes it hard to follow the development of the inner whorls, as extra whorls of organs develop within the first four. The *AGAMOUS (AG)* gene was isolated from a T-DNA tagged line of Arabidopsis and encodes a MADS box transcription factor (Yanofsky *et al.* 1990). The Antirrhinum C function gene, *PLENA (PLE)*, was isolated from a mutant line containing a transposon insertion in the gene, and the proteins encoded by *PLE* and *AG* were shown to be 64% identical (Bradley *et al.* 1993). However, it has become clear that *AG* and *PLE* are non-orthologous genes derived from a duplication in a common ancestor (see below).

Initial activation of *AG* expression requires the activities of LFY and WUS. Ectopic expression of *LFY* causes ectopic expression of *AG* (Parcy *et al.* 1998), and the second intron of *AG* has been shown to contain sequences which are bound *in vitro* by LFY (Busch *et al.* 1999). This second intron had previously been shown to be essential for correct regulation of *AG* expression (Sieburth and Meyerowitz 1997). The WUS homeodomain transcription factor is active very early in flower development in whorls 3 and 4, and also binds to the second intron of *AG* to activate its expression (Lohmann *et al.* 2001). AG then downregulates *WUS* expression to ensure that the floral meristem remains determinate (Lohmann *et al.* 2001). An additional cofactor, PERIANTHIA (PAN), has recently been shown to act with LFY and WUS in the activation of *AG* expression, and also binds to the second intron of *AG* (Das *et al.* 2009; Maier *et al.* 2009). In addition, gibberellin has been shown to activate *AG* expression through inactivation of DELLA proteins that inhibit B and C function gene expression (Yu *et al.* 2004).

The expression of *AG* in whorls 3 and 4 of the developing flower is regulated in part by antagonism with *AP2*, the A function gene, as in *ap2* mutants *AG* expression spreads into whorls 1 and 2 (Drews *et al.* 1991). However, in Antirrhinum, *PLE* expression is not regulated by the *LIP1* and *LIP2* genes, as its expression is restricted to whorls 3 and 4 even in a double *lip1/lip2* mutant (Keck *et al.* 2003). Conversely, AG itself represses *AP1* expression, which is initially throughout the floral meristem but is restricted to whorls 1 and 2 following the expression of *AG* (Gustafson-Brown *et al.* 1994). It is clear that the postulated mutual A/C repression of the original ABC model operates through both A function genes. AP2 represses *AG* in whorls 1 and 2, but is not itself influenced by AG activity, instead being restricted in its zone of action by miR172. AG represses *AP1* in whorls 3 and 4, but is not itself influenced by AP1 activity. C function represses one of the A function genes, and is itself repressed by the other one (see Fig. 10.3). The expression domain of the C function genes is also determined by a number of other genes, the cadastral genes (which are discussed in Section 10.4).

Although *AG* and *PLE* are very similar in sequence and play the same C function role in Arabidopsis and Antirrhinum, it has become clear that they are non-orthologous genes derived from a duplication in a common ancestor. This duplication led to two lineages of closely related MADS box genes. In Arabidopsis, these lineages are represented by *AG* and *SHATTERPROOF*. In Antirrhinum, *PLE* is the orthologue of *SHATTERPROOF* while *FARINELLI* is the orthologue of *AG*. It is hypothesized that C function represents the ancestral role of the original, pre-duplication gene, and that this function has been retained by different members of the duplicated gene pair—by the *AG* gene in the Arabidopsis lineage, but by the *SHATTERPROOF*-like gene (*PLE*) in the Antirrhinum lineage (Causier *et al.* 2005). Regulation of the C function gene

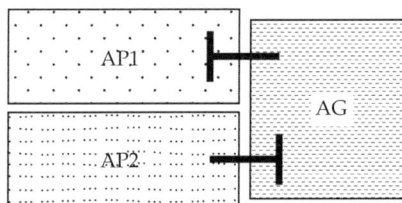

Figure 10.3 The C function gene, *AGAMOUS*, is repressed from whorls 1 and 2 by one of the A function genes, *APETALA 2*, but represses the activity of the other A function gene, *APETALA 1*, in whorls 3 and 4.

apparently works through similar mechanisms, whichever gene lineage is providing the function. Causier *et al.* (2009) used *in planta* mutagenesis to demonstrate that the LFY binding site and other conserved regions of the Antirrhinum *PLE* second intron are essential to proper spatial and temporal expression of the gene, just as they are in Arabidopsis *AG*. However, subsequent mutations have differentiated the ability of the non-C function lineage in each species to fulfill the C function role if ectopically expressed. Although expression of Arabidopsis *SHP* in the third and fourth floral whorls is sufficient to induce carpel and stamen formation (Pinyopich *et al.* 2003), a splice site mutation has introduced an extra amino acid into Antirrhinum *FAR*, and ectopic expression of this gene causes stamen, but not carpel, initiation (Airoldi *et al.* 2010).

10.2 The role of D function genes

An additional class of homeotic genes, the D function genes, was described in *Petunia hybrida* (Solanales) in 1995. The D function genes are required for ovule and placenta development within the carpel, and without their activity the ovules are replaced by carpelloid structures. The D function genes are late acting and work in combination with the C function genes to specify ovule development.

The mature Arabidopsis ovule consists of a haploid embryo sac with two layers of protective integuments, connected to the maternal tissue by a stalk-like structure called the funiculus. The Arabidopsis ovule differentiates from the inner carpel wall, although in petunia the ovules are produced directly from the floral meristem, which remains active after the emergence of the carpel primordia. The development of the gametes themselves will be discussed in Chapter 12, but the ovule represents an appendage of a floral organ, within the carpels, and so is considered in this section.

The D function genes were defined following the discovery that two MADS box proteins in petunia, FLORAL BINDING PROTEIN 7 (FBP7) and FBP11, were necessary for ovule development. Downregulation of both *FBP7* and *FBP11* by co-suppression resulted in the replacement of ovules by carpelloid structures (Angenent *et al.* 1995). The fact that the two genes function redundantly was demonstrated

later when single loss of function mutants in either gene were shown to have wild type ovule development (Vandenbussche *et al.* 2003b). *FBP11* activity is sufficient to induce ovule formation, as ectopic expression of the gene in petunia, using the CaMV 35S promoter, causes the formation of ectopic ovules on the sepals of whorl 1 and the petals of whorl 2 (Colombo *et al.* 1995).

Investigation of *FBP7*- and *FBP11*-like genes in Arabidopsis revealed that the MADS box-encoding gene *SEEDSTICK* (*STK*) is orthologous to both *FBP7* and *FBL11* (Pinyopich *et al.* 2003). *STK* was shown to act redundantly with two other MADS box genes, *SHATTERPROOF 1* (*SHP1*) and *SHP2*, to specify ovule identity in Arabidopsis. Pinyopich *et al.* (2003) showed that the *stk* mutant has defects in funiculus development, but that the ovules are otherwise normal. Similarly, a double mutant between *shp1* and *shp2* has perturbations in seed dehiscence, but ovule development is undisturbed. However, a triple *shp1/shp2/stk* mutant has arrested ovule development, indicating that at least one of the three genes must be active to specify the ovule fate. *SHP1* and *SHP2* encode MADS box proteins that are more similar to the C function AG than to the D function STK (they are the gene lineage that gives rise to the C function gene *PLENA* in Antirrhinum). The fact that each of the *STK*, *SHP1*, and *SHP2* genes is also sufficient for ovule development was shown by Favaro *et al.* (2003). Ectopic expression of *STK*, *SHP1*, or *SHP2* was sufficient to induce the formation of ectopic ovules on sepals and petals. However, it was also shown that ectopic expression of *STK* induced expression of *SHP1* and *SHP2* in outer whorls of the flower, implying that the three genes regulate each other's expression and function together to specify ovule identity. The downstream targets of the ovule-specifying complex are currently under analysis. Matias-Hernandez *et al.* (2010) identified the *VERDANDI* (*VDD*) gene as a direct target of the complex that encodes a transcription factor necessary for correct specification of cell identities within the ovule.

The D function genes are thought to act together with the C function gene *AG* to specify ovule identity. Ectopic expression of *STK* in Arabidopsis induces ectopic *AG* expression (Favaro *et al.* 2003). Recent studies of transposon-induced mutants in petunia further support this model of combinatorial action.

Heijmans *et al.* (2012b) found that the petunia D function genes *FBP7* and *FBP11* not only specified ovule development, but also played a role in floral determinacy, and that their functions in ovule development are at least partially redundant with the same functions specified by the petunia C function genes. It is perhaps not surprising that carpel and ovule development, which is essential for female reproductive success, is apparently specified by an interacting and partially redundant network of MADS box transcription factors.

10.3 The role of E function genes

The E function genes were the latest addition to the ABC model, and their identification finally showed how combinations of transcription factors could be sufficient to induce floral organ identity. Although the ABC genes are each necessary to specify particular floral organ identities, they are not sufficient. Ectopic expression of the B function genes *AP3* and *PI*, for example, is sufficient to convert all floral organs to petals or stamens, but has no effect on leaves (Krizek and Meyerowitz 1996). The B function genes are sufficient *within the flower* to modify organ fate, but insufficient to convert leaves into petals. Addition of the E function genes into the equation, however, is sufficient to convert vegetative organs to floral ones. The E function genes are necessary for the development of all floral organs except sepals.

SEPALLATA 1 (*SEP1*), *SEP2*, and *SEP3* all encode MADS box transcription factors, and they are all expressed in whorls 2, 3, and 4 of the developing flower. *SEP1* and *SEP2* are also expressed in whorl 1 (Flanagan and Ma 1994; Savidge *et al.* 1995; Mandel and Yanofsky 1998). This led Pelaz *et al.* (2000) to develop the hypothesis that the *SEP* genes might be involved in floral organ development. Transposon insertion mutants were identified in each of the three genes, and each single mutant was shown to be phenotypically wild type, as were double mutant combinations. However, in a triple mutant of all three genes the petals, stamens, and carpels were all converted to sepals, so the flower develops just like a B and C function double mutant (see Fig. 10.4a). The E function genes therefore act redundantly, but are necessary for the B and C function genes to do

their job, and therefore necessary for petal, stamen, and carpel development (Pelaz *et al.* 2000). The triple mutant also had reduced determinacy of the floral meristem, suggesting that the E function genes act with *AGAMOUS* to maintain the determinate state of the floral meristem.

A fourth E function gene was later identified in Arabidopsis. *SEP4* encodes a protein with very strong sequence similarity to the other SEP proteins, and with an initially similar expression pattern in all four whorls of the young flower. This expression later resolves into a strong signal in whorl 1, with weaker expression in the rest of the flower. Ditta *et al.* (2004) identified a mutant line with a T-DNA insertion in *SEP4*, and found the plant to be phenotypically wild type. However, when they constructed a *sep1/sep2/sep3/sep4* quadruple mutant they discovered that the flowers produced four whorls of leaf-like organs. The conversion of floral organs to leaves (rather than sepals, as in the *sep1/sep2/sep3* triple mutant) was apparent from the larger size of the organs, the amoeboid shape of their epidermal cells, and the branched nature of their trichomes (Arabidopsis sepal trichomes do not branch). These data indicate that *SEP4* is necessary for sepal development, whereas the other three *SEP* genes are only necessary for the development of petals, stamens, and carpels.

Transgenic plants ectopically expressing any one of the *SEP* genes using the CaMV 35S promoter are not phenotypically very different to wild type, showing only mild alterations to the sepals (see Fig. 10.4b). However, combination of the E function genes with the ABC genes is sufficient to convert leaves into floral organs. Specifically, Honma and Goto (2001) showed that ectopic expression of *SEP3*, *SEP2*, *AP3*, *PI*, and *AP1* (i.e. A, B, and E) was sufficient to convert all leaves to petals. They also showed that ectopic expression of *SEP3*, *AG*, *PI*, and *AP3* (i.e. B, C, and E) was sufficient to convert all cauline leaves (but not rosette leaves) to stamens (see Fig. 10.4c). These experiments demonstrated that the ABC genes are sufficient in combination with the E function genes to convert vegetative organs into floral ones.

The interactions between the ABC and E genes are based on physical contact and complex formation, rather than on regulation of one another.

Figure 10.4 E function genes. (a) A triple mutant between *sep1/sep2/sep3* has four whorls of sepals and an indeterminate meristem. Image kindly supplied by Gary Ditta and Marty Yanofsky (UCSD). (b) A plant ectopically expressing *SEP3* shows only slight alterations to the sepals. (c) A plant ectopically expressing *AP3*, *PI*, *AG*, and *SEP3* has conversion of leaves to stamens. Images (b) and (c) kindly supplied by Koji Goto (Research Institute for Biological Sciences, Okayama).

Pelaz *et al.* (2000) showed that the ABC genes have normal expression patterns in *sep1/sep2/sep3* triple mutant plants, and it had previously been shown that the *SEP* genes are normally expressed in ABC mutants (Flanagan and Ma 1994; Savidge *et al.* 1995). However, yeast two-hybrid studies have revealed physical interactions between ABC and E proteins. For example, AG has been shown to bind SEP1, SEP2, and SEP3 (Fan *et al.* 1997), and the AP3–PI heterodimer has been shown to associate with SEP3 and AP1 (Honma and Goto 2001). These physical interactions are the basis of floral organ identity, providing complexes that function together to activate transcription of downstream genes involved in the detailed development of each organ.

The *DEFH84*, *DEFH200*, and *DEFH72* genes function as the orthologues of *SEP1*, *SEP2*, and *SEP3* in Antirrhinum (reviewed by Jack 2001).

10.4 The role of cadastral genes

The ABCDE genes rely on precisely defined expression patterns to ensure the production of appropriate floral organs in the correct places. While some of these expression patterns are determined at least in part by mutual inhibition, such as the boundary between the A and C function genes, other expression patterns are determined by additional factors. A gene that delimits whorls and organ primordia territories but has no obvious direct role of its own is called a cadastral gene. Cadastral genes may operate through the spatial regulation of the expression patterns of other genes or through the inhibition of cell proliferation in regions delimiting organogenic territories (Breuil-Broyer *et al.* 2004). Analysis of cadastral genes determining floral organ positioning has lagged some way behind the analysis of the floral organ identity genes, but information is beginning to emerge on the ways in which the boundaries of ABCDE gene expression are set.

10.4.1 Regulating the B function expression domain

The best studied floral cadastral gene is the *SUPERMAN* (*SUP*) gene of Arabidopsis, which negatively regulates B function gene expression in whorl 4, restricting it to whorls 2 and 3. The *sup* mutant produces stamens in whorl 4, as a result of ectopic expression of the B function genes *AP3* and *PI* in those organ primordia (see Fig. 10.5a; Bowman *et al.* 1992). *SUP* encodes a complex transcription factor protein with a zinc-finger domain, a serine/proline-rich domain, a basic domain, and a leucine-zipper motif (Sakai *et al.* 1995). *SUP* is not expressed until after the first detection of the B function genes, when its transcript appears in the stamen primordia, adjacent to the boundary between the stamens and the carpels. This expression pattern has led to the suggestion that *SUP* is initially activated by AP3 and PI, and then acts in a negative feedback loop to inhibit their extension into whorl 4. The SUP protein has been shown to work at least in part through the inhibition of cell proliferation at the boundary between the third and fourth whorls. Bereterbide *et al.* (2001) showed that ectopic expression of *SUP* in tobacco reduced cell number in many organs, with the result that individual plants were much shorter than wild type. Given such an effect, regulation of *SUP* expression is particularly important. The precise expression pattern of *SUP* depends in part on epigenetic effects, and relies on a complex

mix of elements. Several motifs in the *SUP* promoter have been shown to be responsible for aspects of the temporal and spatial expression pattern, along with regions of the coding sequence that are likely to be regulated by methylation (Ito *et al.* 2003).

10.4.2 Regulating the C function expression domain

A number of cadastral genes control the extent of C function gene expression, marking the boundary between the second and third whorls and preventing the conversion of the petals into reproductive organs. In Arabidopsis, the repression of *AG* expression from whorls 1 and 2 is under the control of AP2, as discussed earlier, along with several other proteins. The genes encoding LEUNIG (LUG), SEUSS (SEU), and STERILE APETALA (SAP) all interact with AP2 to repress *AG*. In single mutants with lesions in any of these three genes, *AG* transcript is found in whorls 1 and 2 (Liu and Meyerowitz 1995; Byzova *et al.* 1999; Franks *et al.* 2002). *LUG* encodes a transcriptional co-repressor protein with several WD repeat domains, while SEU encodes a glutamine-rich protein with a conserved dimerization domain (Conner and Liu 2000; Franks *et al.* 2002). It has been hypothesized that SEU acts as a bridge to mediate interaction between LUG and a protein complex composed of ABCE model transcription factors in whorls 1 and 2 that is targeted to the second intron of *AG* by the specific DNA-binding activity of AP2 and represses *AG* transcription (Jack 2004; discussed by Liu and Mara 2010). Although *AP2*, *LUG*, and *SEU* are all expressed in all four whorls of the developing flower, their

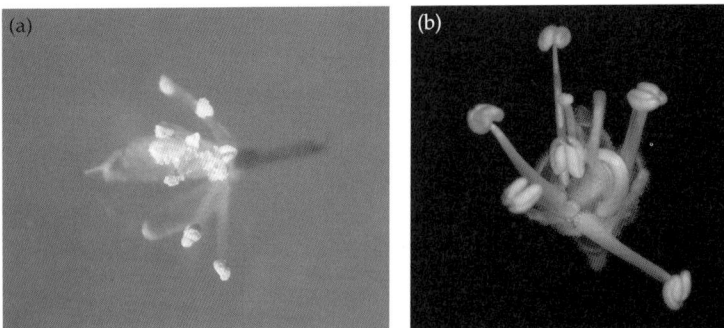

Figure 10.5 Cadastral genes control the activity patterns of other genes. (a) The *superman* mutant has inappropriate B function in whorl 4, generating extra stamens (sepals and petals removed). (b) The *stylosa/fistulata* double mutant has inappropriate C function in whorls 1 and 2, generating stamens in whorl 2 and carpelloid sepals in whorl 1. Image kindly provided by the much missed Zsuzsanna Schwarz-Sommer (MPIZ, Cologne).

repression of *AG* only occurs in whorls 1 and 2. This is likely to be because the post-translational repression of AP2 in whorls 3 and 4 means that the protein complex cannot be targeted to the *AG* sequence in cells in those whorls (Chen 2004). However, a recent report suggests that LUG and SEU play an additional role in whorls 1 and 2, acting also to repress *miR172* and thus to ensure that AP2 activity is not inhibited in these outer whorls (Grigorova *et al.* 2011). This conclusion was based on a combination of mutant analysis, chromatin binding experiments, and yeast two-hybrid assays, and indicates a role for LUG and SEU in the maintenance of A function activity in whorls 1 and 2, as well as in the repression of C function from those whorls. SEU and LUG may play additional, AG-independent roles in the development of petals and sepals, including regulation of cell number, vasculature development, and organ polarity (Franks *et al.* 2006).

The RABBIT EARS (RBE) protein plays a similar role to that of SEU and LUG in the repression of AG activity in whorls 1 and 2. The *rbe* mutant has ectopic expression of *AG* in whorl 2, and consequent partial conversion of petals to stamens. *RBE* encodes a zinc-finger protein with many similarities to SUPERMAN (Takeda *et al.* 2003). The *RBE* gene is required not only to specify the boundary between whorls 2 and 3, but also to specify the boundaries between organs within a whorl, as some *rbe* mutant alleles have fusion of sepals and/or petals (Krizek *et al.* 2006). It has recently been shown that RBE acts to regulate the activity of another microRNA, miR164. The latter post-transcriptionally regulates the activity of a set of NAC domain transcription factors, *CUC1* and *CUC2*, which together results in correct organ number and organ boundary development in the Arabidopsis flower (Baker *et al.* 2005; Mallory *et al.* 2004; Sieber *et al.* 2007). Huang *et al.* (2012) demonstrated that RBE is a transcriptional repressor of *miR164*, and so influences organ whorl boundaries through a downstream effect on *CUC1* and *CUC2* expression.

Several cadastral genes also act to control the region of C function activity in Antirrhinum. *STYLOSA* (*STY*) is the orthologue of *LUG* (Navarro *et al.* 2004), and *sty* mutants have ectopic expression of the C function gene *PLE* in whorl 2, and, as a consequence, their petals are staminoid (Motte *et al.* 1998).

The *fistulata* (*fis*) mutant also has ectopic *PLE* expression in whorl 2 and develops staminoid petals (McSteen *et al.* 1998). *FIS* encodes a miRNA (miRFIS), related to the miR169 family (Cartolano *et al.* 2007). A double mutant between *sty* and *fis* produces carpelloid sepals in whorl 1 and stamens in whorl 2, essentially developing with an A function mutant phenotype (see Fig. 10.5b). This confirms that *STY* and *FIS* act as cadastral genes to ensure repression of C function in whorls 1 and 2, allowing A function genes to specify correct sepal and petal development (Motte *et al.* 1998).

10.5 The quartet model of organ identity

The combined data on floral organ identity led to the model that tetramers of MADS box proteins function to specify the identity of each floral organ (Theissen and Saedler 2001). Many MADS box proteins have been shown to associate with other MADS box proteins in yeast two-hybrid and *in vitro* assays, and we know that some MADS box proteins (such as the B function genes) operate as a heterodimer. Theissen and Saedler (2001) therefore hypothesized that two dimers of MADS box proteins bind together and act as a complex to specify the activity of downstream target genes in each whorl. The situation is clearest in whorls 2 and 3, where the tetramers are proposed to consist of AP1, PI, AP3, and a SEP protein, and PI, AP3, AG, and a SEP protein,

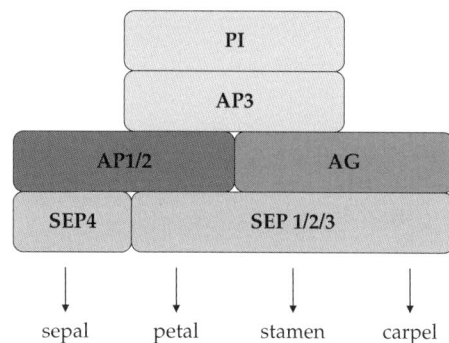

Figure 10.6 The revised ABCE model in Arabidopsis. Tetramers of proteins specify the development of each whorl of organs. In whorls 1 and 4, multiple copies of the SEP and A/C function proteins are believed to make up the tetramers.

respectively. In whorl 4 it is postulated that two AG molecules and two SEP proteins interact, while in whorl 1 it is possible that two AP1 proteins interact with two SEP4 molecules (see Fig. 10.6.). This model was initially tested by *in vitro* DNA-binding assays and analysis of protein combination in protoplasts. More compelling evidence in support of floral quartets was provided by Smaczniak *et al.* (2012), who expressed the ABCE genes fused to GFP from their native promoters, and then isolated protein complexes from floral tissues by immunoprecipitation using anti-GFP antibodies. The proteins in these complexes were identified biochemically, revealing that complexes of AP3 and PI (B function) interacted with SEP3 (E function), AP1 (A function),

and AG (C function). Perhaps not surprisingly, the E function protein SEP3 was most abundant in their assays, indicating a core supporting role in all of the quartet combinations. Further experiments reported in the same paper confirmed that these higher-order transcriptional complexes bound together to target DNA sequences.

When the data described in this chapter are taken together, it is clear that the floral meristem identity genes activate the expression of suites of floral organ identity genes, their domains of expression bound by cadastral genes, and that these floral organ identity genes function as transcriptional activators to induce expression of the structural genes required to build the various floral organs.

CHAPTER 11

The ABC model and the diversity of plant reproductive structures

In Chapter 10 we discussed floral organ development in Arabidopsis, and the functions of a set of transcription factors in activating the structural genes necessary for correct organ development in the correct position. The majority of these ABC genes encode MADS box transcription factors, a very ancient family of DNA-binding proteins. By tracing the evolutionary history of this gene family we can investigate the molecular processes that underlie the evolution of flowers, and by tracing the variation in regulation and function of this gene family we can explore the basis of floral morphological variation. In addition, this chapter considers in more detail the status of the A function genes as an integral part of the ABC model. Since flowers are one of the key distinguishing features of the angiosperms, and have been argued to be responsible for their astonishing radiation into the most species-rich plant division by far, the evolution of these transcription factors is of particular interest in reconstructing plant evolutionary history.

11.1 Evolutionary history of MADS box transcription factors

MADS box genes are defined by a highly conserved 180-base-pair DNA sequence, the MADS box, which encodes the DNA-binding domain of the MADS box protein. This MADS domain folds into an N terminal extension followed by an antiparallel coiled coil of α-helices that lies flat on the minor groove of DNA (Pellegrini et al. 1995). The basic structural similarity of all MADS domains means that all MADS domain proteins bind to similar DNA sequences, with a consensus motif of $CC(A/T)_6GG$,

known as a CarG box (Theissen et al. 2002). It has recently been shown that the MADS domain itself evolved by duplication of the DNA binding domain of subunit A of a prokaryotic topoisomerase IIA, probably in an early ancestor of the eukaryotic lineage (Gramzow et al. 2010).

MADS box genes can be divided on the basis of sequence homology and structural properties of the proteins they encode into two types, known as type I and type II. Both types are present in plants, animals, and fungi, suggesting not only that a MADS box gene was present in the single-celled eukaryote that was the last common ancestor of these three great multicellular kingdoms, but also that a gene duplication event had already occurred in that lineage, giving rise to at least two MADS box genes at least 1 billion years ago (Alvarez-Buylla et al. 2000). In a more detailed analysis, Gramzow et al. (2010) aligned MADS box genes from all available major eukaryotic lineages. They identified a number of instances of loss of either type I or type II MADS box genes, and a surprising absence of MADS box sequences in the excavate lineage (Trichomonas vaginalis and Giardia lamblia), but found that the distribution of type I and type II genes was consistent with a duplication having occurred in or before the most recent common ancestor of the extant eukaryotic lineages.

In animals, type I MADS boxes include the human SERUM RESPONSE FACTOR (SRF) gene family, a highly conserved group of genes with roles in response to growth factors and subsequent development. Fungal type I MADS boxes include MINICHROMOSOME MAINTENANCE 1 (MCM1) of yeast, which regulates the cell cycle and cell growth. The plant type I MADS boxes have been less well studied than the plant type II MADS boxes. They are defined by their similarity to the MADS domain of animal SRF, but show a number of clear differences from the type II genes. Plant type I MADS box genes are generally shorter than type II genes, are usually formed of a single exon, and are focused in a few regions of the genome where recent duplications seem to have occurred. Parenicova *et al.* (2003) divided the 61 Arabidopsis type I MADS box genes into three subfamilies with distinct structural elements, the Mα, Mβ, and Mγ subgroups. Dimerization within subgroups has been shown to be rare, with type I MADS boxes acting either as heterodimers between subgroups or as single proteins (Immink *et al.* 2009). The great variability and low expression levels of the Arabidopsis type I MADS box genes have contributed to the lack of generalizations about their function. However, a number of recent analyses suggest that they might frequently play roles in male and female gametophyte development, and in embryo and endosperm function (reviewed by Masiero *et al.* 2011).

The animal type II MADS box genes are involved in muscle-specific gene regulation, while the small number of fungal type II MADS boxes (two in yeast) have yet to be fully characterized (reviewed by Becker and Theissen 2003). However, the plant type II MADS box proteins have been extensively studied, and include the ABC genes discussed in Chapter 10.

There are 46 type II MADS box genes in the Arabidopsis genome, and all plant type II MADS boxes share a common domain structure, which has led to them also being known as MIKC MADS boxes (see Fig. 11.1; Munster *et al.* 1997; Heijmans *et al.* 2012a). In addition to the MADS domain, described above, plant type II MADS boxes contain an I (Intervening) domain, a K (Keratin-like) domain, and a C (C terminal) domain. The MADS domain is responsible for DNA binding but also plays a role in dimerization. The I domain is only weakly conserved, consists of around 30 amino acids, and is involved in the selective formation of DNA-binding dimers (Reichmann *et al.* 1996). Although dimers can potentially form between any combination of MADS box proteins, only certain dimer combinations bind DNA. Without the I domain, the specificity of the interaction between dimers and DNA is lost. The K domain contains hydrophobic amino acids at regular intervals, generating an amphipathic helix involved in protein–protein interactions (reviewed by Kaufmann *et al.* 2005). The K domain is absent from type II MADS boxes in animals and fungi, and seems to be a plant-specific innovation, perhaps enhancing flexibility of transcriptional regulation by allowing multiple complexes to form from different combinations of MADS domain proteins (Alvarez-Buylla *et al.* 2000). The C terminus of the plant type II MADS boxes is the most variable region of the protein, and can fulfil a number of functions. It may be involved in transcriptional activation, through interaction with components of the cell's core transcriptional machinery, or it may be necessary for the formation of multiprotein complexes which themselves are necessary for transcriptional activation (Egea-Cortines *et al.* 1999).

The plant type II MADS box genes can be divided into two types, known as the MIKC^c and MIKC* types (Henschel *et al.* 2002). There are 40 MIKC^c genes in the Arabidopsis genome, and only 6 MIKC* genes (Kwantes *et al.* 2012). The differences between these two groups of proteins include much longer I domains in MIKC* MADS boxes and divergent K domains. Both types of gene are present in the moss *Physcomitrella*, as well as in the lycophyte *Selaginella*, suggesting that they are derived from a gene duplication event that occurred early in land plant evolution. This gene duplication can be dated to before the divergence of the bryophyte groups from the vascular plants, around 450 million years ago (Henschel *et al.* 2002). The MIKC* genes have only recently been the subject of molecular genetic analyses. These suggest important roles in gametophyte development, and analysis of the gene family across the land plant lineage suggests that this role

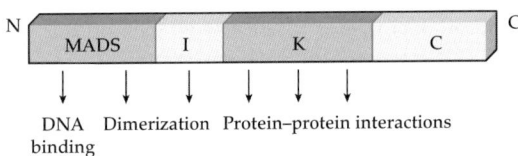

Figure 11.1 The structural domains of plant type II MADS box proteins, with their functions indicated underneath.

might be a conserved one that originated in the earliest land plant groups (Kwantes *et al.* 2012).

The MIKC^c type II MADS box genes are the better studied subgroup. They can be further subdivided into clades, the number of which depends on the method of analysis used. Becker and Theissen (2003) divided the group into 13 clades, each described by the name of the first clade member to be identified. Functional analysis of the genes in these clades has led to hypotheses about the functions of the ancestral gene in each group. Analysis of the groups in different plant species has determined that at least six distinct MIKC^c MADS box genes were present in the last common ancestor of the gymnosperms and the angiosperms, which diverged around 300 million years ago.

11.2 ABC genes in gymnosperms

Among the clades of MADS box proteins identified in both angiosperms and gymnosperms are the groups containing the B function genes and the C and D function genes. Analysis of the function of these genes in plants that do not produce flowers has helped us to understand the evolution of the angiosperm flower we see today. Most molecular work on gymnosperms is conducted on conifers, and a lesser amount on cycads and gnetophytes, particularly the tree *Gnetum gnemon* (see Chapter 1 for a reminder of the relationships between the different gymnosperms). The reproductive structures of conifers differ from those of angiosperms in several major respects, and are described in detail in Chapter 1, along with a discussion of the homology between gymnosperm cones and angiosperm reproductive structures. Briefly, conifers do not produce hermaphroditic reproductive structures, but two types of cones (see Fig. 11.2a). The male cones are usually quite small, and release pollen. The female cones contain the ovules, which in turn contain the female gametes, on individual scales. Notably, the ovules are not entirely enclosed within maternal tissue as they are in the angiosperms. Another big difference between gymnosperm and angiosperm reproductive structures is the absence of a perianth in gymnosperms, resulting in reproductive structures that rarely attract animals to transfer gametes.

Figure 11.2 Gymnosperms and the ABC genes. (a) Male (left) and female (right) pine cones. (b) B and C function genes are sufficient to specify male cone development, while C function genes alone specify female cone development.

Several homologues of Arabidopsis B function genes have been isolated from conifers, notably from *Pinus radiata* (Monterey pine) and *Picea abies* (Norway spruce), and similar B function-like genes have also been isolated from *Gnetum*. The gymnosperm B function-like genes always resolve as the most ancestral of the B function genes in phylogenetic analyses (Zahn *et al.* 2005). *In situ* hybridization studies have shown that all of these genes are expressed in the pollen cones, but not in the female cones, and usually in no other plant tissue at all. This strongly suggests that the gymnosperm B function genes have a role in male reproductive structure development, just as the B function genes have a role in producing stamens in angiosperm flowers. Interestingly, it appears that the formation of heterodimers which are obligate for functionality is a recent trait in the B function genes. Winter *et al.* (2002) showed that one of the two *Gnetum* B function proteins, GGM2, preferentially formed homodimers *in vitro*, rather than heterodimerizing with AP3, DEF, or GLO. This result was replicated in a yeast two-hybrid analysis, and it was further shown that the GGM2 homodimers were able to bind DNA in a perfectly functional manner. Obligate heterodimerization, as shown by the Arabidopsis and Antirrhinum B function proteins, may therefore have evolved later in B function history, following further gene duplication events. In many other respects the gymnosperm B function genes appear very similar to angiosperm ones. Indeed, Sundstrom and Engstrom (2002) showed that ectopic expression of the B function genes *DAL11* and *DAL12* from *Picea abies* in Arabidopsis resulted in the same phenotype as that observed when the endogenous B function gene *PI* was ectopically expressed, even though the spruce proteins also form homodimers.

Genes that fall into the C/D function clade have been isolated from all seed plant groups, including the gymnosperms *Cycas*, *Ginkgo*, *Gnetum*, and *Picea* (a conifer). Each of these C function genes has been shown to be expressed in both the seed cones and the pollen cones, indicating that C function is active in all reproductive structures in gymnosperms. This mirrors the situation in angiosperms where C function genes are expressed in both whorls 3 and 4, specifying the reproductive identity of organs in those whorls. In one of the earliest reports of gymnosperm C function, Rutledge *et al.* (1998) showed that the *Picea mariana* (black spruce) *SAG1* gene was expressed in the pollen and seed cones, and that ectopic expression of the gene in Arabidopsis phenocopied ectopic expression of *AGAMOUS* itself, generating first whorl carpels and second whorl stamens. In a very similar paper, published concurrently, Tandre *et al.* (1998) showed that the *Picea abies* (Norway spruce) C function gene *DAL2* was also expressed in reproductive organs and could mimic *AGAMOUS* activity when ectopically expressed in Arabidopsis. More recently, Zhang *et al.* (2004) showed that the C function gene from the cycad *Cycas edentata* could complement the Arabidopsis *agamous* mutant when expressed using the Arabidopsis *AGAMOUS* promoter. These data confirm the strong conservation of C function proteins over 300 million years of evolutionary history.

Thus it appears that the last common ancestor of gymnosperms and angiosperms had a functional set of B and C/D function genes. Expression of the C function genes, just as in Arabidopsis, confers a reproductive identity on an organ. It seems to be the case in gymnosperms that it is expression of the B function gene that determines which sort of reproductive structure is produced. Expression of B function genes as well as C function genes results in the development of a male reproductive structure, a pollen cone. Expression of just C function genes alone results in the development of a female reproductive structure, a seed cone (see Fig. 11.2b). This ability of B function genes to specify male identity to a reproductive structure also fits with the role of B function genes in producing stamens in the third whorl of the angiosperm flower. On the basis of these observations it has been proposed that the development of the classic hermaphroditic reproductive structure of the angiosperms involved a change in B function activity in an axis of a gymnosperm that produced only male or only female cones. Loss of B function gene expression from the upper regions of a male axis would result in the production of female organs within whorls of male organs. Similarly, gain of B function gene expression in the lower regions of a female axis would result in the production of male organs outside whorls of female organs (Theissen *et al.* 2002). These scenarios are discussed further in Chapter 1.

The presence of E function genes in the gymnosperms is currently uncertain. The *SEP* genes of Arabidopsis form a distinct clade in MADS box phylogenies, and no member of that clade has been isolated from a gymnosperm. However, the SEP clade clusters in a superclade with the *AGAMOUS LIKE 6* (*AGL6*) genes, and members of this group have been identified in *Pinus*, *Picea*, and *Gnetum*. The function of *AGL6* in Arabidopsis has not been well defined, but several recent reports in petunia (Solanales), wheat, and rice (both Poales) have independently suggested that AGL6 clade genes have a role as E function genes redundant with the SEP clade genes (Ohmori *et al.* 2009; Rijpkema *et al.* 2009; Thompson *et al.* 2009). The work in petunia found that an *agl6* mutant allele, when combined with a *sep* mutant, results in a greater degree of floral organ perturbation than is seen in a *sep* mutant alone (Rijpkema *et al.* 2009). In maize, the positional cloning of a mutant allele perturbing floral organ and floral meristem development identified another

AGL6 gene (Thompson *et al.* 2009). The rice mutant has a similar phenotype to the maize mutant, and combination of the *agl6* mutant allele with a *sep* mutant allele increased the severity of the phenotype (Ohmori *et al.* 2009). Although mutant data are not available from gymnosperms, the expression profiles of the *AGL6* genes isolated from these species at least correlate with a role in reproductive development. The most detailed analysis of a gymnosperm *AGL6* gene was conducted by Carlsbecker *et al.* (2004), who showed that *DAL1* from *Picea abies* was expressed at low levels in vegetative branches but at higher levels in reproductive branches, with no expression detected before the 3–5 year mark, when these trees make the transition to the reproductive stage. Similarly, *AGL6*-like sequences from *Pinus* and *Gnetum* are also most strongly expressed in reproductive tissues, perhaps suggesting that an E function-like role had originated before the divergence of gymnosperms and angiosperms (reviewed by Melzer *et al.* 2010).

However, it appears that the last common ancestor of the angiosperms and gymnosperms did not have A function genes. The clade of MADS box genes containing the A function genes is known as the *SQUA-like* genes, and contains many genes from a diverse range of plants, including *AP1* of Arabidopsis (described in Chapter 10). *SQUA*-like genes have been isolated from monocots (maize, rice, barley, and wheat, all Poales) and early diverging angiosperms (including *Nuphar*, Nymphaeales), as well as from several eudicot species. This clade had therefore apparently been established at the base of the angiosperm phylogenetic tree. However, no gene in this clade has yet been isolated from any gymnosperm species, suggesting that the clade is younger than the divergence of the angiosperms and the gymnosperms (Becker and Theissen 2003; Melzer *et al.* 2010). The absence of A function genes from the gymnosperm lineage is not entirely surprising, given the variable function of genes from this lineage in angiosperms (see Chapter 10 and discussion in Section 11.7 of this chapter). In many angiosperm species, including Antirrhinum, the *SQUA*-like genes are primarily involved in floral meristem function, not specifically in sepal and petal identity. Since studies in gymnosperms suggest that the potential E function *AGL6*-like genes

specify reproductive identity, just as the related *SEP* genes do in angiosperms, there may be no need for a separate *SQUA*-like lineage in the gymnosperms. However, the duplication event that led to the establishment and subsequent radiation of this clade may be in part responsible for some of the radiation of angiosperm floral form.

11.3 ABC genes in early diverging angiosperms

In Chapter 1 (Section 1.5) we noted that the greatest diversity of floral body plans occurs in the earliest diverging groups of angiosperms, while the more recently evolved monocots and eudicots have relatively constrained *basic* floral morphology, despite enormous diversity of morphological *detail*. In particular, the early diverging angiosperms are characterized by an undifferentiated perianth, spirals rather than whorls of organs, variable numbers of floral organs, and transitional organs present as one set of organs grades into the next. It is likely that this less constrained floral body plan is retained from an ancestral angiosperm, and so it is particularly interesting to understand how it is produced with reference to the eudicot ABC model. A number of authors have discussed the idea of a 'fading borders' model, where the ABC genes do not occupy discrete zones of the floral meristem but instead have focal regions with weaker expression either side (e.g. Soltis *et al.* 2007, 2009). Such a model could certainly explain the transitional organs observed, and the apparent flexibility of organ number. It is supported by gene expression studies, although the non-model nature of early diverging angiosperms means that no genetic or transgenic evidence is yet available. Kim *et al.* (2005) used quantitative RT-PCR and *in situ* hybridization approaches to determine the expression patterns of A, B, C/D, and E class genes from *Amborella* (Amborellales), *Nuphar* (Nymphaeales), and *Illicium* (Austrobaileyales). Their data indicate that A function genes (from the *SQUA*-like clade) play no specific role in floral organ identity. The *Nuphar SQUA*-like gene was expressed throughout the plant, most strongly in the leaves and in the carpels. B function genes from all three basal angiosperm species had floral-specific expression patterns, but these patterns were broader than those seen in eudicots. The *Amborella* B function genes were expressed in whorls 2, 3, and 4, the *Nuphar* genes in all four whorls of the flower, and the *Illicium* genes in whorls 1, 2, and 3. However, C function genes showed a more restricted expression pattern, being mostly confined to whorls 3 and 4. *SEP* and *AGL6* genes were generally expressed throughout the flower, consistent with an E function-like role in floral meristem identity. These data support the establishment of a basic 'ABC-like' model early in the angiosperm lineage, even if that model was more flexible in the position and number of organs than the eudicot ABC model is today.

In the final section of Chapter 1 it was noted that the flowers of many species in the magnoliids also share characters in common with the early diverging angiosperms. However, the position of the magnoliids relative to the eudicots and monocots is unclear, with many phylogenies representing the three groups as a polytomy. Accordingly, we cannot be certain whether the apparently ancestral features of magnoliid flowers are conserved or derived. For this reason I shall not discuss them in detail, but will simply mention that various authors have provided evidence to suggest that the fading borders model might also apply to flowers within the magnoliids (Kim *et al.* 2005; Chanderbali *et al.* 2009).

11.4 ABC genes in monocots

Within the angiosperms, ABC genes have been most extensively studied in the core eudicots (where Arabidopsis and Antirrhinum are placed) and the monocots. Within the monocots, the economically important grasses have been the subject of most work. Among the clades of MIKC MADS box genes identified in the monocots are the groups containing all of the A, B, C/D, and E function genes.

There are two main floral forms in monocots, each usually producing whorls of floral organs containing three organs per whorl. The petaloid, or animal-attracting, monocots produce flowers that look superficially like eudicot flowers. This group includes common garden flowers such as tulips and lilies (both Liliales), daffodils and freesias, as well as the orchids (all Asparagales). However, closer inspection of these flowers reveals that their reproductive organs are usually surrounded by two whorls

of petaloid perianth organs, and that green sepals are not normally present. The two outer whorls of brightly coloured organs are often referred to as tepals. The wind-pollinated monocots, including the grasses such as maize, rice, and wheat (all Poales), produce very different floral structures again. Most grasses produce separate male and female flowers, and those flowers do not produce petals or sepals. Instead each flower has a lemma and a palea, which function in a similar way to sepals to protect the developing flower. Within these, in most derived grasses, is a pair of small globular structures called the lodicules, nominally in whorl 2 of the flower, which expand on flower opening and force the palea and lemma away from the reproductive structures, allowing access to the wind. Inside the lodicules are the stamens or carpels (see Fig. 11.3a).

Figure 11.3 Grasses and the ABC genes. (a) Two flowers of maize, enclosed within a pair of glumes. The entire structure is known as a spikelet. Each flower has a lemma and a palea in the outer whorl, lodicules in the second whorl (barely visible), and stamens in the third whorl. Maize is monoecious and these are male flowers, so have no fourth whorl carpels. Image kindly provided by Bob Schmidt (UCSD). (b) B, C, E, and possibly A function genes specify organ identity in grass flowers.

Molecular analysis of ABC genes in grasses has revealed that the lemma and palea are homologous to sepals, and that the lodicules are homologous to the petals of eudicots (see Fig. 11.3b).

11.4.1 Grasses

Despite the obvious morphological differences between the second whorl organs of grasses and species with petaloid flowers, the B function genes, which have been particularly well characterized in the grasses, are apparently functionally almost identical to their eudicot orthologues, differing only in their downstream targets. Studies have analysed the functions of both *AP3/DEF*-like genes and *PI/GLO*-like genes, and have used genetic and transgenic approaches as well as expression analysis. For example, in rice, loss of function of *OsMADS4*, a *GLO/PI* orthologue, has been shown to result in homeotic transformation of the second whorl lodicules into paleas and lemmas, as is normally found in the first whorl. Stamens were also converted into carpels. This phenotype is identical to the sepal/sepal/carpel/carpel phenotype of a eudicot B function mutant, confirming both the orthology of the genes and the homology (at least from a developmental genetic perspective) of the lodicules with petals (Kang *et al.* 1998). Similarly, loss of function of *OsMADS16*, an *AP3* homologue, also results in a classic B function mutant phenotype (Nagasawa *et al.* 2003). Analysis of the rice genome sequence revealed the presence of only the single *AP3/DEF*-like gene, *OsMADS16,* but of two *PI/GLO*-like genes, *OsMADS2* and *OsMADS4* (reviewed by Yoshida 2012).

Detailed studies of the B function genes have also been conducted in maize. Ambrose *et al.* (2000) described the phenotype of the *silky1* mutant, lacking a functional version of the single maize *AP3/DEF*-like gene. The mutant flowers produced paleas and lemmas in the second whorl and carpels in the third whorl, again identical to a eudicot B function mutant phenotype. The *SI1* gene was shown to be expressed in developing organ primordia in whorls 2 and 3 of the flower. Whipple *et al.* (2004) extended this work by investigating the relationship between SI1 and the ZMM16 protein, encoded by one of the three maize *GLO/PI*-like genes (*ZMM16, ZMM18,* and *ZMM29*). The two B function proteins were

shown to heterodimerize *in vitro* and bind DNA as a heterodimer, just as the Arabidopsis proteins do. Remarkably, ZMM16 was also able to bind DNA as a heterodimer with Arabidopsis AP3 (the SI1 orthologue), but not with Arabidopsis PI (its own orthologue). Similarly, SI1 was able to bind DNA as a heterodimer with Arabidopsis PI but could not interact with Arabidopsis AP3. Expression of *SI1* from the *AP3* promoter in an *ap3* mutant line of Arabidopsis resulted in partial rescue of the mutant phenotype, with both petal and stamen development being restored (although neither organ was completely normal). Similarly, *ZMM16* was partially able to rescue a *pi* mutant (Whipple *et al.* 2004). These data reveal the remarkable degree of conservation between monocot and eudicot B function genes, despite the over 100 million years of evolution that separate them and the disparate second whorl structures they induce. The fact that those structures are truly homologous was demonstrated by analysis of B function activity in an early diverging grass without lodicules. *Streptochaeta* does not produce lodicules, but has second whorl bracts, and phylogenetic trees indicate that it is basal to the derived grasses which do produce lodicules. Whipple *et al.* (2007) reasoned that, if B function genes had been recruited separately to determine lodicule development and petal development, then they would not be expressed in the bracts of *Streptochaeta*. *In situ* hybridization confirmed their expression in these second whorl bracts, indicating that the role of B function genes in second whorl organ development is ancient, and that the various morphologies of those organs are derived secondarily.

As with both the gymnosperms and the eudicots, C function genes similar to *AG* function in the development of both types of reproductive structure in the grasses. Both maize and rice MADS box genes which fall into the C/D function clade have been shown to be expressed in the third and fourth whorls of the developing flowers (Goto *et al.* 2001; Yamaguchi *et al.* 2006). There appears to have been a major gene duplication of *AGAMOUS*-like genes within the grass family, with at least four *AG*-like genes present in the rice genome and six in maize, representing two C-lineage clades (rice *OsMADS58* and *OsMADS3*, and their maize orthologues *ZmZAG1* and *ZmZMM2/ZmZMM23*) and two D-lineage clades (*OsMADS13* and *OsMADS21*, and their maize orthologues *ZmZAG2*/*ZmZMM1* and *ZmZMM25*) (Ciaffi *et al.* 2011). Subfunctionalization of these genes has occurred, with no single gene fulfilling the complete role of *AG* (i.e. specifying stamen and carpel development, inhibiting A function in whorls 3 and 4, and maintaining meristem determinacy). Genetic loss of *OsMADS3* function resulted in the conversion of stamens to lodicules but did not affect carpels, indicating that *OsMADS3* is required for stamen development in whorl 3 but not for carpel development. However, *OsMADS58* RNAi knockdown lines produced flower-within-a-flower phenotypes with perturbed carpels, indicating that this C function gene is required to maintain floral meristem determinacy as well as to induce carpel formation (Yamaguchi *et al.* 2006). Such subfunctionalization and subsequent complexity are likely to be found in many species with genomes more complex than that of Arabidopsis, not just in grasses.

E function genes also appear to play roles in grass flower development. In the maize genome, at least 10 MADS box genes falling into the greater E function clade (eight *SEP* and two *AGL6* genes) have been identified (seven in rice), and reports suggest that the expression patterns of each are distinct but overlapping, and that together they cover a broad range. These expression patterns suggest that the genes play a number of partially redundant roles in flower development, including a role in the maintenance of floral meristem determinacy (Theissen *et al.* 2000). Within rice, the *SEP*-like gene *OsMADS1* has been shown to be necessary for correct development of all floral organs, rather like the orthologous E function *SEP1* gene of Arabidopsis. The *Osmads1* mutant has reduced determinacy of the floral meristem, but also produces flowers with palea- and lemma-like lodicules, a reduced number of stamens, and an increased number of carpels (Jeon *et al.* 2000). Similarly, mutation of *OsMADS6*, an *AGL6*-like gene, results in perturbations to lodicules, paleas, stamens, and ovules, as well as reduced meristem determinacy (Ohmori *et al.* 2009). As with B and C function, it is apparent that E function as it is understood in eudicots is also present in grass flowers.

The A function genes have not yet been well characterized in the grasses, although several members of the *SQUA*-like clade have been isolated from monocots. In maize, five genes fall into this clade, and there are three in rice, clustering as three subclades derived from a duplication event at the base of the monocots and a second duplication near the base of the grasses (Preston and Kellogg 2006; Ciaffi *et al.* 2011). Analysis of expression patterns suggests that these genes play a range of roles, most probably in floral meristem determination as well as vegetative development (Becker and Theissen 2003). There is no evidence to suggest an Arabidopsis-style A function role in specifying development of the outer two whorls for the grass *SQUA*-like genes. The other A function gene in the Arabidopsis model is the non-MADS box *AP2* (see Chapter 10). A recent report from barley does suggest that grass *AP2* genes play roles in development at least of whorl 2 floral organs (Nair *et al.* 2010). *HvAP2* is necessary for lodicule development, and mutation of this locus causes production of small lodicules which fail to open the flower fully. The gene is regulated by a microRNA, miR172, just as *AP2* is in Arabidopsis (Chapter 10). Although there is no apparent effect of *HvAP2* on first whorl organ development, the conservation of this microRNA-mediated regulation of second whorl development does imply that some aspects of A function had developed prior to the divergence of the monocots and the eudicots. Considering the complexities of A function gene activity in Arabidopsis and Antirrhinum, it is not surprising that this aspect of the grass ABC model should be the least clear. However, the presence of these genes in grasses, as in early diverging angiosperms, compared with their absence in the gymnosperms, may provide some of the genetic inspiration underlying the evolution of perianth organs.

11.4.2 Petaloid monocots

ABC genes have been identified from a variety of petaloid monocots, including tulip, lily, and *Agapanthus* (all Liliales), and asparagus, crocus, and various orchids (all Asparagales). In most cases the gene expression patterns have led to the hypothesis that the B, C, and E function parts of the ABC model apply to these flowers just as they do to the eudicots and grasses, with one key difference—in those petaloid monocots with two whorls of coloured tepals, B function gene expression is not confined to

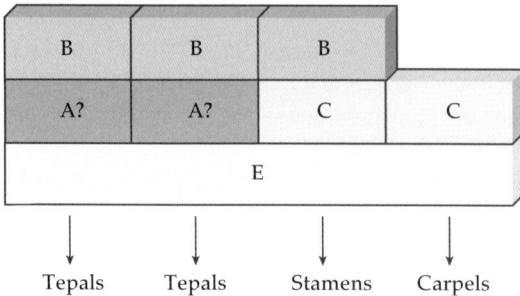

Figure 11.4 Petaloid monocots and the ABC genes. (a) Lilies have two whorls of coloured tepals, an inner whorl of three and an outer whorl of three. (b) B, C, E, and possibly A function genes specify organ identity in petaloid monocot flowers, with the B function genes being expressed in whorls 1, 2, and 3.

whorls 2 and 3, but is also found in whorl 1. The two (often, but not always, identical) whorls of coloured tepals are therefore both developing under the control of petal identity genes (see Fig. 11.4). Where mutant or transgenic experiments have been possible with these species, they have confirmed this model. However, expression analyses suggest that the situation might be more complex in the orchid family. Analysis of A, B, C/D, and E function gene expression in *Dendrobium* indicated that the E function *SEP*-like gene was expressed throughout the flower, as expected, but that so too were the C function gene and two of the three B function genes identified (Xu *et al.* 2006). In contrast, Mondragon-Palomino and Theissen (2011) showed that variation in expression levels of the four duplicate *DEF* genes found in orchids was in principle sufficient to explain the differentiation of three tepal types

(outer tepals, inner lateral tepals, and the showy inner labellum). In *Vanilla planifolia, Phragmipedium longifolium*, and *Phalaenopsis* they found that outer tepals expressed only clade 1 and 2 genes, inner lateral tepals expressed high levels of clade 1 and 2 genes but low levels of clade 3 and 4 genes, and the labellum expressed low levels of clade 1 and 2 genes but high levels of clade 3 and 4 genes. More detailed analyses of gene function are necessary to resolve the role of the ABC genes in development of the complex orchid flower.

One surprising aspect of B function in petaloid monocots is that, unlike the experiments described in Section 11.4.1, which confirm that the B function proteins in maize operate as a heterodimer, the petaloid monocot B function proteins appear to homodimerize. Kanno *et al.* (2003) isolated two *AP3*-like genes and one *PI*-like gene from *Tulipa gesneriana*. All three genes are expressed in whorls 1, 2, and 3, confirming their role in the development of both whorls of tepals and the stamens. Each of the AP3-like proteins was shown to heterodimerize with the PI-like protein and bind DNA as a heterodimer. However, the PI-like protein was also able to bind DNA as a homodimer, an ability not shown by the eudicot B function proteins. These data suggest that the obligate nature of heterodimer formation in the eudicot and grass B function proteins might be a convergent evolutionary phenomenon. Similar results have also been reported in *Lilium regale*, where AP3-like proteins heterodimerize with a PI-like protein, but the PI-like protein can also bind DNA as a homodimer (Winter *et al.* 2002). These data suggest that the evolution of obligate B function heterodimerization was a two-step process, with the AP3-like descendents of a gene duplication in the eudicots losing the ability to homodimerize first, followed by the PI-like descendents (see Section 11.5).

A function genes have not been well characterized in petaloid monocots, but the data available suggest that they function in floral meristem development rather than in organ specification, just as in grasses. Chen *et al.* (2007) identified two *SQUA*-like genes from the orchid *Phalaenopsis* and showed them to be expressed throughout the plant, most strongly in the developing floral bud. Heterologous expression in tobacco (Solanales) induced an early floral transition, supporting an *AP1*-like role in meristem specification. Similarly to grasses, members of the other A function gene family, *AP2*, can also be found in petaloid monocots. For example, Tsaftaris *et al.* (2012) recently reported the identification of three *AP2*-like genes in saffron crocus (Asparagales), and noted that they were expressed throughout the plant. However, without an indication of function or of regulation by microRNAs it is not possible to say whether such genes play a role in specification of outer whorl organs in the flower.

11.5 ABC genes in the basal eudicots

It is apparent from the previous sections that the B, C/D, and E function parts of the ABC model were established before the divergence of the monocots and eudicots, and that some aspects of A function were also present. A function is discussed in more detail in Section 11.7. However, there is still one major difference in floral developmental genetics between the basal orders of eudicots and the core eudicots that must be discussed. This difference concerns the B function genes. While the duplication event that gave rise to the *DEF/AP3* and *GLO/PI* lineages arose before the divergence of gymnosperms and angiosperms, a further—and crucial—gene duplication event occurred in the *AP3* lineage in the basal eudicots. Kramer *et al.* (2006) date this duplication event to after the divergence of the Ranunculales, but before the base of the core eudicots, suggesting that it occurred around the position of the Buxales and Trochodendrales. The ancestral *AP3* gene contains a C terminal motif known as the paleoAP3 motif, and in one of the lineages derived from this duplication event, the *TM6* lineage, this motif is retained. However, in the second lineage derived from the duplication, the *euAP3* lineage, either an eight nucleotide insertion or a single nucleotide deletion, each of which would cause a frameshift, has generated a novel C terminal motif, the euAP3 motif (Vandenbussche *et al.* 2003a; Kramer *et al.* 2006). The *TM6* lineage has been lost from both Arabidopsis and Antirrhinum, but in those species where it is retained it is thought that B function is partitioned semi-redundantly between the *TM6* and *euAP3* genes. Most notably, the euAP3 C terminal domain facilitates heterodimerization between B function proteins (although it is not itself thought to directly interact with other proteins). The presence of the paleoAP3 motif in petaloid monocot B function genes can then explain their ability to homodimerize, when the Arabidopsis and Antirrhinum proteins cannot (Kramer *et al.* 2006).

Within the basal eudicots that pre-date the paleoAP3/euAP3 duplication event, lineage-specific gene duplications have also occurred. For instance, the Ranunculales have experienced two separate duplication events within the *AP3* gene lineage, giving rise to three distinct *AP3*-like genes, each with the paleoAP3 C terminal motif. Detailed analysis of this gene family in *Aquilegia* has revealed that the proteins can function as heterodimers with the PI protein (Kramer *et al.* 2007), despite lacking the euAP3 domain. RNAi studies have shown that the duplicate genes have apparently experienced both subfunctionalization and neofunctionalization. *AqAP3-1* has acquired a new role in the development of sterile staminodes, *AqAP3-2* specifies stamen formation, and *AqAP3-3* is required for petal development (Sharma *et al.* 2011; Sharma and Kramer 2013).

11.6 Variations on the ABC model

The discovery that the petaloid monocots use an AB/AB/BC/C variation on the ABC model to specify their distinctive floral structure illustrates the inherent flexibility of the system. Simply by modifying the expression pattern of one or more of the ABC genes, through mutations to their promoter sequences or to upstream factors, it is possible to produce a flower with altered positions of floral organs. It is likely that the varying floral structures of many species can be explained through shifts to the expression patterns of their ABC genes. For example, the monocots are not the only plants to produce brightly coloured first whorl organs, and the expansion of B function gene expression into whorl 1 may turn out to explain the bright blue sepals of *Delphinium* (Ranunculales), the bright pink sepals of *Fuchsia* (Myrtales), and many other instances of petaloid sepals. Similarly, the dogwood *Cornus* (Cornales) is one of many examples that attract pollinators through brightly coloured bracts, which may in some species be the result of ectopic B (and possibly A) function expression in those organs (Feng *et al.* 2012).

The case of the sterile staminodes of the Ranunculales, induced by modified B function activity, was described in Section 11.5. Another novel organ that has been shown to develop through a variation on the basic ABC model consists of the pappus hairs of the composite inflorescence of daisies (Asterales). All members of the Asteraceae family produce a composite inflorescence, with the sex of flowers varying across the capitulum. In *Gerbera hybrida*, female flowers are produced in the outer whorls and hermaphrodite flowers in the inner whorls. The flowers do not have sepals, which would be awkwardly placed in a condensed inflorescence. Instead, structures called pappus hairs form in the first whorl, where they act as parachutes for the seed later in development (see Fig. 11.5). B, C/D, and E genes from *Gerbera* have been shown by transgenic experiments and expression analyses to fulfil essentially similar roles to their Arabidopsis counterparts (although note that both *TM6* and *euAP3* lineages of *AP3* are present, so B function involves some separation of functions). In particular, analysis of the B function genes has confirmed that the pappus hairs, like the palea and lemma of the grass flower, are simply variations on the theme of a sepal. Loss of function of *Gerbera AP3* or *PI* genes through antisense downregulation results in the conversion of the petals to pappus hairs and the stamens to carpels. Similarly, ectopic expression of *PI* converts the pappus hairs in whorl 1 to petals and the carpels to

stamens (reviewed by Teeri *et al.* 2002). However, it appears that at least one of the E function genes operates slightly differently in *Gerbera* from their known mode of action in Arabidopsis. The *Gerbera* gene *GERBERA REGULATOR OF CAPITULUM DEVELOPMENT 1* (*GRCD1*) is orthologous to the E function *SEP* genes of Arabidopsis, and, like the *SEP* genes, is expressed in all floral organs. However, downregulation of *GRCD1* function through an antisense approach resulted in the conversion of the sterile stamens (found in female flowers) into a second whorl of petals. These data suggest that this particular *Gerbera* E function gene has non-redundant whorl-specific activity, operating with the B and C function genes to specify stamen development (Kotilainen *et al.* 2000). This contrasts markedly with the position in Arabidopsis, where the E function genes are redundant, only resulting in a phenotypic change when all three are inactivated together, and influence the development of all floral organs, rather than just one whorl. However, other E function genes in *Gerbera* do fulfil the usual role, with *GRCD4* and *GRCD5* operating in a similar way to the Arabidopsis *SEP* genes (Rukolainen *et al.* 2010).

The production of unisexual flowers is not unique to the Asteraceae, occurring also in monoecious systems such as maize, and in dioecious systems such as *Silene dioica* (red campion, Caryophyllales). These will be discussed in more detail in Chapter 13. In some species that produce unisexual flowers, the abortion of one type of reproductive organ has been shown to be correlated with the downregulation of the appropriate B or C function genes, but it is not yet clear how the relationship between floral gender and ABC gene function is controlled. These and many other variations on the basic ABC model remain to be analysed, but the current depth of knowledge on the control of floral organ identity in Arabidopsis serves as an excellent starting point for such studies.

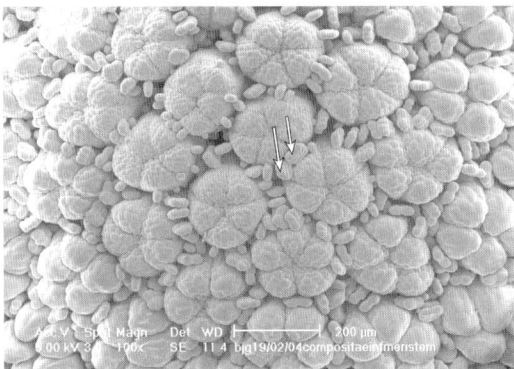

Figure 11.5 Pappus hairs (arrowed) develop in the first whorl position around each floral meristem in the capitulum of a daisy.

11.7 Is A function unique to the Brassicaceae?

In Section 10.1.1 of Chapter 10 we discussed A function, the specification of sepal and petal identity, in

Arabidopsis and Antirrhinum. In Arabidopsis it is provided by the MADS box gene *AP1*, which also plays a role in floral meristem specification, and by the non-MADS box transcription factor AP2. These proteins interact with the C function protein AG and a microRNA (miR172) to specify sepal and petal development in whorls 1 and 2. However, the situation in Antirrhinum is more complex, with no identified A function mutant. Reverse genetic approaches revealed that the two *AP2* orthologues in Antirrhinum acted redundantly to specify some aspects of petal development, but could not be said to provide classical A function (Keck *et al.* 2003). The *AP1* orthologue in Antirrhinum, *SQUA*, is necessary for floral meristem identity, but has no role in floral organ development (Huijser *et al.* 1992).

The observation that A function does not seem to be provided in Antirrhinum by the *AP2* or *AP1* genes could reflect loss of the module in Antirrhinum, or its gain in Arabidopsis. In this chapter we have seen that the early diverging angiosperms lack conventional A function, since *Nuphar SQUA* is expressed throughout the plant (see Section 11.3), and that the monocot A function genes also seem to play more general roles in floral meristem conversion. Litt (2007) reviewed the evidence on function of the gene lineages containing *AP2* and *AP1* from across the angiosperms. In addition to the data presented above, she reported evidence from tobacco and petunia (both Solanales) to suggest that *AP2*-like genes do not generally play a role in sepal and petal specification. More extensive data were available for *AP1*-like genes, but again evidence from pea (*Pisum sativum*, Fabales), petunia, tomato, and tobacco (all Solanales), apple (Rosales), *Silene latifolia* (Caryophyllales), birch (*Betula pendula*, Fagales), and various daisies (Asterales) suggested that the conserved role of *AP1*-like genes was in floral meristem identity, not in floral organ specification. It is possible that conventional A function activity, involving the specification of sepals and petals and repression of C function activity, is unique to the Brassicaceae, where its evolution may have been facilitated by the recent gene duplication event that gave rise to *AP1* and *CAULIFLOWER* (see Chapter 9). Without A function it is still possible to initiate four distinct organs on a florally defined meristem—if the default situation is to produce sepals, addition of B function in whorl 2 could generate petals, B + C function in whorl 3 could generate stamens, and C function in whorl 4 could generate carpels. The evidence to date suggests that this two gene model might more accurately reflect the diversity of angiosperm flowers.

CHAPTER 12

Function and development of gametophytes

In previous chapters we have considered the induction and development of flowers, without paying much attention to the reason for their existence. The role of flowers is to ensure the transfer of a small but precious package, the pollen grain, from one plant to another. That pollen grain contains the sperm, which fertilizes an egg cell retained within the tissues of another flower, resulting in the formation of a zygote and, in due course, a new generation of plants that will produce flowers in their turn. Although the details of seed and fruit development are beyond the scope of this book, it is necessary to provide a brief summary of the events that occur within the male and female reproductive organs of the flower. Here, meiosis and then *haploid mitosis* produce an independent generation, the gametophyte, including the gametes. In this chapter we discuss the development of male and female gametophytes, and then briefly consider the events that occur when they are brought together.

12.1 Alternation of generations in multicellular organisms

The alternation of generations in plants is a topic often considered deeply confusing. In general, this confusion stems from a failure to understand the life cycles of all multicellular organisms, including animals. Put simply, in multicellular eukaryotes there are three types of life cycle, shown in Fig. 12.1. The existence of three alternative forms of life cycle is most probably a simple consequence of multicellularity evolving multiple times, each independently. There are several possible ways of developing a multicellular sexual life cycle from a unicellular one, and it seems that animals, plants, and fungi have each adopted different systems. These different life cycles have different genetic effects and so have different evolutionary consequences, although these differences should not be thought of as driving the evolution of the different forms, which almost certainly arose through simple chance. The different life cycles are defined on the basis of where ordinary mitotic cell divisions occur relative to the process of meiosis.

The animal life cycle (and that of many brown algae) is based on gametic meiosis (see Fig. 12.1a). In organisms that have gametic meiosis, most of the life cycle is spent in the diploid stage, and accordingly animal bodies are composed of diploid cells. When reproduction occurs, those diploid cells undergo meiosis to produce gametes directly, and those gametes immediately fuse to make a diploid zygote. Thus animal cells only undergo mitosis in the diploid state, and the haploid cells of animals have one job and one job alone—to fuse together to make a new diploid cell.

The fungi, and some of the green algae, have a life cycle that operates in exactly the opposite way (see Fig. 12.1b). The zygote is diploid, formed by the fusion of two haploid gametes, but instead of undergoing mitosis to produce a body, it first undergoes meiosis and becomes haploid again. These haploid cells then undergo mitosis to produce the fungal body, which is still haploid. Gametes are produced by this haploid body through more mitotic divisions. Thus, in contrast to the animal system, fungal cells only undergo mitosis in the haploid state, and the diploid cells of fungi have one job and one job

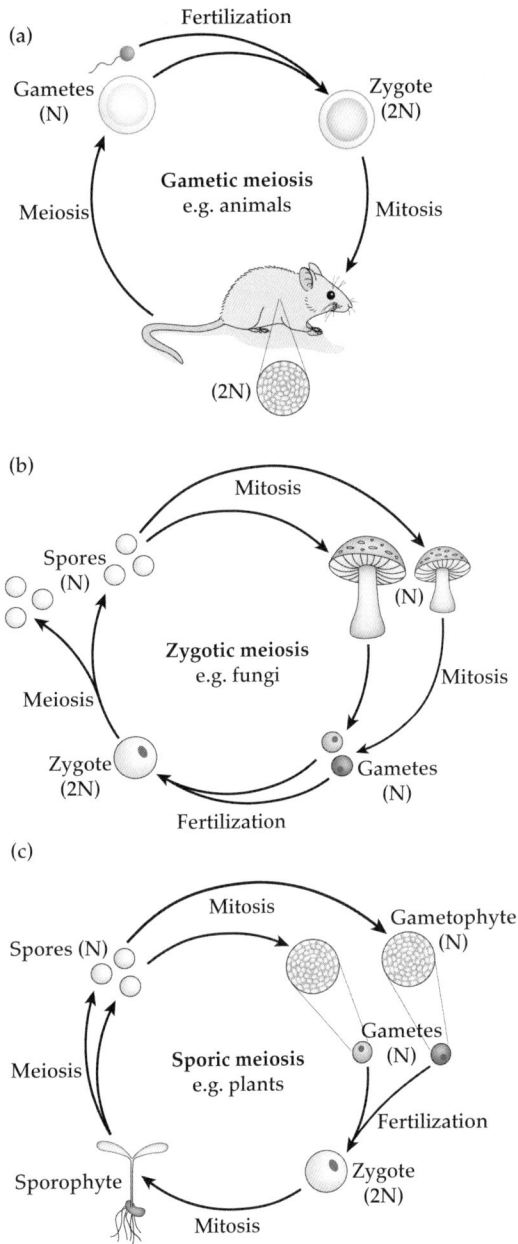

Figure 12.1 Different multicellular life cycles. (a) Gametic meiosis, as seen in animals. (b) Zygotic meiosis, as found in fungi. (c) Sporic meiosis, found in plants.

alone—to undergo meiosis to make a new haploid cell. This system is known as zygotic meiosis.

The third type of life cycle is the one that some green algae and all land plants use, and is known as sporic meiosis (see Fig. 12.1c). In an organism with sporic meiosis, mitosis occurs in both the haploid and the diploid stages of the life cycle. The diploid zygote undergoes mitosis to produce a diploid body, just like an animal. This diploid body undergoes meiosis to produce haploid cells, which are not gametes but are called spores. The spores do not fuse with other haploid cells, but instead behave like the haploid cells of a fungus—they undergo mitosis to produce a haploid body. This haploid body then undergoes further mitosis to produce the haploid gametes, which fuse to make a diploid zygote again. Thus there are essentially two types of body—a haploid type and a diploid type. The haploid body is known as the gametophyte, because it produces gametes, and the diploid body is called the sporophyte, because it produces spores. All land plants have both diploid and haploid phases in their life cycles. Although all terrestrial plants show sporic meiosis, they vary greatly in the length of time they spend in the haploid and diploid states, and also in the independence or interdependence of the two phases.

Different life cycles carry different genetic consequences. Undergoing mitosis in the haploid state, as fungi and the plant gametophyte do, places enormous selective pressure on expressed genes. A gene expressed in the haploid stage cannot be masked by a corresponding dominant allele, so all alleles of genes expressed in the haploid stage are directly exposed to selection. Since the haploid gametophyte undergoes the complex processes of cell growth, cell division, and cell differentiation, many genes are expressed during this stage. In animals, the haploid phase is limited to a single cell, the gamete, which does not have to do much—no mitosis, no growth, and no development of complex organs. Therefore lethal recessive alleles can pass unnoticed in the animal haploid stage and then be masked by dominant alleles in the diploid stage. However, for a plant, those lethal alleles will never get through the gametophyte stage. Accordingly, plants cannot carry lethal recessive alleles across generations, and so have a much lower genetic load than animals do (although note that mutation within the lifetime of a long-lived plant can result in a high genetic load within individuals). Plant populations have generally healthier genetic make-ups than animal populations and are less subject to inbreeding depression.

12.2 Diversity of gametophyte form

As mentioned in Section 12.1, all land plants have sporic meiosis, indicating that it is an ancestral state, typical of the first land plants. However, within the constraints of an alternating haploid/diploid multicellular form, plants vary greatly in the length of time they spend in the haploid and diploid states, and also in the relative independence of the two phases. Over the course of their evolution, land plants have undergone a transition from haploid or gametophyte dominance, with a dependent sporophytic phase, to dominance of a free-living diploid or sporophytic phase. It is possible that the increasing dominance of the plant life cycle by the diploid sporophyte through evolutionary time reflects the high mortality of the gametophyte stage. By reducing the gametophyte stage to only a few cell divisions, angiosperms may have succeeded in minimizing this mortality while retaining the ability to lose catastrophic mutations before they enter a diploid phase. An alternative explanation is that all land plants retain the egg attached to the female gametophyte, and so can protect and nourish both the egg and, after fertilization, the zygote and developing embryo (sporophyte). This opportunity to provide early support to the developing sporophyte might have led to the development of larger and more complex sporophyte forms.

In early diverging land plants, such as mosses and liverworts, the gametophyte stage of the life cycle is dominant. Thus the familiar tufted green moss plant is the haploid gametophyte, as is the green thallus of liverworts found in damp places. In both of these groups of bryophytes, the diploid sporophyte is parasitic and retained attached to the gametophyte body throughout its development. In moss plants, the little brown stalked head which grows on top of the tufted green body is the sporophyte.

The life cycle of all vascular plants is sporophyte dominant. For example, the typical fronded fern plant is diploid, producing haploid spores on the undersides of the leaves. These spores germinate to produce the gametophyte stage, which is a flat green blob of tissue, very like liverwort thallus in appearance. The eggs that are produced and retained on this gametophyte are fertilized by sperm from the same or a nearby gametophyte, and the new diploid (sporophyte) plant grows from the gametophyte tissue.

In the seed plants, the size and duration of the gametophyte stage have been reduced even further, to only a few cell divisions. The female gametophyte is retained wholly or partly within the sporophyte body, as the egg sac (within the carpel of angiosperms). The male gametophyte is simply the pollen grain, built within the sporophyte tissue but released into the wind or on to an animal's body. When the angiosperm male gametophyte reaches a stigma it germinates, eventually releasing the sperm cells into the egg sac to fertilize the egg. The detailed anatomy of gametophytes (egg sacs and pollen grains) is somewhat variable among the angiosperms, but that variability is on a very different scale from the variation in gametophyte form across the green plant lineage.

12.3 The angiosperm female gametophyte

The female gametophyte is also known as the megagametophyte. In angiosperms it is found within the ovule, within the carpel of the flower (see Fig. 12.2a). Several female gametophyte forms are known in flowering plants, consisting of a variety of cell and nuclei numbers. The most common form (and the one found in Arabidopsis), called the *Polygonum* form, consists of seven haploid cells, one of which contains two haploid nuclei (so there are eight nuclei in total). These seven cells are of four types. At the top of the embryo sac, furthest from

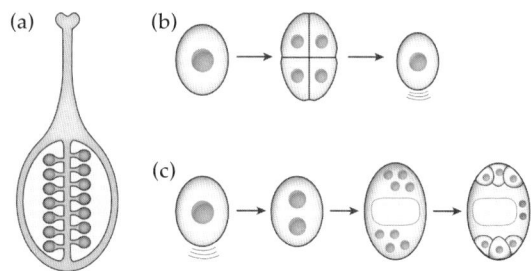

Figure 12.2 The angiosperm female gametophyte. (a) Position of ovules within the Arabidopsis carpel. (b) Megasporogenesis in Arabidopsis. (c) Megagametogenesis in Arabidopsis.

the point at which sperm will later enter (the micropyle), are three antipodal cells. Below these is the central cell, which contains two nuclei, and will later be fertilized by a single sperm to generate the triploid nutritive endosperm. Below this is a single egg cell (the gamete) flanked by two synergid cells which play roles in attracting pollen tubes. The female gametophyte is retained at all times within the tissues of the sporophyte parent, but from there it plays a crucial role in directing the growth of the incoming pollen tube. It also expresses genes that control the initiation of seed development after fertilization, and later control the development of both the embryo and the endosperm (Yadegari and Drews 2004; Drews *et al.* 2011).

The development of the megagametophyte is a two-stage process (reviewed by Beger and Twell 2011). First, a cell within the sporophyte tissue, known as the megaspore mother cell, undergoes meiosis and gives rise to four haploid nuclei. This process is called megasporogenesis. Usually three of these nuclei degrade, and the fourth, the megaspore, then undergoes mitosis to produce the eight haploid nuclei of the mature gametophyte. This process is called megagametogenesis.

12.3.1 Megasporogenesis

Megasporogenesis is a relatively simple process, during which a single meiotic event gives rise to four haploid nuclei. Ma (2005) provides an excellent summary of chromosome behaviour during meiosis in plants. In different plant species the four hapoid nuclei may be packaged into one, two, or four cells. Most commonly, in plants which produce a *Polygonum*-type megagametophyte, the megaspore mother cell within the ovule produces four haploid cells, as cell division occurs after both meiosis I and meiosis II. Three of these haploid cells undergo cell death, leaving a single haploid megaspore (see Fig. 12.2b). However, in other species, cell division occurs after only one, or even neither, of the meiotic divisions. In species with bisporic megasporogenesis, cell division occurs only after meiosis I, so meiosis II results in the production of two cells, each containing two haploid nuclei. One of these survives and enters megagametogenesis to produce a final total of six haploid nuclei. In species with

tetrasporic megasporogenesis, cell division does not occur after either meiotic division. The final megaspore has four haploid nuclei, which undergo mitosis to produce around 16 nuclei in the mature megagametophyte.

Since the development of the megaspore is a phase of sporophyte development, it is genes expressed in the sporophyte that control megasporogenesis. Mutations in such genes often manifest themselves as female infertility, due to failure of the megaspore mother cell to develop, or failure of meiosis to occur, or incorrect positioning or polarity of the embryo sac within the ovule (reviewed by Yadegari and Drews 2004). A recent report by Makkena *et al.* (2012) implicated the MYB transcription factors *FOURLIPS* (*FLP*) and *MYB88* in controlling the entry of the megaspore mother cell into meiosis. In *flp* mutants and *flp/myb88* double mutants the megaspore mother cell either fails to enter meiosis or undergoes abnormal meiosis. FLP and MYB88 were first described for their role in stomatal development, where they restrict the guard mother cell to a single symmetric division through the regulation of genes involved in cell cycle progression. It is likely that they are regulating similar pathways during megasporogenesis. In contrast, reports have implicated small RNA pathways operating in the sporophyte cells around the ovule in limiting the number of megaspores generated (Olmedo-Monfil *et al.* 2010). Arabidopsis lines that lack a functional copy of *ARGONAUTE 9* (*AGO9*), a protein that binds small interfering RNAs and so is crucial in regulating gene silencing, produce excess numbers of megaspores, indicating that both positive and negative factors regulate the entry into meiosis.

12.3.2 Megagametogenesis

Megagametogenesis consists of a number of mitotic cell divisions, which result in the mature megagametophyte. In the most common system this simply involves the single haploid spore undergoing three rounds of mitosis to produce eight haploid nuclei (see Fig. 12.2c). As with the entry into meiosis in megasporogenesis, the entry into mitosis is partially regulated by factors expressed in the surrounding sporophyte tissues, particularly small interfering RNA pathway components such as *ARGONAUTE 5*

(*AGO5*) (Tucker *et al.* 2012). Genes expressed in the megaspore are also important. In Arabidopsis, the mitotic divisions require the activity of CHROMATIN REMODELLING PROTEIN 11 (CHR11), as transgenic lines with *CHR11* silenced are unable to proliferate nuclei and so fail to undergo megagametogenesis (Huanca-Mamani *et al.* 2005).

Megagametogenesis is the whole of the gametophyte's life, and it is genes expressed by the gametophyte that determine correct gametogenesis. Mutations that perturb cell division or any essential metabolic function will be lethal at this stage, and can usually be identified because of perturbed segregation ratios in the progeny of a plant. Classically, a megagametophytic mutation will result in an absence of 50% of the seeds in a pod or silique. For example, Ebel *et al.* (2004) showed that the plant retinoblastoma protein, RBR1, is essential for correct megagametogenesis in Arabidopsis. The *rbr1* mutants could only be maintained as heterozygotes, as the megagametophyte failed to develop normally if it carried the *rbr1* mutant allele. The RBR1 protein is involved in cell cycle control in plants and animals, and the *rbr1* mutant megagametophytes underwent excessive mitotic divisions, resulting in a non-viable embryo sac. This mutant phenotype also supports the suggestion that the ancestral angiosperm megagametophyte contained more cells, but that genes such as *RBR1* have been recruited to reduce cell division and minimize the gametophyte lifespan. In theory, mutations in genes expressed in the sporophyte may also affect megagametogenesis, by altering the environment surrounding the developing gametophyte or inhibiting its access to water or essential nutrients.

Yadegari and Drews (2004) reported the identification of over 150 mutants with perturbed megagametogenesis, and described their classification into mutations affecting different stages of gametophyte formation. Detailed characterization of the lesions occurring in these mutants is complex, as visualization of gametogenesis is not simple, given that it occurs within living tissues of the sporophyte. However, molecular characterization of these mutated genes will greatly enhance our understanding of how the angiosperm female gametophyte develops and functions. Similarly, Pagnussat *et al.* (2005) described 130 gametophytic mutants in Arabidopsis, generated using transposon insertions perturbing segregation ratios in the next generation. These mutants affect all processes from the first mitotic division of megagametogenesis, through megagametophyte development, to later embryo sac functions including post-fertilization processes such as embryo development. The authors were able to amplify sequence flanking the inserted transposons, and catalogue the genes perturbed in the mutants. The proportions of proteins of different types involved at different stages varied quite significantly. During megagametogenesis itself, genes with roles in secondary metabolism seemed to be particularly important, along with genes involved in protein degradation. During the later fertilization process, genes related to energy metabolism and stress responses seemed to be more important.

Cell walls do not usually form in the megagametophyte until all divisions have occurred, allowing migration of nuclei during development. Specifically, one nucleus from the top of the embryo sac, and one from the bottom (the micropylar end, where the pollen tube will arrive) migrate to the centre of the structure and fuse, either immediately or at fertilization, to generate the diploid central cell which will be fertilized to give rise to the triploid endosperm. This fusion requires the RPL21M protein, a mitochondrial 50S ribosomal subunit, and in a mutant defective in the production of this protein the nuclear membranes of the two central cell nuclei fail to merge (Portereiko *et al.* 2006). In different species other gametophyte cells may also undergo different differentiation events. For example, in Arabidopsis, the three antipodal cells (those furthest from the micropyle) undergo cell death shortly before fertilization occurs.

Pattern formation within the embryo sac has recently been shown to depend on gradients of auxin. The micropyle acts as a souce of auxin, actively expressing genes required for auxin synthesis, and a gradient exists from high auxin concentration at the micropylar end to low auxin concentration at the antipodals. Pagnussat *et al.* (2009) showed that this gradient was the result of auxin synthesis at the micropyle, rather than of auxin transport within the embryo sac. They further showed that perturbing auxin levels resulted in interconversion of the cell types within the megagametophyte.

Little is known about the evolutionary history of this simple eight-celled megagametophyte. Other lineages of land plants, including gymnosperms, inevitably produce megagametophytes containing many more cells. Reports of a nine-celled megagametophyte in *Amborella trichopoda* (Amborellales), the sister to all other extant angiosperms, suggested that early angiosperms might also have produced larger megagametophytes (Friedman 2006; Rudall 2006). However, a megagametophyte with four cells and four nuclei is found in other early diverging angiosperms, in the Nymphaeales and Austrobaileyales (reviewed by Friedman and Ryerson 2009). It has therefore been hypothesized that the ancestral state for the angiosperm megagametophyte might have been a four cell, four nuclei structure, which then experienced a duplication to generate an eight nuclei structure (with an additional division occurring in *Amborella*). Pagnussat *et al.* (2009) postulated that, if such a duplication occurred in the context of an auxin gradient, the duplicated module furthest from the micropyle (and thus with least auxin) would produce sterile antipodals, rather than an egg. One of the four nuclei arising from this duplication might then have migrated to the central cell, facilitating the evolution of the triploid endosperm characteristic of angiosperms.

12.4 The angiosperm male gametophyte

The male gametophyte, also known as the microgametophyte, begins life within the stamens of the flower, and completes its development by germinating to produce a pollen tube—usually within the style of a (hopefully different) flower. It is even more reduced than the female gametophyte, and so has less scope for variability in structure between different plants. Essentially, the microgametophytes are pollen grains, and the locules of the stamens of most plants are filled with many hundreds or even thousands of them. The mature microgametophyte consists of three haploid cells. A larger vegetative cell contains two sperm cells, which go on to fertilize the egg cell and the diploid central cell of the female gametophyte. The vegetative nucleus itself also travels down the developing pollen tube, alongside the sperm nuclei. The whole microgametophyte is coated with a specialized outer layer

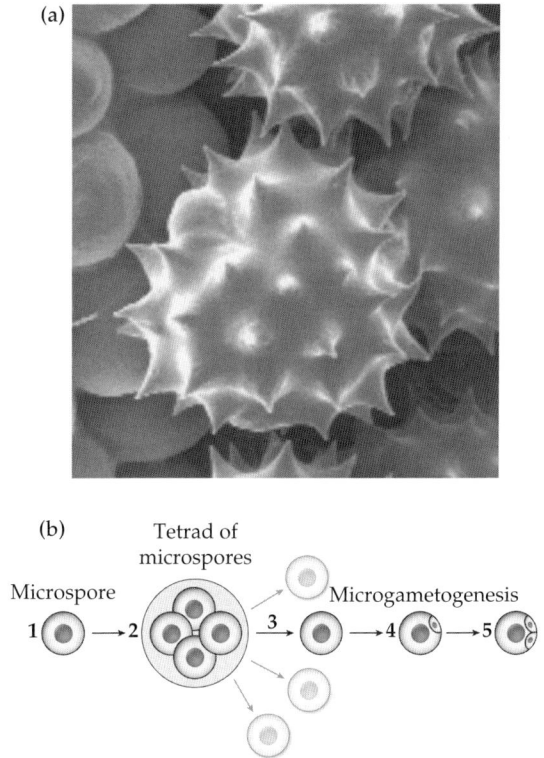

Figure 12.3 The angiosperm male gametophyte. (a) Mature microgametophytes are coated with elaborate sporopollenin. (b) Microsporogenesis and microgametogenesis in Arabidopsis. The tetrad of microspores present at stage 2 disassociate before undergoing microgametogenesis.

of sporopollenin, which protects the contents from dehydration and mechanical damage as they are transferred between plants (see Fig. 12.3a). This pollen coat also serves to mediate interactions between the microgametophyte and the female sporophyte tissue that it lands on, and so plays an important role in mating success. The pollen coat is mainly derived from sporophytic tissue on the inside of the stamen locules.

The development of the microgametophyte is a two-stage process (see Fig. 12.3b). First, a cell within the sporophyte tissue (within the stamen), known as the microspore mother cell, undergoes meiosis and gives rise to a tetrad of four haploid cells (for a summary of the events that occur during meiosis, see Ma 2005). This process is called microsporogenesis. The tetrads are initially enclosed in a thick callose wall, which is broken down by an enzyme secreted from the stamen wall, releasing the microspores. These microspores then enlarge and undergo an asymmetric mitotic division to produce a large vegetative cell and a small generative cell, which is retained within the vegetative cell. This process is called microgametogenesis. The generative cell divides again to produce two sperm cells, although this division may not occur until after pollination has been achieved (McCormick 2004).

12.4.1 Microsporogenesis

Microsporogenesis occurs within the stamen. Specifically, on top of the stamen filament an anther is formed, which usually consists of four cell layers enclosing the microspore mother cells. The innermost of these four layers, the tapetum, has a key role in pollen grain development. When the microspore mother cells initiate meiosis, callose is deposited on their outer surfaces. Thus the tetrads of microspores are surrounded by callose. These tetrads then disassociate from the tapetum, generating a space called the locule. Within the locule, callase released by the tapetum degrades the callose outer walls and frees the four microspores from each tetrad (Ma 2005). They are then free to enlarge and begin mitotic divisions to produce the microgametophytes.

As with megasporogenesis, microsporogenesis is controlled by genes expressed by the sporophyte itself (that is, the diploid parental tissue). Such genes must ensure the correct formation of the crucial cell layers in the anther, alongside their role in regulating the onset of meiosis. For example, a feedback loop operating between the BARELY ANY MERISTEM 1/2 (BAM1/2) leucine-rich receptor-like kinases and the SPOROCYTELESS (SPL) transcription factor controls the relative production of somatic cells of the anther walls and microspore mother cells (Hord *et al.* 2006; reviewed by Chang *et al.* 2011). Perturbation of *SPL* results in a reduction in microspore mother cell formation, while perturbation of *BAM1/2* results in loss of the inner three cell layers of the anther, replaced by microspore mother cells. Similarly, the *EXCESS MALE SPOROCYTES 1* (*EMS1*) gene is necessary for the normal development of both the tapetum cells in the innermost layer of the anther and the microspore mother cells

attached to them (Zhao *et al.* 2002). In the *ems1* mutant, excess microspore mother cells develop at the expense specifically of the tapetal layer. However, since the tapetum is necessary for pollen development, this mutant is male sterile. It seems that the EMS1 protein is also essential for the division of cells following meiosis, as the microspore mother cells in the *ems1* mutant undergo meiotic separation of chromosomes and chromatids but do not form new cell membranes around the new nuclei. These incomplete microspores are then incapable of undergoing mitosis. *EMS1* encodes a leucine-rich receptor-like protein that localizes to cell membranes and is hypothesized to be present in the membranes of tapetal precursor cells (Zhao *et al.* 2002). It may be bound by the TAPETUM DETERMINANT 1 (TPD1) protein, which is a very small, possibly secreted protein, encoded by a gene expressed in the developing microspore mother cells (Yang *et al.* 2003a). Mutation of *TPD1* results in a very similar phenotype to that of the *ems1* mutant, namely an increase in microspore mother cells at the expense of tapetum. TPD1 and EMS have been shown to physically interact (Jia *et al.* 2008), supporting the hypothesis that these two proteins interact to regulate the respective numbers of tapetal cells and microspore mother cells. Other leucine-rich repeat receptor kinases are also involved in distinguishing tapetal and microspore mother cells, possibly through the same pathway. Colcombet *et al.* (2005) reported that a double mutant between the *SOMATIC EMBRYOGENESIS RECEPTOR KINASE 1* and *2* genes (*SERK1* and *SERK2*) showed a phenotype similar to that of the *ems1* mutant, with an absence of tapetum and an excess of microspore mother cells. It has been hypothesized that EMS1, SERK1, and SERK2 form a heteromeric receptor complex that is bound by the TPD1 ligand (Jia *et al.* 2008). It is certainly likely that a number of receptors and ligands are involved in this feedback loop, to provide multiple checkpoints in this essential process. Following this establishment of cell fate and cell number, a collection of downstream genes, including several encoding transcriptional regulators, act together to ensure correct differentiation of microspore mother cells (reviewed by Ma 2005 and by Chang *et al.* 2011). Mutation of any of these genes results in a failure of microsporogenesis and subsequent male sterility.

Once cell identity is established, the microspores themselves develop in response to sporophyte-expressed genes. These genes specifically control the meiotic divisions that produce the microspores. Many essential genes are required for correct pairing and separation of chromosomes during meiosis (Ma 2005). Later, the *TETRASPORE* (*TES*) gene is essential for the formation and expansion of cell plates between the meiotically produced haploid nuclei. The *tes* mutants produce large microspores containing four haploid nuclei in a single cell. The TES protein is a kinesin which may determine microtubule positioning within the cell, a necessary component of cell division. The absence of the TES protein is correlated with an absence of the radial arrays of microtubules usually seen when the microspores undergo cell division (Yang *et al.* 2003b; Oh *et al.* 2008).

Detailed analysis of microsporogenesis has been used in the dissection of evolutionary relationships between flowering plants. One key character is the timing of the cell divisions in relation to the nuclear divisions, while another is the orientation of formation of successive cell plates during meiosis. Analysis of these traits in early diverging angiosperms has revealed a surprising degree of variability between taxa, and suggests that the details of microsporogenesis are relatively labile in evolutionary time (Furness *et al.* 2002).

12.4.2 Microgametogenesis

Following microsporogenesis, the anther locules contain free microspores. These become heavily vacuolized, pushing the nucleus to the edge of the cell. The microspores then undergo a stereotyped set of cell divisions (one asymmetric, one symmetric), to produce the three nuclei of the microgametophyte. The first mitosis is asymmetric because the nucleus is at the periphery of the cell. The large daughter cell is the vegetative cell, and contains most of the cytoplasm from the microspore. The smaller cell, the generative cell, is entirely engulfed by the vegetative cell. It then undergoes a further, symmetric division to produce two sperm cells. This final division occurs within the developing pollen grain in some species (those with trinucleate pollen), but in other species (those with binucleate

pollen) it occurs during pollen tube growth after pollination (Ma 2005). The developing pollen grain is coated with two protective layers—the exine and the intine. The exine is made of highly elaborate sporopollenin, which mediates interactions with stigmatic cells, whereas the intine is made from cellulose, pectin, and proteins. Both layers appear to be mainly derived from the anther tapetum. The mature pollen grain is then dehydrated ready for release into the outside world.

As with megagametogenesis, microgametogenesis is regulated by genes expressed by the gametophyte nuclei themselves. However, because the tapetum contributes to pollen coat development, there is also a clear role for sporophytically expressed genes in microgametogenesis. For example, the genes described earlier that are necessary to correctly specify tapetal and microspore fate also influence gametophyte development. Those microspores which survive in mutants defective in these genes fail to undergo correct gametogenesis due to the abnormal tapetal layer surrounding them. Similarly, the parental sporophyte tissue expresses many genes involved in building the pollen coat. Plants carrying mutations in these genes are usually male sterile, as a result of perturbed exine and intine development. For example, the *male sterile 2* mutant of Arabidopsis carries a lesion in a gene encoding a protein that promotes the synthesis of long-chain molecules, such as sporopollenin (cross-linked carotenoids). The mutant produces non-viable pollen grains with thin cell walls lacking an exine (Aarts *et al.* 1997). A number of such mutants are reviewed by Ma (2005).

Male gametophytic mutants influence segregation ratios in the progeny of plants, usually resulting in abortion of 50% of pollen grains. However, unlike the situation with female gametophytic mutants, where this results in a 50% reduction in seed set, the vast excess of pollen relative to egg cells means that male gametophytic mutants can be difficult to identify using conventional screens. Instead, perturbations in the segregation ratios of other markers are used to pick out male gametophyte mutations. Since many thousands of genes are likely to be expressed in the male gametophyte, it is not surprising that many mutants have been described. Indeed, microarray analysis of genes

Plate 1 The Gnetophytes. (a) *Ephedra distachya* subsp. *monostachya* (male). Photo by Le.Loup.Gris (Wikimedia Commons). (b) *Welwitschia mirabilis* (male). Photo by Franzfoto (Wikimedia Commons). (c) *Gnetum latifolium* var. *funiculare*. Photo by Vinayaraj (Wikimedia Commons). See also Figure 1.2.

Plate 2 The flower of *Amborella trichopoda*. Photograph kindly supplied by Sangtae Kim and Pam Soltis (University of Florida). See also Figure 1.4.

Plate 3 Species in which the floral transition has been studied. (a) *Pisum sativum* (Fabales). Photo by Rasbak. (b) *Triticum* species (Poales). Photo by Optograph. (c) *Populus* species (Malpighiales). Photo by Matt Lavin. (d) *Arabis alpina* (Brassicales). Photo by Franz Xaver. (e) *Oryza sativa* (Poales). Photo by C.T. Johansson. (f) *Beta vulgaris* (Caryophyllales). Photo by Forest and Kim Starr. All images from Wikimedia Commons. See also Figure 8.1.

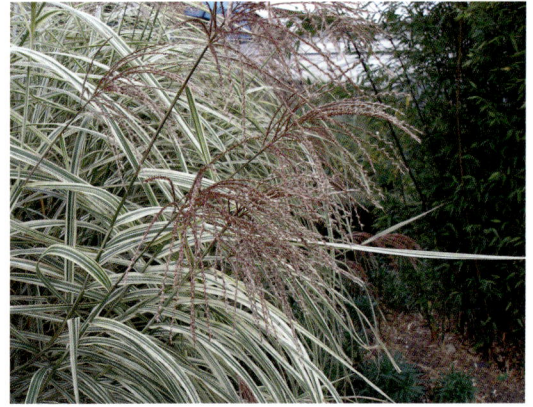

Plate 6 The long tasselled flowers of a wind-pollinated grass hang far from the main body of the plant. See also Figure 14.1.

Plate 4 Early floral meristem identity mutants. (a) The Antirrhinum *flo* mutant (left) has inflorescence shoots produced in place of the flowers found in the axils of wild type bracts (right). Image kindly supplied by Enrico Coen (John Innes Centre). (b) The Arabidopsis *ap1* mutant is slightly better converted to the floral form, with indeterminate floral structures arising from the meristem. See also Figure 9.2.

Plate 5 The original ABC mutants. (a) *apetala 1* (A function in Arabidopsis). (b) *ovulata* (A function in Antirrhinum) (right) compared with wild type (left). (c) *apetala 3* (B function in Arabidopsis). (d) *deficiens* (B function in Antirrhinum) (right) compared with wild type (left). (e) *agamous* (C function in Arabidopsis). (f) *plena* (C function in Antirrhinum) (right) compared with wild type (left). Antirrhinum images kindly supplied by Enrico Coen (John Innes Centre). Image (e) provided by Ian Furner (University of Cambridge). See also Figure 10.2.

Plate 7 Insect-pollinated flowers. (a) *Magnolia* (Magnoliales) flowers are beetle-pollinated. (b) The fly-pollinated flowers of *Fatsia japonica* (Apiales). (c) Bumblebee entering a *Hebe* flower (Lamiales). (d) Many daisies (Asterales) are butterfly-pollinated. (e) The flowers of *Angraecum sesquipedale* (Asparagales) have very long nectar spurs and are pollinated by extremely long-tongued moths. Photographs (a), (d), and (e) kindly supplied by Cambridge University Botanic Garden and H. Rice. See also Figure 14.3.

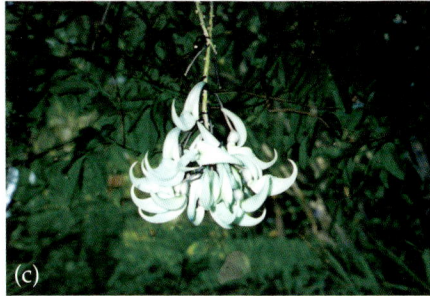

Plate 8 Vertebrate-pollinated flowers. (a) The pendant form of *Fuchsia* flowers (Myrtales) is ideal for hovering hummingbirds. (b) Bird of paradise (*Strelitzia regina*, Zingiberales) flowers provide a sturdy landing platform for non-hovering birds. Photograph kindly supplied by Cambridge University Botanic Garden. (c) The flowers of *Strongylodon macrobotrys*, the jade vine (Fabales), hang far below the foliage, making them readily accessible to bats. See also Figure 14.4.

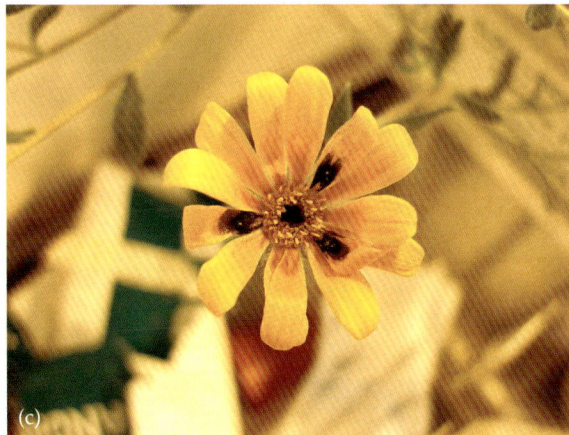

Plate 9 Floral mimicry. (a) The titan arum (*Amorphophallus titanium*, Alismatales) attracts pollinators by releasing a strong scent reminiscent of rotting flesh. Image kindly provided by Cambridge University Botanic Garden. (b) *Ophrys episcopalis* (Asparagales), which mimics female insects to achieve pollination through pseudocopulation. Image kindly provided by Richard Bateman (Royal Botanic Garden, Kew). (c) The composite inflorescence of *Gorteria diffusa* (Asterales) mimics its pollinating flies. See also Figure 14.5.

Plate 12 The petals of *Aquilegia formosa* (Ranunculales) are heavily modified to produce nectar spurs. Image kindly provided by Scott Hodges (UCSB). See also Figure 15.4.

Plate 10 Zygomorphy and actinomorphy. (a) Many flowers are radially symmetrical, or actinomorphic. (b) The flowers of *Antirrhinum* species are bilaterally symmetrical, or zygomorphic. See also Figure 15.2.

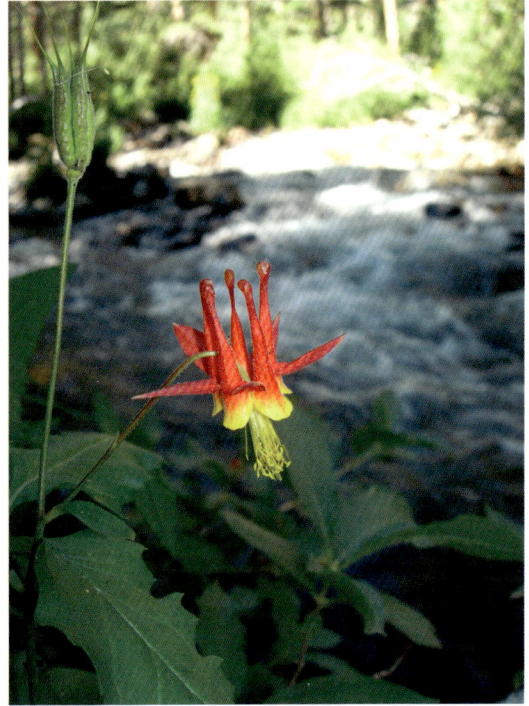

Plate 11 The enantiostylous flower of *Solanum heterodoxum* (Solanales). The style is the filamentous pink organ. Image kindly provided by Sandy Knapp (Natural History Museum, London). See also Figure 15.3.

Plate 13 Composite inflorescences in the daisy family. (a) The capitulum of *Gerbera hybrida*, with zygomorphic outer florets (each with one large petal) and actinomorphic inner florets. (b) The capitulum of *Senecio vulgaris* usually contains only actinomorphic disc florets. See also Figure 15.5.

Plate 14 Carotenoids give the yellow and orange colours to (a) *Freesia* (Asparagales), (b) *Gerbera hybrida* (Asterales), and (c) lilies (Liliales). The photo in (b) is a modified version of a photo by Mauro Girotto (Wikimedia Commons). See also Figure 16.2.

Plate 15 Anthocyanins give the purple, magenta, and pink colours to (a) *Petunia hybrida* (Solanales, delphinidin and petunidin), (b) *Antirrhinum majus* (Lamiales, cyanidin), and (c) *Pelargonium* (Geraniales, pelargonidin). The photo in (c) is adapted from a photo by Rameshng (Wikimedia Commons). See also Figure 16.4.

Plate 16 Betalains give the yellow, purple, and pink colours to (a) *Portulaca oleracea*, (b) *Mirabilis jalapa*, and (c) *Sesuvium portulacastrum* (all Caryophyllales). All images kindly provided by Sam Brockington (Cambridge). See also Figure 16.6.

Plate 17 Pigment regulation. (a) *Viola cornuta* 'Yesterday, Today and Tomorrow' is fully purple 5–8 days after pollination (left), but opens as a white flower (middle) in which pigmentation steadily increases (right). Image kindly provided by Martha Weiss (Georgetown University, Washington, DC). (b) The *delila* mutant of Antirrhinum lacks pigmentation in the tube as a result of loss of activity of a bHLH transcription factor. (c) The *Venosa* locus produces pigmentation over the petal veins in a pale Antirrhinum flower. *VENOSA* encodes a MYB transcription factor. (d) The *an11* mutant of petunia lacks pigmentation as a result of loss of activity of a WD40 protein. The transposon in the *AN11* locus excises somatically, generating patches of wild type red tissue. Image kindly provided by Ronald Koes (Vrije Universiteit, Amsterdam). See also Figure 17.1.

Plate 18 Metals and pH both affect flower colour. (a) The Himalayan blue poppy owes its blue colour to an interaction between anthocyanin and iron. Photograph kindly supplied by Cambridge University Botanic Garden. (b) Hydrangea flowers can be blue or pink, depending on the metal ions present in the soil. (c) Morning glory flowers have a high vacuolar pH. Image kindly provided by Felix Jaffe. (d) An unstable *pH4* mutant of petunia, with revertant wild type red (acidic) sectors on a mutant bluish-pink (more alkaline) background. Image kindly provided by Ronald Koes (Vrije Universiteit, Amsterdam). See also Figure 17.2.

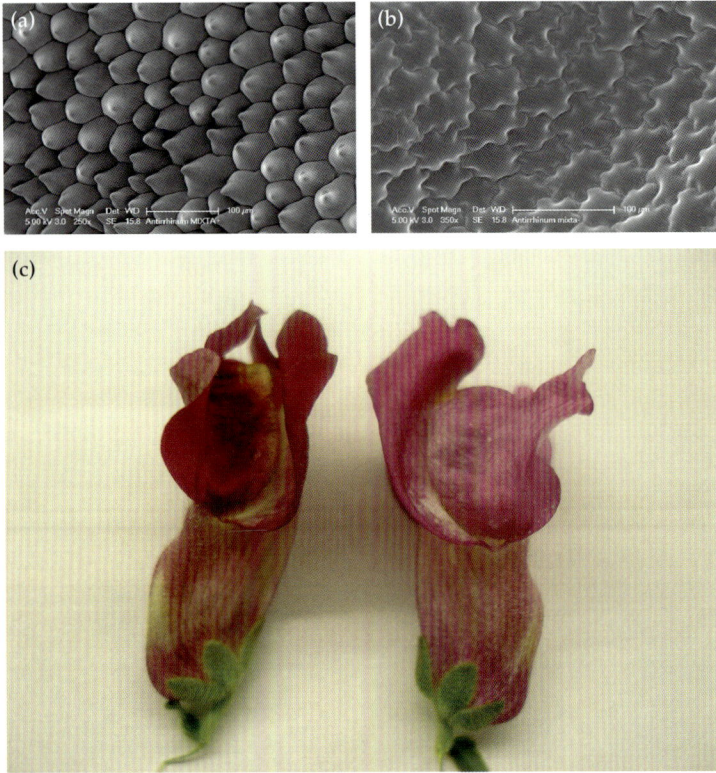

Plate 19 Petal cell shape affects flower colour. (a) Wild type Antirrhinum petal epidermis, composed of conical cells. (b) *mixta* mutant petal epidermis, composed of flat cells. (c) Wild type (left) and *mixta* mutant (right) flowers, showing the difference in colour attributable to the cell shape. See also Figure 17.3.

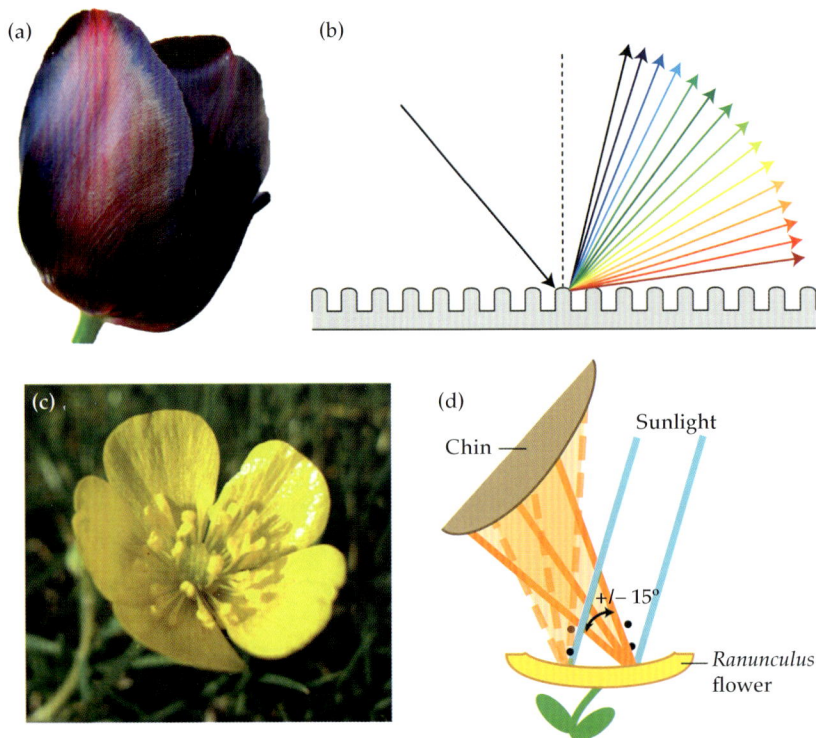

Plate 20 Structural colour. (a) Tulip 'Queen of the Night' has iridescent rainbow colours on top of purple pigmentation. (b) This iridescent effect is caused by a diffraction grating. (c) The bright yellow buttercup reflects yellow light very strongly. (d) The buttercup acts as a double mirror, reflecting yellow and white light together on to nearby surfaces such as a child's chin. See also Figure 17.4.

Plate 21 Variation in zygomorphy in the Antirrhineae. (a) Highly zygomorphic *Antirrhinum majus*. (b) Moderately zygomorphic *Maurandya scandens*. (c) Slightly zygomorphic *Mabrya acerifolia*. (d) Almost actinomorphic *Rhodochiton atrosanguineum*. All scale bars 1 cm. See also Figure 18.1.

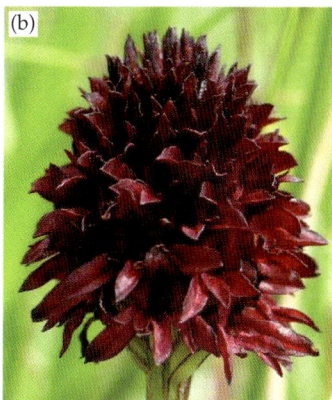

Plate 22 Variation in nectar spur length. (a) *Gymnadenia conopsea* flowers have very long nectar spurs. (b) *Gymnadenia rhellicani* flowers have almost no nectar spur. Images kindly provided by Matt Box (Sainsbury Laboratory, Cambridge University). See also Figure 18.3.

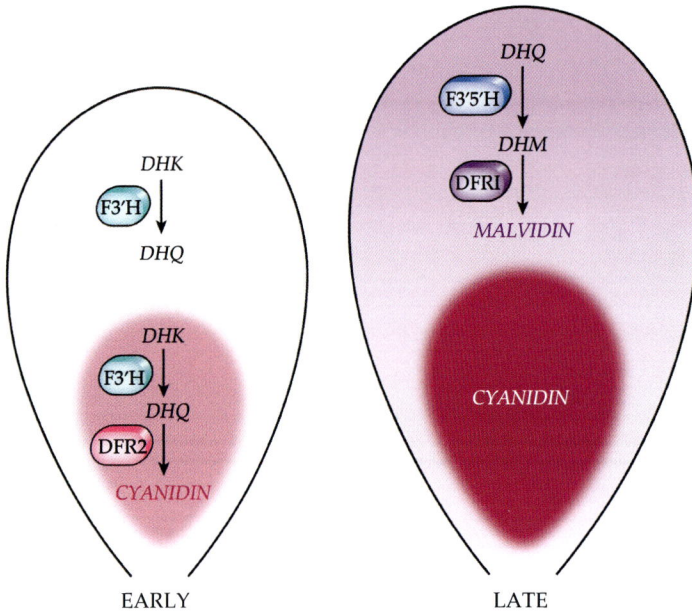

Plate 23 Development of the *Clarkia gracilis* petal spot. Early expression of *DFR2* in the spot region, in the presence of F3'H but not F3'5'H, results in red cyanidin pigment. Later expression of *DFR1* throughout the petal, in the presence of F3'5'H, results in mauve malvidin production. See also Figure 18.4.

Plate 24 Flowers for which pollinators have been shown to discriminate between colour morphs. (a) *Mimulus lewisii*. (b) *Mimulus cardinalis*. Images (a) and (b) kindly provided by Toby Bradshaw (Washington State University). (c) *Antirrhinum majus* wild type and *incolorata* lines. (d) *Antirrhinum majus* wild type and *sulfurea* lines. Images (c) and (d) kindly provided by Enrico Coen (John Innes Centre). See also Figure 20.2.

Plate 25 Flowers with significant nectar guides. (a) *Delphinium nelsonii*. Image kindly supplied by Nick Waser (University of California – Riverside). (b) *Clarkia xantiana* subsp. *xantiana*. Image kindly provided by Vince Eckhart (Grinnell College, USA). See also Figure 20.3.

Plate 26 Flowers for which pollinators have been shown to discriminate on the basis of size. *Ipomopsis aggregata*. Image kindly provided by Nick Waser (University of California – Riverside). See also Figure 20.4.

Plate 27 Plants for which the main pollinator type can be accurately predicted from floral morphology. (a) *Penstemon centranthifolius* (hummingbird-pollinated). (b) *Penstemon heterophyllus* (bee-pollinated). Images (a) and (b) kindly supplied by Scott Armbruster (University of Portsmouth). See also Figure 21.1.

Plate 28 Plants that have cast doubt on the pollination syndrome concept. (a) *Viola cazorlensis.* Image kindly supplied by Carlos Herrera (Seville). (b) *Microloma sagittatum* with its sunbird pollinator. Image kindly supplied by Anton Pauw (Stellenbosch). See also Figure 21.2.

expressed in the pollen grain suggested a likely total of around 4000 pollen-expressed genes across the Arabidopsis genome. The expressed genes included a high proportion of essential genes with roles in signal transduction and cell wall biogenesis (Honys and Twell 2003). Many of the mutants that have been described so far underline the importance of the asymmetry of the first mitotic division. For example, the *gemini pollen 1* mutant produces pollen grains with apparently randomly orientated first divisions, which then arrest after that first division (Park *et al.* 1998). The cells produced by this first division express vegetative cell markers, indicating that the vegetative cell is the default state in the absence of asymmetry. The GEM1 protein binds to microtubules and appears to be involved in microtubule positioning, apparently essential for the appropriate asymmetry of the first mitotic division (Twell *et al.* 2002). Similarly, the *two-in-one* (*tio*) mutant microsporophyte completes asymmetric nuclear division, but fails to produce cell membranes as a result of failed cell plate formation, resulting in binucleate pollen that does not produce sperm cells (Oh *et al.* 2005). Other mutants are affected in the number of mitotic divisions that occur, and therefore the number of sperm cells produced, and in the positioning of the vegetative and generative cells with respect to one another (all reviewed by Twell 2011). As with the female gametophyte, analysis of small RNA pathways indicates that these are also present and active in microgametophyte development (Grant-Downton *et al.* 2009). It is clear that many genes are involved in microgametogenesis, and it should therefore be noted that the reduced size and duration of the male gametophyte stage of the angiosperm life cycle is still quite sufficient to ensure that few lethal mutations are retained within the genome.

12.5 Events following pollination

Mature pollen arrives at the stigmatic surface at the distal end of a carpel through any number of routes, although the important role of animal pollinators will be discussed in detail in Section III of this book. Once at the stigma of a suitable flower of the same species, the pollen grain germinates. The developing pollen tube grows through the carpel tissue, in a special tissue called the transmitting tract. The pollen tube itself is a remarkable structure, thought to have evolved from an earlier spore-derived structure with a filamentous or branching form adapted for uptake of nutrients from surrounding tissues (discussed in Rudall and Bateman 2007). Indeed, in the gymnosperm *Ginkgo biloba* the pollen tube branches extensively within the female tissues, extracting nutrients over a period of several months before releasing motile sperm. In angiosperms, genes expressed in the stigma and style (that is, female sporophytic genes) play important roles in guiding the germinating pollen tube towards the embryo sac. The vegetative nucleus and the sperm nuclei migrate with the pollen tube, staying close to the growing tip. The pollen tube is attracted to the micropylar end of an ovule by a number of female gametophytic chemical messengers, and there bursts within a synergid, releasing the two sperm cells into the entrance to the embryo sac. The synergids die, allowing the sperm cells to fuse with the embryo sac proper. One sperm nucleus enters mitosis with the egg cell, to produce a diploid embryo, while the other enters mitosis with the diploid central cell to produce a triploid tissue, the endosperm. Analysis of a number of Arabidopsis mutants has shown that the female gametophyte guides the pollen tube to the micropyle and from there into the synergids, arrests pollen tube growth, causes pollen tube bursting, and prevents further pollen tubes from entering the embryo sac (reviewed by Dresselhaus 2006). Mechanisms are also in place to ensure that the male and female gametes are at the same stage of the cell cycle when fertilization occurs. The appropriate cell cycle stage varies between plant species, but is consistent within a single species (reviewed by Weterings and Russell 2004). The endosperm grows and divides to provide the nutritive tissue in the seed, while the embryo undergoes cell division and development to produce the next sporophyte generation. The outer layers of the ovule, the integuments, become the seed coat that protects the developing embryo. Development of this new generation is beyond the scope of this book.

CHAPTER 13

Outcrossing and self-fertilization

This section of the book has focused on the development of the angiosperm flower. The flower that has been discussed so far is a 'perfect' (hermaphrodite) flower, containing both male and female reproductive structures, and producing, through sporogenesis and gametogenesis, both male and female gametes. Hermaphroditism carries both advantages and disadvantages. Perfect flowers can, assuming no other constraints, self-pollinate and fertilize their own ovules. This sexual reproduction without the need for another individual of the same species gives the plant a certain set of advantages. The most important of these is guaranteed reproduction, irrespective of the availability of likely mates, and without relying solely on asexual reproduction (such as the production of runners or tubers). Since self-fertilization is a sexual process, involving independent assortment of chromosomes at meiosis and the possibility of recombination between chromosomes, it does not result in the same lack of genetic variability that is the consequence of asexual reproduction. This guaranteed sexual reproduction gives self-pollinating plants the ability to colonize new habitats, and it is a common trait in weedy species. A second advantage of self-fertilization is the saving in energy and nutrients that would otherwise be spent producing costly rewards for pollinators. It is perhaps not surprising, then, that around 20% of angiosperm species are predominantly self-pollinating (Barrett 2002). In the final section of this chapter we shall discuss the evolution of traits that promote self-pollination as a breeding system. However, self-fertilization does carry a disadvantage relative to outcrossing, which is that the genetic variability produced, although greater than in an asexual population, is considerably less than that seen in an outbreeding population. The reduction in viability

of inbred progeny compared with outcrossed progeny is known as inbreeding depression, and is recognized as the primary selective pressure resulting in strategies to avoid self-fertilization (Barrett 2002). The balance between the relative importance of assured reproduction and genetic variability differs in different species, largely as a result of their habitats, life cycles, and the niches that they occupy (Charlesworth 2006). For the majority of this chapter we shall consider the developmental and biochemical ways in which self-fertilization can be reduced or prevented.

13.1 Reducing self-pollination in a hermaphroditic flower

Self-pollination can occur within a flower, or between flowers on the same plant. Pollination within a single flower is called autogamy, and is an obvious consequence of producing hermaphrodite flowers. Autogamy may occur simply as a result of the positioning of stamens and anthers, or it may be facilitated by visiting animals. Weedy plants that self-pollinate (such as Arabidopsis) do not usually require an animal pollinator, as pollen is transferred directly from the stamens on to the stigma, which are in very close proximity. Autogamy can be most simply reduced by separating the stamens and stigma physically within the flower (herkogamy) or by separating the timing of their development (dichogamy).

13.1.1 Herkogamy

Herkogamous flowers have their anthers and stigma positioned such that pollen cannot be passively

Understanding Flowers and Flowering. Second Edition. Beverley Glover.
© Beverley Glover 2014. Published 2014 by Oxford University Press.

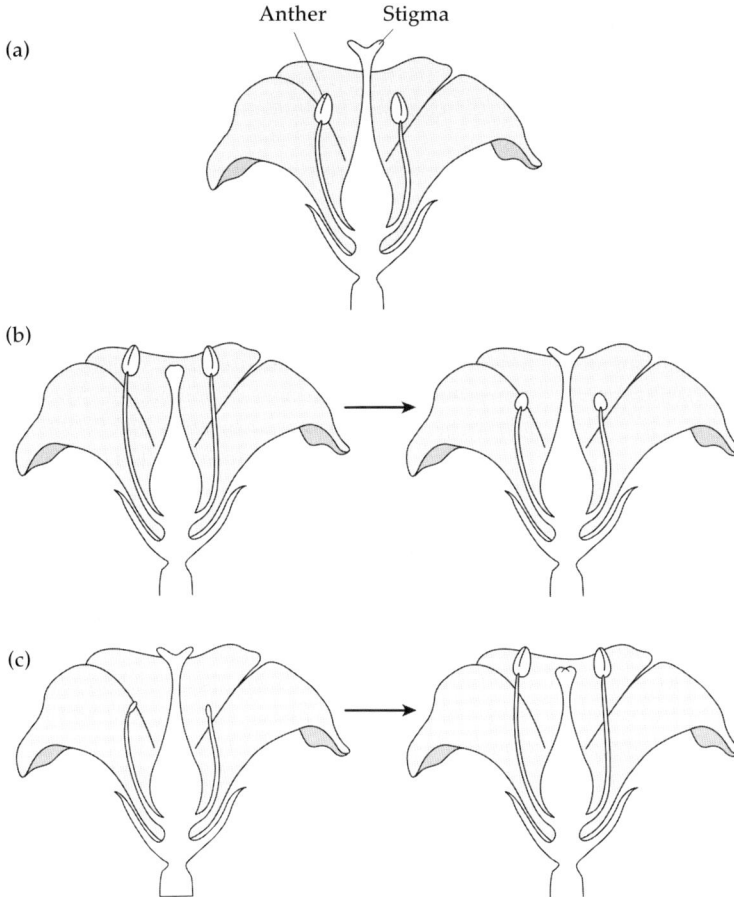

Figure 13.1 Reducing self pollination in a hermaphrodite flower. (a) Herkogamy. (b) Dichogamy—protandry. (c) Dichogamy—protogyny.

transferred to the stigma (see Fig. 13.1a). This trait acts to prevent self pollen reaching the stigma, unless it is transferred by an animal. Even when self pollen is transferred by a foraging animal, the animal is likely to carry pollen from other individuals as well, which means that a herkogamous flower should receive a mixed load of self and non-self pollen. Takebayashi *et al.* (2006) investigated within-population variability in herkogamy in *Gilia achilleifolia* (Ericales), a small blue-flowered Californian phlox which has the stigma extended some way out of the flower, beyond the stamens. They found that a greater degree of herkogamy (measured as the distance between the stigma and the top of the stamen) was correlated with a greater degree of outcrossing, confirming the hypothesis that herkogamy reduces self-pollination. Other studies

have reported similar results in a range of systems. However, not all studies have found correlations between herkogamy and outcrossing. Medrano *et al.* (2005) found no increase in outcrossing associated with increased herkogamy in six populations of *Narcissus longispathus* (Asparagales). This study highlights a number of problems with herkogamy as a method of reducing outcrossing. Herkogamy is dependent on pollinator foraging behaviour to ensure that the non-autogamous pollen deposited on the stigma is collected from a different *individual*, not from a different flower on the same plant. It is also reliant on the pollinator not brushing by the stigma when it leaves the flower, a problem in the *N. longispathus* system as the pollinator often grooms on the edge of the flower, near the stigma, after feeding.

Surprisingly, a recent report indicates that herkogamy varies in different populations of Arabidopsis. Traditional lab lines show slight reverse herkogamy, where the anthers are positioned slightly above the stigma, shedding pollen directly on to it. However, in two different wild populations, Luo and Widmer (2013) recorded the occurrence of approach herkogamy, in which the stigma is positioned above the anthers. Plants from these populations had considerably reduced seed set in insect-free greenhouses, compared with Col-0, indicating that herkogamy is also a system arising in natural populations of a predominant selfer that reduces self-fertilization and presumably allows enhanced outcrossing.

13.1.2 Dichogamy

Dichogamy is a slightly more reliable method of preventing autogamous pollination than herkogamy. Dichogamous flowers have the timing of anther dehiscence separated from the timing of stigma receptivity. Dichogamy is extremely common, with as many as 87% of angiosperm species showing some degree of temporal separation of male and female function (Bertin and Newman 1993). The great advantage of dichogamy over herkogamy is that it is not possible for an animal pollinator to accidentally transfer pollen from within the flower on to the receptive stigma. However, the same difficulty occurs with foraging behaviour—if pollinators forage primarily within an inflorescence then the pollen they transfer to receptive stigmas will still be self pollen, just from different flowers on the same plant. An impressive adaptation to reduce this particular problem is shown by some species in which the flowers of an individual exhibit synchronous dichogamy, all maturing as males at one time point and as females at another.

There are two variations of dichogamy, depending on which reproductive structure matures first. In protandrous flowers the anthers dehisce and the pollen is shed several days before the stigma is receptive (see Fig. 13.1b). In protogynous flowers, which are less common, the stigma is receptive several days before the anthers dehisce and then ceases to be receptive before the pollen is actually released (see Fig. 13.1c). A possible explanation for the relative infrequency of protogyny is that it requires a reversal of the usual order of development of the reproductive organs. As we have seen in previous chapters, the fourth whorl organs, the carpels, usually develop last in floral meristems. Remarkably, some plant species are heterodichogamous, producing flowers that are protandrous and flowers that are protogynous. Wang *et al.* (2012) showed that the two flower types were produced in a 1:1 ratio by the basal eudicot *Kingdonia uniflora* (Ranunculales).

Protogyny is predicted to be a more efficient inhibitor of autogamy than protandry. In protandrous flowers, pollen may persist until after the stigma becomes receptive, causing self-pollination. Another advantage of protogynous flowers is that they have the ability to 'top up' any ovules which have not received outcross pollen with their own pollen when it is released. Since protogyny is more effective than protandry, but protandry is more common, it has been proposed that the main benefit of protandry is in preventing interference between stamens and stigmas, by separating them, rather than in reducing inbreeding depression. Such interference results in the wastage of pollen on self stigmas, when it could contribute to male fitness if properly dispersed. Sargent *et al.* (2006) modelled the evolution of dichogamy, and found that both avoiding inbreeding depression and avoiding interference between stigma and stamens played important roles in the appearance of dichogamy under different conditions. This may go some way towards explaining the surprising prevalence of dichogamy (particularly protandry) in biochemically self-incompatible species (which should not need additional mechanisms to prevent self-fertilization; Bertin 1993).

Comparative studies have demonstrated a correlation between pollinator type and dichogamy. Protandrous species are often pollinated by bees or flies, whereas protogynous species are often wind- or beetle-pollinated (Sargent and Otto 2004). These correlations suggest that, as we saw with herkogamy, pollinator foraging behaviour can have significant consequences for the effectiveness of strategies to minimize self-pollination.

13.2 Monoecy

The only developmental mechanism that completely ensures that autogamy does not occur is the separation of male and female functions into different flowers. Monoecious plants are hermaphroditic, as they produce both male and female gametophytes (and thus gametes), but the individual flowers on a monoecious plant are of a single sex. This strategy still allows self-pollination between different flowers of the same individual (geitonogamy) to occur, and so is not a very effective method of preventing self-pollination. However, it is quite common, with around 5% of angiosperm species being entirely monoecious, and another 4.5% producing both monoecious and hermpahroditic flowers.

Monoecy has often been associated with inefficient pollination systems. It is common in grasses and wind-pollinated trees, and has recently been found to be particularly frequent in tropical Australian tree species (Gross 2005). This may reflect the infrequency or absence of specialized pollinators such as birds, bats, and large bees in the Australian forests, and the prevalence of generalist pollinators such as beetles, flies, small bees, and thrips. When pollination is unreliable or inefficient, monoecy may be selectively advantageous over dioecy (see below) because pollen does not have to travel between individuals, just between flowers, to achieve fertilization. It may also be advantageous to advertise strongly to such inefficient pollinators by providing 'spare' male flowers with plenty of rewarding pollen.

A second major advantage of monoecy is that it allows the plant to vary the number of male flowers and the number of female flowers it produces according to its age, health, or environment. For instance, in many monoecious trees, young plants have almost entirely male flowers, whereas older trees have mostly female flowers. Older trees have usually built up more resources and can afford to support fruit, whereas younger trees benefit from the ability to contribute to the next generation without the investment in costly fruit and seed. Similarly, drought-stressed monoecious plants in dry habitats tend to produce mostly male flowers while those in more moderate habitats tend to produce more female flowers (Freeman et al. 1981). Light, nutrients, and water have all been shown to increase female production in various monoecious species. This ability to selectively allocate resources to different reproductive functions is often considered

a much bigger advantage of monoecy than any effect on outcrossing rate, which is seen as a secondary consequence of responding to selective pressure to maximize reproductive output under different conditions.

Monoecy is believed to be a derived condition (Richards 1997). A number of plant species produce both unisexual and hermaphroditic flowers, and these are believed to represent evolutionary intermediates. Andromonoecious plants, such as *Cucumis melo* (melon, Cucurbitales), produce both male and hermaphroditic flowers, whereas gynomonoecious plants, such as various *Poa* species (Poales), produce both female and hermaphroditic flowers. It has been observed that andromonoecious plants with few resources produce a greater proportion of male flowers than their better resourced contemporaries. These data suggest that the ability to vary male and female output, rather than avoidance of autogamy, is the driving force behind the evolution of monoecy.

The development of monoecy from hermaphroditic flowers requires male sterility and female sterility to be expressed developmentally and spatially within the same individual plant. It is likely that this involves specific loss of B or C function gene activity in whorl 3 to make female flowers, and specific loss of C function gene activity in whorl 4 to make male flowers (see Chapter 10). It is usually the case that there is a distinct pattern of arrangement of male and female flowers within a monoecious plant. For instance, most grasses have female flowers at the bottom of the inflorescence and male flowers at the top (see Fig. 13.2). This architectural separation of reproductive function is probably easier to produce developmentally than a more integrated arrangement, as promoter activity of B and C function genes can be tied to developmental stage. However, it should be remembered that the B and C function genes are only necessary to differentiate male and female function, but that other genes may actually specify that function. Perl-Treves et al. (1998) examined three *AGAMOUS*-like (C function) genes in *Cucumis sativus* (cucumber, Cucurbitales), and found no changes in expression correlated with hormone treatments that alter floral sex. Boualem et al. (2009) found that variation in ethylene biosynthesis was correlated with

Figure 13.2 Monoecy. Flowers of each sex are often developmentally separated in the plant.

development of monoecious or andromonoecious individuals in *Cucumis*. Another plant growth regulator, gibberellin, has also been shown to influence floral sex, and some of the maize (Poales) mutants with disruptions in monoecy are thought to be perturbed in gibberellin synthesis or perception (Dellaporta and Calderon-Urrea 1994). Of course, the pleiotropic nature of plant growth regulator function makes it likely that perturbation of synthesis or perception of a plant growth regulator will have an influence, even if it is non-specific, on floral organ development. Maize has so far proved the best system in which to explore the development of monoecy, with a number of mutants identified. The *tasselseed* group of mutants produce hermaphrodite flowers in their male inflorescences. In addition to mutants with lesions in the production of enzymes involved in gibberellin and jasmonic acid production and degradation, the *tasselseed* mutants have pointed to a role for ABC genes in unisexual flower development. In particular, the *TS4* locus encodes miR172, which restricts activity of the A function gene *APETALA2* to whorls 1 and 2 in Arabidopsis, and *TS6* encodes an *AP2*-like gene that is targeted by this microRNA (Chuck *et al.* 2007).

13.3 Dioecy

Dioecy is a relatively common solution to the problem of self-fertilization, with around 6% of angiosperm species showing complete separation of the sexes (Barrett 2002). Dioecy is scattered throughout the angiosperm phylogenetic tree, suggesting that it has evolved independently many times. As many as 160 plant families include dioecious species, and it has been predicted that the trait has evolved around 100 times independently. However, there are a few plant families where dioecy is particularly prevalent. These include the Euphorbiaceae (in the Malpighiales), the Cucurbitaceae (Cucurbitales), and the Urticaceae (Rosales) (Ainsworth 2000).

Dioecy is often viewed as the logical end point of mechanisms to separate sexual function in space or time. However, although it does effectively eliminate self-fertilization, it also carries a great risk of reproduction failing entirely. Dioecy would clearly be a big disadvantage to weedy species and those that regularly colonize new habitats. Attempts to model the evolution of dioecy have concluded that avoiding inbreeding is one selective pressure that may be responsible for its appearance, but sexual specialization may be equally important (Freeman *et al.* 1997). Sexual specialization involves the development of characters that enhance male or female reproductive success. For example, the enlarged feathery stigmas of female wind-pollinated flowers would interfere with activity of stamens, and so retaining the two organs in separate flowers or on separate plants is advantageous. Similarly, female flowers must be borne on branches strong enough to support developing fruit, while male flowers may benefit from thinner, more flexible branches that encourage pollen dispersal. This differential development to maximize a single sexual function may, like the differential allocation of resources to sexual functions in monoecious plants, actually represent the driving force behind the evolution of dioecy, with the elimination of self-pollination simply being a fortuitous consequence.

Dioecy can evolve through monoecy, and dioecy has also been shown to revert to monoecy—for example, in the genus *Momordica* (Cucurbitales; Schaefer and Renner 2010). An alternative evolutionary route to dioecy is through one of the intermediate states of androdioecy or gynodioecy. Androdioecy (male individuals and hermaphrodite individuals) is extremely rare, and is thought to occur when selection favours the reappearance of hermaphroditism in a dioecious population (Pannell 2002). Gynodioecy (female individuals and hermaphrodite individuals) is much more common, and is the more likely route from hermaphroditism to dioecy. Females would be at an advantage in an otherwise hermaphrodite population if their increased rate of outcrossing reduced inbreeding depression. Once such a gynodioecious population is established there may be selective advantage to the hermaphrodites in maximizing pollen output, establishing males and a fully dioecious population. The common European weed *Plantago coronopus* (Lamiales) is a familiar example of a gynodioecious species (Koelewijn and van Damme 1996).

Sex determination in dioecious plants is complex, largely because the multiple origins of the trait seem to be associated with multiple developmental mechanisms. In some species, plant growth regulators play key roles and sex determination is environment dependent. In 15 angiosperm families it has been shown that sex chromosomes, nonrecombining chromosomes, or chromosomal regions in which female sterility/fertility and male sterility/fertility loci are tightly linked control the development of dioecy (Ming *et al.* 2011). These families are widely distributed in the angiosperm phylogeny, and are not related by any obvious features of habitat or life history. Both between and within these families distinct sex determining mechanisms are found, possibly representing different stages in sex chromosome evolution. *Silene latifolia* (campion, Caryophyllales) contains sex chromosomes that operate under the same active Y system as is seen in mammals, with XY individuals developing as males and XX individuals developing as females (Grant *et al.* 1994). In contrast, *Rumex acetosa* (sorrel, also Caryophyllales) uses a dosage-dependent system similar to that of *Drosophila melanogaster*, with an X:autosome ratio of more than 1.0 resulting in female development and a ratio of less than 0.5 resulting in male development (Ainsworth 2000). The *Rumex acetosa* system may represent loss of a degenerated Y chromosome in a system that was originally XY, as other *Rumex* species (such as *R. acetosella*) use an XY sex determining mechanism, the most common sex chromosome system in plants.

The differentiation of male or female flowers on a dioecious individual requires the same developmental modifications as those already discussed for monoecious plants. The timing of abortion of male or female organs varies greatly across different species. It has been suggested that dividing unisexual flowers into two groups—type I, in which the flowers initiate as hermaphrodite but become unisexual by termination of development of one set of organs, and type II, in which the flowers initiate as unisexual—would assist efforts to understand the different developmental processes at work (Diggle *et al.* 2011). Interestingly, male flowers and female flowers of both dioecious and monoecious plants tend to arrest at similar developmental stages within a single species.

The hypothesis that ABC function genes are at least involved in sexual differentiation in dioecious plants has been proved correct for *Rumex acetosa*. B function genes in this species are expressed only in male flowers, and only in the stamen whorl. They are not expressed in the sepaloid perianth organs of either male or female flowers. C function genes are expressed in both sexes, initially in both the stamen and carpel whorls. However, as the 'inappropriate' set of organs aborts, transcript of the C function gene becomes undetectable in those organs. Thus in the female flowers C function expression is lost from the developing stamens at the time when their development ceases, and in the male flowers C function expression is lost from the carpel at the time when its development ceases (Ainsworth *et al.* 1995). However, further reports of C function activity being maintained in aborted carpel primordia in a mutant *Rumex* plant producing both flower types suggest that loss of C function activity cannot cause the cessation of development in this species, but may itself be a consequence of abortion of organ development in individuals of different sex (Ainsworth *et al.* 2005). A related mechanism for female-specific flower development was demonstrated in *Silene latifolia*, where an orthologue of the Arabidopsis *SUPERMAN* gene (see Chapter 10) was found to be expressed solely in female flowers. In Arabidopsis, SUP represses activity of the B function genes in whorl 4, allowing carpel development, a role which it might also be playing in this dioecious system (Kazama *et al.* 2009).

It is currently not clear how sex determination through sex chromosomes is linked to sexual differentiation at the level of floral organ development. It has been suggested that the basal eudicot *Thalictrum dioicum*, a member of the Ranunculales, might provide a system in which to link these two stages of dioecious development. *T. dioicum* flowers produce type II unisexual flowers—no carpel primordia are found in male plants and no stamen primordia are found in females. B and C function gene expression has been analysed in this species, and although the presence of multiple copies of both types of genes makes the analysis complex, there is clear segregation of expression patterns with plant gender. This suggests that sex determination methods in this species operate through complete repression of the transcription of organ identity genes (Di Stilio *et al.* 2005).

13.4 Self-incompatibility (SI)

It is clear that developmental separation of male and female reproductive functions occurs for complex reasons, only some of which may be related to preventing self-fertilization. Such separation can vary greatly in its effectiveness, and all mechanisms except dioecy do allow some self-fertilization. However, in around 90 angiosperm families, and up to 50% of angiosperm species, it has been reported that some form of biochemical self-incompatibility exists (McClure and Franklin-Tong 2006). Biochemical self-incompatibility results in the failure of self pollen to fertilize an ovule of the same individual, whether the pollen grain was produced on the same flower or on a different flower. In those families where self-incompatibility has been studied, the control of this block to self-fertilization can be attributed to a single genetic locus, the S locus. The S locus does not necessarily represent one single gene, but is a closely linked region of the chromosome containing several genes which together control self-incompatibility. The S locus exists in multiple forms, known as haplotypes (rather than alleles, since multiple genes are present, although the term 'S allele' is often used, too), providing different

'mating types.' An incompatible pollination occurs when one or both of the S haplotypes in the male parent is the same as one or both of the S haplotypes in the female parent.

Self-incompatibility can be divided into two main forms, according to whether the male determinant of the system is produced by haploid pollen grains or by the diploid parental anther tissue. Within each of these two divisions, multiple molecular mechanisms may be at work in different species. In gametophytic self-incompatibility it is the genotype of the pollen grain, the gametophyte, that is key. Since pollen grains are haploid, they carry only one S haplotype (see Fig. 13.3). In sporophytic self-incompatibility it is the genotype of the parent plant from which the pollen grain came—the diploid sporophyte—that matters. Since sporophytes are diploid, the pollen grain will carry the imprint of 2 S haplotypes (Fig. 13.4). Differences in the mechanism of preventing fertilization also separate the best studied gametophytic and sporophytic systems. In gametophytic self-incompatibility in the Solanales, the growth of the pollen tube is inhibited within the tissue of the style. In sporophytic self-incompatibility in the Brassicales, the germination of the pollen grain is inhibited on the surface of the stigma and no pollen tube enters the style. Of the 90 families of angiosperms that have self-incompatibility, no family consists entirely of self-incompatible species. It is also the case that sporophytic and gametophytic self-incompatibility systems are not found within a single family. Gametophytic self-incompatibility is thought to be the more common of the two types, but examples of sporophytic self-incompatibility have been more thoroughly characterized.

13.5 Sporophytic self-incompatibility (SSI)

The best studied sporophytic self-incompatibility system is that found in the Brassicaceae (Brassicales), because of the importance of this family in crop production. However, sporophytic self-incompatibility is also found in other large families, including the Asteraceae (Asterales), the Convolvulaceae (Solanales), and the Caryophyllaceae (Caryophyllales). It is not yet clear whether similar molecular systems

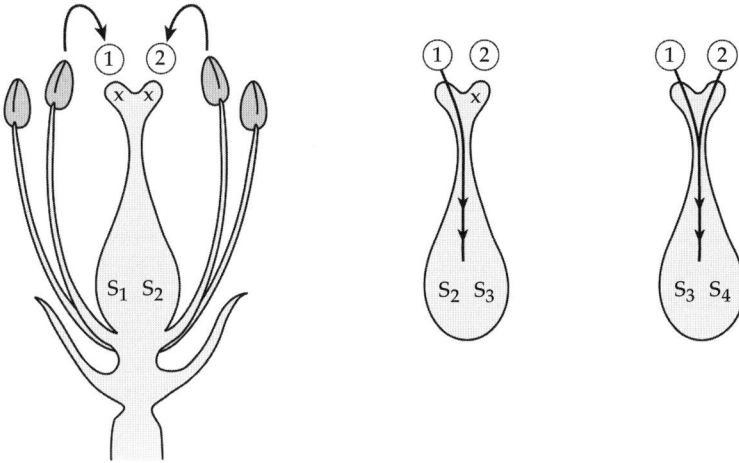

Figure 13.3 Gametophytic self-incompatibility. Pollen grains carry only a single mating type, according to their own haploid genotype.

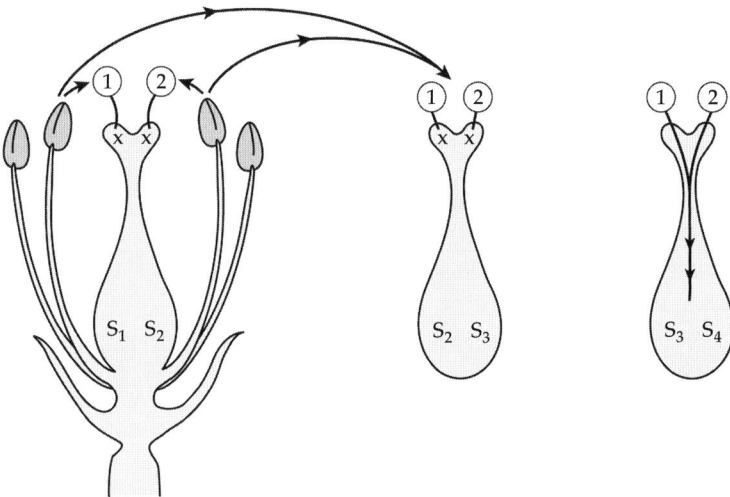

Figure 13.4 Sporophytic self-incompatibility. Pollen grains carry the imprint of the parent plant's diploid genotype.

are at work in each of these families, but reports suggest that not all of the components of the *Brassica* system are present in other examples of SSI (reviewed by Iwano and Takayam 2012). Sporophytic self-incompatibility can be genetically complex, because different S locus haplotypes may be dominant, recessive, or co-dominant, and these relationships may differ between the male components of the system and the female components. Thus a particular S locus haplotype may be dominant in the stigma but co-dominant in the pollen, making the analysis of relationships between

haplotypes rather complex (Hiscock and McInnis 2003).

The *Brassica* self-incompatibility reaction can be divided into two stages. Firstly, self pollen, or at least pollen with the appropriate genotype to trigger the response, is recognized by the presence of the same S haplotype in the male and female parents. Secondly, this recognition stimulates a signal transduction cascade that leads to pollen rejection. The self-incompatibility response in *Brassica* is controlled by a single S locus, at which up to 80 different S haplotypes have been identified. The locus itself contains as many as 14 expressed genes (Watanabe *et al.* 2000), and the genes determining the male and female components of the SI response are distinct, although closely linked. The tight linkage ensures that male and female self-incompatibility alleles are not brought together in inappropriate combinations through meiotic recombination. In contrast, the genes involved in the signal transduction cascade that causes pollen rejection appear to be entirely unlinked and are scattered throughout the genome.

13.5.1 The female determinant of SSI in *Brassica*

The S locus contains multiple genes, so identifying the female and male components of self-incompatibility involved analysis of their expression patterns and functions. Two *Brassica* S locus genes encode proteins only detected in the stigma. These two highly polymorphic genes are called the *S locus receptor kinase (SRK)* and the *S locus glycoprotein (SLG)*. *SLG* encodes an abundant glycoprotein, which is secreted into the cell wall matrix of the stigmatic papillae (Nasrallah *et al.* 1985; Kandasamy *et al.* 1989), while *SRK* encodes a membrane-spanning receptor kinase, with an extracellular domain, a transmembrane domain, and a cytoplasmic domain with serine/threonine kinase activity (Stein *et al.* 1991). The extracellular domain is structurally very similar to the SLG protein. Because of this similarity it has been hypothesized that *SLG* was produced by a partial duplication of *SRK*. Stein *et al.* (1996) confirmed that the SRK protein is localized to the plasma membrane of stigmatic papillae.

Analysis of mutant lines of *Brassica* in which self-compatibility has been lost confirmed the importance of the *SRK* gene in determining female self-incompatibility. Several such lines have been shown to contain mutations within *SRK*. Transgenic work also supports this conclusion. Introduction of the *SRK* gene from S locus haplotype S28 into a *Brassica rapa* line containing the S60 haplotype resulted in the production of plants that rejected S28 pollen. These data confirmed that the *SRK* gene is sufficient to determine which S type the female reproductive organs function as. The transgenics continued to reject S60 pollen, so the *SRK28* transgene had simply added to their female self-incompatibility type

(Takasaki *et al.* 2000). The haplotype specificity of the SRK protein has been shown, through analysis of variants with substitutions at various positions, to be attributable to a small number of amino acid sites in the extracellular domain of the protein (Boggs *et al.* 2009).

Similar experiments have been conducted with *SLG*, and have concluded that it is not sufficient alone to determine female S locus type, but that it acts with *SRK* and enhances its activity. Introduction of the *SLG28* gene into a heterozygous line of *B. rapa* carrying S52 and S60 failed to confer recognition or rejection of S28 pollen. However, crossing these plants to the transgenic plants described above, which express *SRK28*, resulted in a significant enhancement of rejection of S28 pollen, with much lower seed set than was shown by the plants containing *SRK28* alone. From these data it is apparent that *SRK* is the female specificity determinant in *Brassica*, and that *SLG* enhances the recognition process in the stigma (Takasaki *et al.* 2000). One possibility is that *SLG* determines the strength of the incompatible reaction, with strong incompatibility resulting from particular *SLG* alleles but weak incompatibility resulting from other *SLG* alleles (Hiscock and McInnis 2003).

13.5.2 The male determinant of SSI in *Brassica*

Similar approaches have been taken to establish which genes at the S locus determine the male mating type of a plant. The *S locus cysteine-rich (SCR)* gene (also sometimes known as *S-locus pollen protein 11*, or *SP11*) encodes a small cysteine-rich protein that is present only in the anthers and has a signal peptide for secretion (Schopfer *et al.* 1999; Takayama *et al.* 2000). Schopfer *et al.* (1999) demonstrated the role of the *SCR* gene in male mating type using both loss-of-function and gain-of-function approaches. First, they showed that a self-compatible mutant line of *Brassica oleracea* lacked the *SCR* transcript. Then they introduced the *SCR6* gene transgenically into a line homozygous for the S2 haplotype. Pollen from all 12 resulting transformants shown to express the transgene was rejected by the stigmas of plants homozygous for the S6 haplotype, confirming the importance of the *SCR* gene in the male phenotype.

Further analysis of SCR protein function has revealed how haploid pollen grains can carry the imprint of two S locus haplotypes, resulting in sporophytic self-incompatibility. When the *SCR9* gene was expressed in bacterial cells, and the SCR9 protein was purified, the protein alone resulted in a self-incompatibility response when applied to stigmatic cells of a plant carrying the S9 haplotype, but not of a plant carrying the S8 haplotype (Takayama *et al.* 2000). Within the plant, the *SCR* genes are expressed early in anther development in the tapetal cells that line the inside of the anther, as well as in the developing microspores (Takayama *et al.* 2000). As pollen grains develop within the anther, SCR protein is rubbed from the tapetum on to the surface of the pollen grain. As a result, pollen grains carry SCR proteins encoded by both *SCR* alleles present in the parent plant, even though the haploid gametes within them only carry a single allele at the *SCR* locus.

The dominance of *SCR* alleles is a reflection of their expression patterns, regulated by RNA-directed DNA methylation. Dominant *SCR* alleles are expressed tapetally and gametophytically, whereas recessive alleles are only expressed in the tapetum or are silenced completely in a heterozygous context. Tapetum-specific expression is sufficient to provide a coating of SCR protein to the outside of the pollen grain, maintaining self-incompatibility in a homozygous recessive individual. The silencing of recessive alleles in the presence of a dominant allele has been observed in *B. rapa*, *B. oleracea*, and *Arabidopsis lyrata* (Kusaba *et al.* 2002; Shiba *et al.* 2002). Tarutani *et al.* (2010) showed that silencing of the recessive *SCR* allele is due to methylation, directed by a small non-coding RNA, *SP11 methylation inducer* (*Smi*), encoded by regions of the S locus flanking the *SCR* in dominant haplotypes.

13.5.3 The SSI response in *Brassica*

When a pollen grain lands on a stigma, the pollen coat SCR diffuses into the cell wall of the stigmatic papillae, where it binds to the receptor domain of the SRK. Prior to SCR binding, the SRK is present as loose dimers associated with two thioredoxin-H-like proteins (Mazzurco *et al.* 2001). The thioredoxin-H-like proteins are thought to prevent auto-phosphorylation of the SRK, maintaining it in an inactive state in the absence of the SCR ligand (Cabrillac *et al.* 2001). If the SCR and SRK are of the same haplotype, then the SCR binds to the SRK, and residues in the SRK kinase domain are phosphorylated (Takayama *et al.* 2001). At the same time the SRK dimer is stabilized, and this phosphorylated stable dimer is an active signalling molecule (reviewed by Iwano and Takayama 2012). This change to SRK conformation leads to an intracellular signal transduction cascade that inhibits pollen germination. The signal transduction cascade downstream of SRK acts through degradation of factors essential for pollen tube growth. ARC1, a stigma-specific U-box E3 ubiquitin ligase containing an Armadillo motif, is bound by SRK and phosphorylated (Gu *et al.* 1998; Stone *et al.* 2003). Further experiments indicate that transgenic plants expressing an antisense version of *ARC1* show loss of their SI responses (Stone *et al.* 1999). Yeast two-hybrid studies also suggest that a kinase-associated protein phosphatase, calmodulin, and a nexin all bind to the SRK (Vanoosthuyse *et al.* 2003). Downstream of the immediate SRK-driven signal it is thought that multiple diverse signalling pathways, including auxin signalling, interact to suppress pollen germination.

13.5.4 Loss of SSI in Arabidopsis

One obvious member of the Brassicaceae that does not show SSI is Arabidopsis. Tsuchimatsu *et al.* (2010) demonstrated that the loss of self-incompatibility in European ecotypes of Arabidopsis was attributable to a 213-base-pair inversion of DNA within the *SCR* region of the S locus. Re-inversion of this region in transgenic plants of the Wei-1 ecotype restored self-incompatibility, confirming that the *SRK* region and all downstream responses were still functional. However, in many ecotypes further degradation of the S locus has occurred.

13.5.5 SSI in other plant families

Sporophytic self-incompatibility has been most extensively studied in the Brassicaceae, but recent reports of its control in the Convolvulaceae (Solanales) and the Asteraceae (Asterales) suggest that it has evolved multiple times through different molecular mechanisms. Within the Convolvulaceae, research has focused on *Ipomoea trifida* with ongoing efforts to isolate the genes at the S locus through a variety of approaches. Although little is yet known about what does constitute the S locus in this species, it has been shown that genes with sequence similarity to *SRK* and *SLG* do *not* segregate with the S locus, which suggests that they are not involved in self-incompatibility, and are expressed in all tissues rather than in a stigma-specific manner (Kowyama *et al.* 1996; Rahmann *et al.* 2007). In the Asteraceae, similar conclusions have been drawn from work with *Senecio squalidus*. Hiscock *et al.* (2003) reported that *SRK*-like sequences did not segregate with the S locus and were not specifically expressed in the stigma. Recent work in *Senecio* has focused on genes encoding glycoproteins which are expressed in the stigma and do segregate with the S locus, although their molecular function is as yet unknown (Hiscock and Tabah 2003), and on transcriptome-based analyses of stigma-specific transcripts, which have provided a number of additional candidate genes for further analysis (Allen *et al.* 2011).

13.6 Gametophytic self-incompatibility (GSI)

Gametophytic self-incompatibility has been studied in most detail in the Solanaceae, although it is found in 60–90 families, including the Rosaceae (Rosales), Leguminosae (Fabales), Plantaginaceae (Lamiales), and Papaveraceae (Ranunculales) (Franklin-Tong and Franklin 2003). As with sporophytic self-incompatibility, gametophytic self-incompatibility is controlled by a single S locus which is highly polymorphic and contains the determinants of both male and female function tightly linked. However, since male S type is determined by the genotype of the haploid pollen grain, the gametophyte, there are no issues of dominance with respect to the male function. Gametophytic self-incompatibility in the best studied system, that of the Solanaceae, differs from sporophytic self-incompatibility in that pollen grains do germinate but are inhibited within the transmitting tract of the style, rather than on the stigmatic surface.

13.6.1 The female determinant of GSI in the Solanaceae

A small glycoprotein, present in mature styles, is the female determinant in the gametophytic self-incompatibility system. The protein was initially identified on the basis of its abundance in the style and its segregation with the S locus in *Nicotiana alata* (Anderson *et al.* 1986). The protein was noted to have significant structural and sequence similarity with a group of fungal ribonucleases, or RNases, and was then shown to have ribonuclease activity, and so was named S-RNase. Gray *et al.* (1991) confirmed that S-RNases prevent protein production, both within a pollen tube and in an *in vitro* system, and showed that S-RNases degraded all mRNAs, not specific substrates. Transformation experiments similar to those used to test the female determinant of sporophytic self-incompatibility have shown conclusively that S-RNase is sufficient to determine female mating type in *N. alata*. Introduction of the gene encoding S-RNase from the SA2 haplotype into a different genetic background (a hybrid between *N. alata* and *N. langsdorffii*) resulted in the production of plants that specifically rejected SA2 pollen (but not pollen grains carrying other S haplotypes) (Murfett *et al.* 1994). Similarly, introduction of the S3 S-RNase gene from *Petunia inflata* into an individual carrying the S1S2 haplotype induced the ability to recognize and reject S3 pollen (Lee *et al.* 1994a). To date, it appears that S-RNase is the only protein involved in specifying female mating type in the Solanaceaous gametophytic self-incompatibility system.

13.6.2 The male determinant of GSI in the Solanaceae

F box proteins, which target other proteins for degradation, have been shown to be the male determinants of gametophytic SI in both petunia and Antirrhinum (Plantaginaceae, in the Lamiales). F box proteins in many cellular contexts bind specific proteins to the SCF E3 ubiquitin ligase complex, targeting them for degradation by the 26S proteasome. In GSI it is thought that F box proteins target non-self S-RNases for degradation, allowing pollen tubes to grow. Genes encoding F box proteins were first shown to be linked to the S locus by sequencing of the genomic region surrounding S-RNase in *Antirrhinum hispanicum* (Lai *et al.* 2002). These genes were named *S locus F box* (*SLF*) genes, and were then shown to influence compatibility when introduced transgenically into different backgrounds (Qiao *et al.* 2004; Sijacic *et al.* 2004). Rather surprisingly, there seem to be few alleles of *SLF*. This observation, combined with the difficulty in explaining how a single pollen SLF protein could recognize the potentially infinite variety of non-self S-RNase alleles, led Kubo *et al.* (2010) to look at the male determinant of GSI more closely in petunia. They established that multiple *SLF* genes were linked to the S locus and were specifically expressed in pollen, suggesting that the male determinant is not a single SLF but a variety of them, able to recognize a great diversity of S-RNase molecules, but not the matching one. These data led to a revision of the basic model of GSI, which is now known as a collaborative non-self recognition system.

13.6.3 The GSI response in the Solanaceae

Before the male determinant of gametophytic self-incompatibility was identified, debate centred on its likely role in prohibiting RNase activity. The observation that compatible S-RNases do enter the pollen tube (Luu *et al.* 2000) indicated that the male determinant was likely to inhibit RNase activity, rather than act as a gatekeeper, guarding the pollen tube. The identification of the male determinant as an F box protein seemed to reinforce this model, and SLF has been shown to bind S-RNase, using yeast two-hybrid analysis (Qiao *et al.* 2004). F box proteins are most likely to act within the pollen tube to degrade all the S-RNases entering from the style. In this model, when a matching S-RNase enters the pollen tube it is not recognized and degraded by any of the F box proteins, and is able to inhibit pollen tube growth. Accordingly, the Solanaceous GSI system can be thought of as a non-self recognition system—all the non-matching S-RNase proteins are recognized and degraded by the SLF proteins, but the matching S-RNase escapes. However, some studies have suggested an alternative model. Goldraij *et al.* (2006) reported that non-self S-RNases are not degraded in *Nicotiana alata*, where compatible and incompatible pollen tubes contain

similar quantities of RNase. Instead, those researchers showed that S-RNases were sequestered in a vacuolar compartment in the pollen tube, which broke down after pollination with an incompatible (self) pollen grain. They further showed that a stylar protein called HT-B was necessary for the release of S-RNases from the vacuole. HT-B was one of the first to be identified of a number of modifier proteins that interact in the GSI recognition system. This particular protein is significantly degraded following pollination with an incompatible pollen grain, but less so after a compatible pollination. In transgenic plants expressing an antisense *HT-B* construct, the pollen tube vacuole did not break down after an incompatible pollination. The authors therefore proposed an alternative to the simple RNase degradation model (Goldraij *et al.* 2006; and reviewed by McClure and Franklin-Tong 2006 and McClure *et al.* 2011). These authors suggested that S-RNases might be compartmentalized into pollen tube vacuoles in compatible reactions, and only released to inhibit pollen tube growth in incompatible (self) reactions. The SLF proteins might therefore be involved in targeting the non-matching HT-B proteins for degradation, preventing vacuolar breakdown, rather than in directly degrading the S-RNases themselves.

The relatively recent discovery of the male determinant of gametophytic SI means that our understanding of this system lags considerably behind our understanding of sporophytic self-incompatibility. Much exciting work remains to be done, to uncover the details of the incompatibility reaction and any downstream responses.

13.6.4 GSI in other plant families

Gametophytic self-incompatibility has been studied in several plant groups, and extremely similar mechanisms seem to be at work in the Solanales (*Petunia*, *Nicotiana*), the Lamiales (*Antirrhinum*), and the Rosales (*Prunus*). However, it seems that an entirely different system of gametophytic self-incompatibility operates in poppy (Ranunculales). In *Papaver rhoeas*, small secreted proteins known as Papaver rhoeas style S (PrsS) proteins determine the female mating type (Foote *et al.* 1994), and the incompatible reaction occurs in the pollen grain. Sequencing of the S locus revealed that the male determinant of GSI in this system, *Papaver rhoeas pollen* S (*PrpS*), encodes a transmembrane protein, with an extracellular loop shown to bind to the secreted PrsS protein (Wheeler *et al.* 2009). The interaction between the two proteins results in calcium movement into the germinating pollen tube, triggering a set of downstream responses that result in programmed cell death. PrpS has no sequence homology with proteins of known function from other plant systems, so the details of its mode of function are currently unclear (Wheeler *et al.* 2010).

13.7 Heteromorphic self-incompatibility

In around 20 different families of angiosperms there are species which have a heteromorphic self-incompatibility system, where the biochemical self-incompatibility type is associated with particular morphological traits (Barrett 2002). The common primrose, *Primula vulgaris* (Ericales), is a good example of this type of system. It has sporophytic self-incompatibility with only two S haplotypes, which also control the position of the anthers and stigma within the flower. In 'pin' flowers the style extends above the anthers, and in 'thrum' flowers the style sits below the anthers (see Fig. 13.5). Plants that produce pin flowers are genetically homozygous recessive (ss) at the S locus, and plants

that produce thrum flowers are heterozygous (Ss). Pollen from heterozygous plants shows dominance of the 'thrum' (S) haplotype, allowing it to fertilize homozygote pin (ss) plants. Crosses between the two types produce 50% heterozygous (and therefore thrum) flowers and 50% homozygous (and therefore pin) flowers. A similar genetic system operates in other, unrelated species, such as buckwheat (*Fagopyrum esculentum*, Caryophyllales). This system places pollen on specific parts of the pollinator's body, from where it is most likely to be deposited on a stigma of the opposite flower type. This prevents half of the pollen being wasted on the wrong stigma type, which would reject it biochemically, and thus saves energy and resources as well as reducing blockage of the stigma by incompatible pollen. Molecular genetic approaches are currently being used to dissect the S locus in both primrose and buckwheat (Li *et al.* 2011).

The biochemical self-incompatibility that is present in heteromorphic systems may work in different physiological ways in different floral morphs. For example, Massinga *et al.* (2005) showed that incompatible pollen tubes were inhibited on the stigmatic surface of the long-styled form of *Pentanisia prunelloides* (wild verbena, Gentianales), but within the style of the short-styled form. The style length polymorphism in this species is associated with an anther length polymorphism, and these influence pollination by butterflies, which carry pollen on

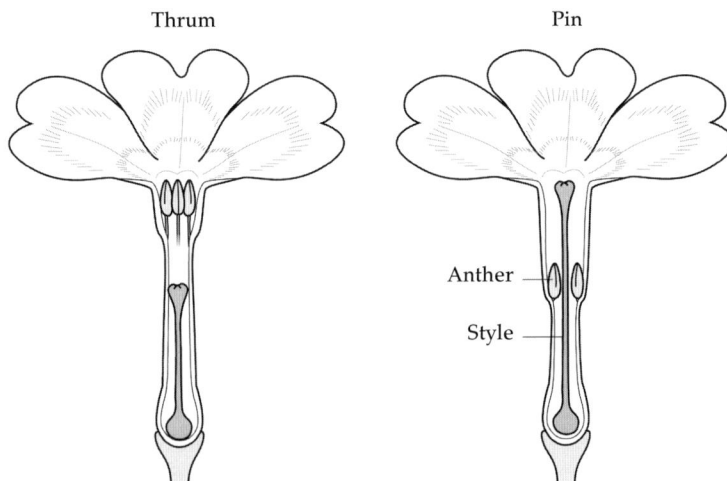

Figure 13.5 Heteromorphic self-incompatibility, a form of reciprocal herkogamy.

their heads or proboscises according to the anther length of the flower from which it came.

In addition to the distylous systems described above, in which two different forms of flower are produced, some species have three different floral forms, with stigmas at short, medium, and tall heights. These flowers produce two whorls of anthers, and within each floral morph the two anther whorls are at the reciprocal heights relative to the stigma (so flowers with short stigmas have medium and long anthers). This system is known as tristyly.

The existence of reciprocal floral types with differences in style and anther height is not always associated with biochemical self-incompatibility. When such reciprocal herkogamy occurs alone it is known as heterostyly, which may be more common than full heteromorphic self-incompatibility. Both heterostyly and heteromorphic self-incompatibility require the linkage of genes controlling a number of traits at a single locus. Research to date has focused on the population dynamics and evolution of heterostyly (discussed by Mast and Conti 2006).

13.8 Ensuring self-pollination

Around 20% of angiosperm species are predominantly self-fertilizing, and the transition from outbreeding to selfing has occurred very frequently in the evolutionary history of the flowering plants (Barrett 2002). It might be expected that selfing is a default state for a plant, and that, in order to achieve it, it is simply necessary not to acquire, or to lose, the various traits described earlier in this chapter. Although hermaphrodite plants (and preferably hermaphrodite flowers) without biochemical self-incompatibility are a prerequisite for a selfing system, many other aspects of floral morphology can become adapted in ways that optimize self-pollination. The transition to selfing is thought to begin in most cases with the loss of the biochemical self-incompatibility system (as described above in Arabidopsis; Tsuchimatsu *et al.* 2010). This loss is often followed by a reduction in flower size, by the loss or reduction of scent and nectar production, by a reduction in herkogamy to position anthers and stigma at the same height, and by a reduction in flower opening. This reduced opening is taken to extremes in cleistogamous flowers, which do not open at all, and instead achieve self-pollination within the closed perianth organs. It is not always possible to predict which morphological traits will be most clearly associated with selfing. In *Collinsia* (Lamiales), a genus that contains both selfing and outcrossing species but no biochemical self-incompatibility, the floral trait most strongly associated with selfing was shown to be loss of dichogamy, rather than floral size or spatial positioning of reproductive organs (Kalisz *et al.* 2012).

Efforts to explore the molecular genetic basis of the transition to selfing have focused on a number of species closely related to genetic models. For example, Sicard *et al.* (2011) demonstrated that the evolution of selfing in *Capsella rubella* (Brassicales) was accompanied by a reduction in petal size attributable to a shorter period of cell division compared with outcrossing relatives.

Those plants that do not allow self-fertilization must ensure that they achieve outcrossing. The interactions of plants with their pollinators, and the adaptations that facilitate those interactions, are the subject of the third section of this book.

Pollination Success: Molecular and Ecological Interactions

How and Why Does Floral Form Vary?

CHAPTER 14

Why are flowers different? Pollination syndromes: the theory

In Chapter 8 we considered the newly emerging data on how different plants make the transition to flowering. However, when focusing on the development of flowers themselves, we have so far largely ignored the concept that the flowers of different species may be very different from one another. However, it is clear from a merely cursory glance around any garden in the summer months that flowers come in an enormous variety of sizes, shapes, colours, and scents. In Section III of this book we shall focus on the differences between flowers, as opposed to the molecular similarities that unite them. In this chapter we begin by considering the different ways in which flowers can be pollinated. It is a basic premise underlying much of floral biology that differences in pollination system explain many of the differences in floral form, as a result of selective pressure applied by pollinator behaviour both on floral traits that optimize animal foraging efficiency and on floral traits that optimize contact between the animal's body and the plant's reproductive organs. The evidence to support this premise is not as compelling as we might like to think, as will be discussed in later chapters (see in particular Chapter 21). To set the stage for those discussions, in this chapter we shall look at the historical concept of the pollination syndrome and the predictions that it makes about floral morphology.

14.1 Cross-pollination

If cross-pollination is to occur within a population of flowering plants, it is necessary for pollen to be transported from the anthers in a flower of one individual to the stigma in a flower of a second

individual. The alternative is self-pollination, when pollen is transferred from the anther of a single flower on to the stigma of the same flower (autogamy), or from the anther of one flower to the stigma of a second flower on the same plant (geitonogamy). Self-pollination is thought of as an adaptation to unreliable pollen vectors, particularly seen in short-lived annual plants and those species that regularly invade new habitats. However, self-pollination is not optimal for genetic recombination, and so in the long term may compromise the genetic diversity and evolutionary lability of a population (see discussion in Chapter 13). Therefore a great many flowering plants have adaptations that maximize their potential for cross-pollination. Since plants are immobile, this transport of pollen from plant to plant must be accomplished with the aid of a pollen vector, which may be abiotic or biotic.

14.2 Abiotic pollen vectors

14.2.1 Wind pollination

Abiotic pollen vectors consist primarily of the wind and water. Of these, wind pollination (anemophily) is by far the more common, being found in 18% of angiosperm families (including the ecologically dominant Poaceae, the grasses) as well as in the conifers (Ackerman 2000). Wind pollination is a secondarily derived state in angiosperm flowers, as the earliest flowers are believed to have been pollinated by beetles (Thien *et al.* 2000). Anemophily is thought to evolve in response to changes in the environment that decrease the efficiency of biotic pollination while enhancing the success of wind pollination, and has arisen around 65 times in the angiosperms (Friedman and Barrett 2008). Examples of environmental changes that might cause a shift towards wind pollination include variations in pollinator abundance, migration into an area with a very dry climate, or the arrival of another plant species which might compete for pollinator attention (Culley *et al.* 2002). Many plant species have been shown to be both wind- and animal-pollinated, either simultaneously or at different times within the same season (Cox 1991). This combination of biotic and abiotic pollination systems is known as ambophily, and may be the transitional route through which full wind pollination most commonly evolves.

Angiosperm species that are wind-pollinated typically show a particular set of characteristics. Most of these characteristics have been shown to enhance the success of wind pollination, so can be thought of as adaptations to a wind-pollinated reproductive system (Culley *et al.* 2002). As wind pollination is a passive process, many of these adaptations are related to the environments and habitats in which anemophilous plants live, rather than to the features of their flowers. To begin with, wind-pollinated species typically inhabit environments with stronger winds but lower humidity than those of biotically pollinated species. Wind-pollinated plants are found in habitats with low rainfall, ensuring that pollen is not washed away. Wind-pollinated species are usually found in moderately high densities within an area, but with low densities of other species surrounding them. This combination of features maximizes the chances of released pollen landing on another individual of the same species.

Besides common features of environment and habitat, wind-pollinated species also share a typical set of floral characteristics. Flowers and inflorescences are commonly held above or away from vegetation, which maximizes access by the wind to reproductive organs. Very often the inflorescences are pendulous, like those of willow trees (Malpighiales). The flowers within these inflorescences are usually unisexual. It would be hard to prevent self pollen clogging the stigmas of flowers exposed to the wind, if both male and female reproductive organs were present in a small area. The outer two whorls of floral organs, the petals and sepals, are usually much reduced or absent in wind-pollinated

Figure 14.1 The long tasselled flowers of a wind-pollinated grass hang far from the main body of the plant. See also Plate 6.

flowers, and, where present, do not usually produce much scent or pigmentation. The loss of a brightly coloured and strongly scented corolla is unsurprising in view of the unconscious nature of the wind as a pollen vector. Similarly, nectar production is usually lost when wind pollination evolves (Friedman and Barrett 2008). The stamen filaments of wind-pollinated flowers are usually long, exposing the locules to the wind and generating an acroclastic release of the pollen as energy is transferred from the wind to the stamen through the long filament (Urzay *et al.* 2009). The pollen grains themselves are smooth-surfaced, small in size, and enormous in quantity. Small size and a smooth surface have been shown to enhance the aerodynamic properties of pollen travelling in wind (Niklas 1985), while large quantities are necessary to increase the chances of wind-blown pollen reaching a stigma. The stigmas themselves are usually long and feathery, and protrude a considerable way beyond the outer edges of the flower, maximizing the surface area available for pollen capture (Culley *et al.* 2002). Wind-pollinated grass flowers are shown in Fig. 14.1.

14.2.2 Water pollination

Water pollination (hydrophily) is a relatively rare system of gamete transfer, but it is employed by a few species of grasses and waterweeds, occurring in only 31 genera of 11 different families, mostly in the monocots (Cox 1988). In most hydrophilous species, pollen is released below the water surface and carried passively by currents to female reproductive structures. This mechanism of water pollination is used by many marine plants, such as the Caribbean turtle grass, *Thalassia testudinum* (Alismatales), which releases pollen grains underwater, bound up in strands of mucilage. These are carried by currents below the water surface, and the mucilage encourages them to stick to the female flowers, which are also underwater.

However, in some species, pollen is released on to the surface of the water and is carried on the surface to female reproductive structures. The pollen of such plants—for example, *Elodea canadensis* (Canadian waterweed, also Alismatales)—is ornamented with many tiny spikes, which trap pockets of air and ensure that the pollen floats. Anther dehiscence is often explosive, scattering the pollen grains widely across the water surface. The female reproductive structures may always be held at surface level in some hydrophilous angiosperms, but in others the female flower closes around the pollen and is then withdrawn below the water surface after pollination. A variation on the theme of releasing pollen on to the water surface is shown by species of *Vallisneria* (tape-grass, also Alismatales), which release entire male flowers on to the water, with the pollen never leaving the anthers until it arrives at a female flower.

The female flowers of many hydrophilous species create depressions in the surface tension of the water, encouraging passing pollen grains to slide down towards them. This system is most effective in relatively still, stable water conditions, and can be easily perturbed by strong winds. Although hydrophilous flowers do produce relatively large quantities of pollen, the use of water surface tension dynamics to facilitate pollination means that less pollen is required than might be predicted (Cox 1988).

14.3 Biotic pollen vectors

Biotic pollen vectors provide some or all of the pollination service for the majority of flowering plants. The frequency with which we observe this interaction leads to a common assumption that plants and animals are cooperating in a mutualistic association. However, animals rarely deliberately pollinate flowers, but instead view flowers as food sources. The reward offered is usually a mixture of nectar, a sugar solution (which may contain various concentrations of several different sugars, along with other nutrients), and pollen itself, which is very rich in amino acids. When collecting this food, animals inadvertently brush against the reproductive organs of the flower, causing pollen to be transferred from stamens to their bodies, and from their bodies on to stigmatic surfaces. Many flowers have structural features that enhance this inadvertent contact between pollinator body and floral reproductive organs, such as narrow corolla tubes, dorsally positioned anthers that deposit pollen on the back or head of the pollinator, or long nectar spurs that force the pollinator to crawl inside the flower to access the reward.

14.4 Principles underlying the pollination syndrome concept

Biotic pollen vectors consist of a range of animals—most commonly insects, as well as birds, bats, and a small number of other vertebrates. In a seminal book published in 1966, Faegri and van der Pijl bravely set out to 'formulate the general principles of pollination ecology, applicable anywhere.' A major device used by those authors was the pollination syndrome, a suite of floral traits associated with the attraction of a particular group of pollinators. The assumption behind the pollination syndrome concept is that coevolution between plant and animal, or else adaptation of the plant to existing animal traits, has led to the acquisition by plants of floral features that maximize their chances of attracting and being pollinated by a particular animal or group of animals.

14.4.1 Do flowers act as specialist or general advertisements?

The pollination syndrome concept explains the differences between different flowers as consequences of the types of animals that they attract. This concept relies on the principle that characteristics such as particular colours, scents, shapes, and markings are specific to flowers pollinated by specific types of animals. The different floral characteristics can be viewed as animal-specific advertisements, adapted to maximize the chances of attracting a particular type of bee or butterfly or beetle.

However, there is an alternative way of explaining the many differences between different flowers. It was Darwin himself who first articulated the reason why flowers are brightly coloured:

Flowers rank amongst the most beautiful productions of nature; but they have been rendered conspicuous on contrast with the green leaves, and in consequence at the same time beautiful, so that they may be easily observed by insects. (Darwin 1859)

At first sight this might be read as consistent with the pollination syndrome concept, in stating that different colours attract different animals. In fact all that this quote means is that colour (and by extension shape and scent) simply provides contrast between flowers and the green vegetation around them. In this case any feature that makes a flower stand out even more against the green backdrop is likely to enhance the success of that flower in attracting pollinators. Accordingly, from this point of view, all the different floral characteristics are simply an enormous number of different ways that plants have hit upon that make their flowers more attractive to a broad spectrum of pollinators. Instead of providing species-specific advertisements, the different floral characteristics might simply represent an almost infinite number of different advertisements, all of which serve to attract animals in general.

It is likely that certain plant species produce advertisements that target particular animal pollinators, and that other plant species produce generalist advertisements. However, we should bear in mind that the pollination syndrome concept rests on the idea of specialization, and may not be compatible with a view of flowers as non-specific advertisers of rewards to generalist pollinators.

14.4.2 Plants and pollinators want different things, making species-specific interactions unlikely

Plants and their pollinators have different requirements of the interaction, and these differences may reduce the likelihood of coevolution occurring between individual species. A plant will experience maximum reproductive fitness if it maximizes pollen dispersal between individuals of the same species, without its stigmas becoming clogged with pollen of other species and without wasting much energy on nectar production. Therefore the optimal pollen vector will be an animal that alights only briefly on each flower, moves rapidly between individuals, is faithful to only one plant species, and does not eat much.

The animal will conserve energy if there is little need to forage, and so should prefer flowers with a large reward, where it can stay for a long time. If the reward is insufficient in one species it should forage on a range of plant species.

In addition, very specific interactions between plant and pollinator carry their own risks. A one-on-one relationship between flower and pollinator should mean that little pollen is wasted, and so its production can be less prolific, saving energy. Female fertility should also be enhanced, as foreign pollen will not be applied to the stigma, where it can block acceptable pollen from germinating. However, such an extreme specialist can expect to set no seed at all if the pollinator is absent in any one season, due to disease or climatic changes, a situation which would be catastrophic for an annual plant and might also be very disadvantageous for a perennial.

It is generally argued that for these reasons the evolution of plants and their pollinators has followed a middle road, with groups of flowers adapting to selective pressures imposed by groups of animals, rather than one-on-one species-specific interactions. It is the broad association between a group of flowers and the group of animals that pollinates them that has come to be called a pollination syndrome (Faegri and van der Pijl 1966).

14.5 The pollination syndromes

A pollination syndrome classically describes a suite of adaptations shown by a flower to a taxonomic order of animals, and by those animals to a particular group of plants, which may not be phylogenetically related to each other. The adaptations shown by the animals may be behavioural or morphological,

	COLOUR	SHAPE	SCENT	REWARD
BEETLE	White/cream	Dish/bowl	Fruity, quite strong	Nectar to lap, some excess pollen
FLY	White/cream/pale yellow	Dish-like or more complex	Minimal	Small amount of nectar
BEE	Blue/yellow/ultraviolet	Deep tube, bilateral symmetry	Minimal	More nectar, some excess pollen
BUTTERFLY	Yellow/red/orange	Deep tubes with landing platforms	Minimal	Only nectar
MOTH	White	Bilateral symmetry, deep tubes	Strong, sweet, at night	Plentiful nectar
BIRD	Red/orange	Pendant or bilateral and upright, tube	Minimal	Much nectar
BAT	White	Large, saucer shaped	Strong, butyric acid	Copious nectar and much excess pollen

Figure 14.2 Table showing the general features of flowers involved in different pollination syndromes.

while the plants can only show morphological adaptations, which may include flower size, structure, colour, reward, and timing of floral induction and opening. A pollination syndrome can result from coevolution, or from adaptation of the plant to pre-existing animal traits. A table summarizing the key floral traits traditionally associated with different pollinators is shown in Fig. 14.2.

14.5.1 Beetle pollination (cantherophily)

Beetle pollination is widely believed to have been the first pollination syndrome, the one used by the first angiosperms, as the Coleoptera (the beetles) constitute one of the oldest orders of insects and were already numerous at the time when the angiosperms came into existence, having themselves arisen around 260–280 million years ago (Ponomarenko 1995). Fossilized pollen found in the digestive tracts of beetles has shown that these insects had already acquired the habit of grazing on the pollen of cycads, conifers, and other gymnosperms, before the angiosperms appeared (Labandeira 1997;

Thien *et al.* 2000). When considering a pollination syndrome, it is useful to begin with the features of the animal's biology that are relevant to its role as an agent of pollen transfer. Beetles have mouth parts positioned parallel to the axis of the body, which limits their ability to manipulate food sources, particularly those with depth. They are also quite big animals, with a moderately high demand for protein as well as carbohydrate. Beetles do not have good colour vision, but do have a strong sense of smell and a particular attraction to fruity smells. Flowers that have become adapted for pollination by beetles might be expected to have characteristics which make them both attractive to beetles and easy to obtain food from once the animal has landed. These features would include the provision of a floral reward consisting of both nectar (for sugar) and pollen (for protein) in a flat structure from which the reward can be lapped. Beetle-pollinated flowers are usually saucer- or bowl-shaped, with nectar secreted into the shallow bowl, and often have excess anthers to produce extra pollen as a reward. Since colour vision is unimportant to beetles, beetle-pollinated flowers might be expected to waste little energy on the production and modification of pigments, but instead to produce a strong scent. Many beetle-pollinated flowers do produce a fruity fragrance, and they are often greenish or off-white in colour. The classic example of a beetle-pollinated flower is magnolia (Magnoliales), which has remained essentially unchanged for 100 million years and is still pollinated by the same sorts of animals (see Fig. 14.3a). Lilies (Liliales), wild roses (Rosales), and some poppies (Ranunculales) are also often beetle-pollinated.

Figure 14.3 Insect-pollinated flowers. (a) *Magnolia* (Magnoliales) flowers are beetle-pollinated. (b) The fly-pollinated flowers of *Fatsia japonica* (Apiales). (c) Bumblebee entering a *Hebe* flower (Lamiales). (d) Many daisies (Asterales) are butterfly-pollinated. (e) The flowers of *Angraecum sesquipedale* (Asparagales) have very long nectar spurs and are pollinated by extremely long-tongued moths. Photographs (a), (d), and (e) kindly supplied by Cambridge University Botanic Garden and H. Rice. See also Plate 7.

14.5.2 Fly pollination (myophily)

The Diptera (the order of true flies) show the greatest variation in methods and habits of pollination of any group of insects. This makes it hard to define the morphological traits that we would expect to see in a fly-pollinated flower. Indeed, some flies behave very like beetles, and pollinate flowers that are morphologically very similar to beetle-pollinated flowers (Thien *et al.* 2000). Other flies may have more in common with wasps, and pollinate flowers with very different structural features. One feature of flies that has been utilized by flowering plants is their persistence throughout all seasons. Flies are one of the few groups of insects that are not strictly periodic, and, as such, many plants that flower under adverse conditions or at odd times of the year can be entirely dependent on flies for their pollination. Another feature common to all flies is that they do not feed their offspring, and are usually lighter-bodied than many other insects. In consequence they do not need much food, and fly-pollinated flowers usually supply only a small quantity of nectar. Although there is much variation within the order, in general flies are more visually acute animals than beetles, and have a positive preference for pale and yellow colours. Fly-pollinated flowers are often coloured cream or yellow, and do not usually have much scent. Classic examples of fly-pollinated flowers include species such as carrot and other members of the Apiaceae (Apiales), as well as species such as groundsel and other daisies (Asterales). A fly-pollinated *Fatsia japonica* flower (Apiales) is shown in Fig. 14.3b.

14.5.3 Bee pollination (melittophily)

Of the Hymenoptera, the bees have specialized most towards a diet of nectar and pollen. Wasps will take nectar to meet their sugar requirements, but do not actively collect either nectar or pollen. There are a few specialized instances of ants feeding on nectar, often from extra-floral nectaries, but for the most part it is the bees that are the great pollinators of the Hymenopteran order. Bees are large animals, and have a substantial energy requirement. They usually forage for nectar for themselves, and pollen to feed to larvae back in their hive or brood chamber. The pollen is carried on their bodies, in specialized structures that range from simply having hairy feet to having pollen baskets on the hind legs, and it is then groomed off back at the hive. In response to these features, bee-pollinated flowers are usually quite large, in order to bear the weight of the animal, and they often have a clear landing platform. They are also often closed, with petals or other structures that must be pushed aside by the bee before it can access the nectar. This type of mechanism prevents nectar robbing by other animals, and helps to keep bees constant, as the supply of nectar is likely to be good. Inside these large flowers there is usually a reasonable volume of fairly concentrated nectar, and a small amount of excess pollen. Bees have the ability to perceive depth, and many species have long tongues. The nectar is often secreted at the base of deep corolla tubes or nectar spurs, again denying access to species without the size and tongue length to access it. Since pollen found on the bee's body is groomed off and fed to the larvae, many bee-pollinated flowers have dorsal anthers that deposit at least some of the pollen on the back of the bee's neck, a position from which it cannot easily be groomed. Bees have good colour vision and can see in ultraviolet, blue, and yellow, but do not have receptors to perceive red, which they see only as a weak green signal. Bee-pollinated flowers are usually brightly coloured, with yellow and blue being considered classic bee colours. However, many bee-pollinated flowers are red or pink. It has been shown that in many of these cases the presence of UV-absorbing pigments modifies the red to a colour which is more highly visible to the bee (see Chapter 20; Chittka and Waser 1997). As quite intelligent animals, bees are sensitive to nectar guides, which enable them to handle the flowers more quickly by directing them straight to the nectar. Many bee-pollinated flowers have nectar guides visible either in the range of the spectrum that humans can see or else in the ultraviolet part. Classic examples of bee-pollinated flowers include the garden snapdragon (*Antirrhinum majus*), rosemary, and foxgloves (all Lamiales), and nettles (Rosales). A bumblebee visiting a *Hebe* (Lamiales) inflorescence is shown in Fig. 14.3c.

14.5.4 Butterfly pollination (psychophily)

In contrast to bees, the other most prominent group of pollinating invertebrates in temperate climates does not feed its offspring and has relatively low energy requirements. Butterflies are relatively light in weight, and usually alight on flowers, which reduces their energy expenditure. They also have long tongues, often 1 to 2 cm in length. Flowers that are pollinated by butterflies usually combine a flat structure to alight on with deep tubes in which the nectar is secreted. This can be achieved either by presenting a tube-shaped flower with a landing rim around it, as in buddleia (Lamiales), or else by clustering a number of small tubes together to produce a larger flat surface, the system used by daisies (Asterales) (see Fig. 14.3d). Butterflies are not known to have much sense of smell at all, and butterfly-pollinated flowers do not usually produce much scent. However, butterflies do have good colour vision, and can see red. Butterfly-pollinated flowers are usually brightly coloured, with reds and yellows often predominant. This prevalence of red does not necessarily reflect a butterfly preference for red, but could simply occur because red is more visible to butterflies than to bees, and red flowers are therefore less likely to have been depleted of nectar by bees, making them attractive to butterflies.

14.5.5 Moth pollination (phalaenophily or sphingophily)

Although moths and butterflies are members of the same insect order, the Lepidoptera, their methods of pollination and the structures of the flowers they visit are completely different. The major reason for this is the difference in their behaviour—many moths are nocturnal, whereas butterflies are diurnal. Moths also prefer not to land on flowers, but to collect their nectar while hovering. Many moths are also much heavier in the body than butterflies. The combination of a heavy body and hovering flight means that moths have very high energy requirements indeed, and moth-pollinated flowers usually produce more nectar than butterfly- or bee-pollinated flowers. To facilitate nectar collection while the moth is hovering, moth-pollinated flowers are usually bilaterally symmetrical, and presented to the pollinator with the petal lobes bent backwards and the corolla tube open for easy access. To prevent theft of this large supply of easily accessible nectar by other animals, it is common for moth-pollinated flowers to close up or fold over during the day, and only open at night. Alternatively, the nectar may be presented in a tube so long that it is inaccessible to other animals. Moth proboscises are longer than those of any other insect, and moth-pollinated flowers often have much longer nectar spurs than any other flower type. The extreme example of this relationship is the story of the Madagascan orchid (*Angraecum sesquipedale*, Asparagales), which has a nectar spur up to 30 cm in length (see Fig. 14.3e). In 1862, on seeing one of these plants in flower at the Royal Botanic Gardens, Kew, Darwin predicted that there must exist a moth with a proboscis long enough to reach the nectar at the bottom of the spur. He was proved correct in 1903 with the identification of the hawkmoth (*Xanthopan morganii* ssp. *praedicta*), which has a tongue of similar length to the nectar spur, and has been shown to pollinate the flower (Wasserthal 1997). Colour vision is often irrelevant to moths, particularly night-flying species, and moth-pollinated flowers are often white or cream. This allows them to stand out against the dark vegetation at night. However, moths do have a good sense of smell, and it is usual for moth-pollinated flowers to release a strong scent, often in the evening. Classic examples of moth-pollinated flowers include gardenia (Gentianales) and some honeysuckles (Dipsacales).

14.5.6 Bird pollination (ornithophily)

Since the first pollinating animals were insects, the later emergence of pollinating vertebrates meant that these animals had access to flowers already adapted for pollination by another animal. There are sufficient similarities between butterfly- and bird-pollinated flowers to suggest that the early pollinating birds fed from flowers which had evolved with butterfly pollinators. There is speculation as to how bird pollination first arose, with some authors suggesting that birds hunting nectar-gathering insects in flowers accidentally discovered nectar. It is certainly true that even hummingbirds, which take almost all of their energy from nectar, still eat the occasional insect to meet their protein requirements. The classic pollinating bird is the hummingbird, which is found only in the Americas, but a variety of other groups also feed on nectar, including the African sunbirds, Australian lorikeets, and American honey-creepers. Different types of birds have different feeding strategies. For example, hummingbirds hover beneath or in front of flowers while feeding, so hummingbird-pollinated flowers tend to be either pendant, like a fuchsia (Myrtales) (see Fig. 14.4a), or stand out with free space in front. Sunbirds, on the other hand, perch while feeding, so sunbird-pollinated flowers have a perch with the nectary facing towards it (see Fig. 14.4b). Bird-pollinated flowers are usually either brush- or tube-shaped, and the nectar is secreted into spurs, which are usually shorter and wider than those on butterfly-pollinated flowers. They must also be quite tough, as a beak is both stronger and harder than a butterfly's tongue. One of the key features of a bird-pollinated flower is the quantity of nectar secreted. Birds are larger animals than the invertebrates discussed previously, and their energy requirements can be very high indeed. They feed from flowers that produce very large quantities of quite concentrated nectar, so much so that it will actually drip from the flowers at certain times of the year. Birds have good colour vision (see Chapter 20), and the flowers they pollinate are usually red, often with contrasting yellow marks to act as nectar guides. However, scent is not important in bird pollination. Classic examples of bird-pollinated flowers include red columbine (Ranunculales), poinsettia and passion flower (both Malpighiales), eucalyptus (Myrtales), and hibiscus (Malvales).

14.5.7 Bat pollination (chiropterophily)

A quarter of all bat species use flowers for food to some extent, and a few species rely on flowers for all of their nutritional requirements. There are even some crop plants that are dependent on bat pollination, such as durian (Malvales; Bumrungsri *et al.* 2009). Bats are large, heavy animals which sometimes land on the flowers they feed from. Those flowers tend to be large and robust, and usually saucer-shaped for ease of lapping nectar. Because

Figure 14.4 Vertebrate-pollinated flowers. (a) The pendant form of *Fuchsia* flowers (Myrtales) is ideal for hovering hummingbirds. (b) Bird of paradise (*Strelitzia regina*, Zingiberales) flowers provide a sturdy landing platform for non-hovering birds. Photograph kindly supplied by Cambridge University Botanic Garden. (c) The flowers of *Strongylodon macrobotrys*, the jade vine (Fabales), hang far below the foliage, making them readily accessible to bats. See also Plate 8.

bats must find the flowers at night, they are not usually within the foliage, but either hang below it for easy access (see Fig. 14.4c) or actually develop on the trunk of the plant itself. Bat-pollinated flowers produce more nectar than any other flower type, with as much as 15 ml sometimes recorded from a single flower. Those bat species that have become entirely dependent on flowers eat pollen as their only protein source. The flowers they visit usually have greatly enlarged anthers, and sometimes a very great number of them. The flowers of some bat-pollinated species commonly develop over 2000 anthers per flower. Those anthers may only open at night, to prevent pollen robbing by beetles. Since bats are nocturnal, the flowers themselves may sometimes only open at night and may only last one night before senescing. Bats are colour-blind, so flower colour is irrelevant in attracting them. Bat-pollinated flowers are usually white to cream or sometimes a greenish pink colour. The main attractant used by flowers to attract bats is scent. Bat-pollinated flowers generate a very strong scent, often containing butyric acid. Bats that are specialized to feed on nectar often have a nasal cavity larger than that of insectivorous bats, suggesting that scent is more important to flower bats than to insectivorous bats. In contrast, the sonar apparatus is sometimes reduced, as little time is spent hunting insects. Flower bats have longer tongues and narrower snouts than insectivores, and the tongue often has papillae on the end for lapping up the nectar. Classic examples of bat-pollinated flowers include many cacti (Caryophyllales), as well as members of the Bignoniaceae (trumpet creepers, Lamiales) and the Bombacaceae (Malvales).

14.5.8 Deceit pollination

The term 'deceit pollination' covers such an enormous range of relationships that it would be impossible to describe them all in this section. Instead we shall consider three examples of deceit pollination, each involving increasingly specialized flowers that appear to have evolved very specifically with particular pollinating animals. Although the relationships considered in this section are more complex than the basic pollination syndromes described earlier, they still represent recognizable syndromes with clearly apparent matches between floral morphology and animal behaviour.

The simplest form of deceit pollination is a version of Batesian mimicry, a phenomenon that occurs when one organism mimics another and has particular features of the model attributed to it, not necessarily correctly. For example, the yellow and black banding of hoverflies is a form of Batesian mimicry, causing predatory birds to avoid the hoverfly in case it has the same poisonous sting as the wasps and bees that it mimics. A similar situation exists where species of plants mimic the flowers of other species, but do not provide a reward. This sort of pollination system requires the mimic to produce flowers of the appropriate colour, shape, and scent to match the model, but that do not produce nectar. The match does not have to be perfect, as the visual acuity of many pollinators is insufficient to distinguish between generally similar flowers. For instance, in the Asparagales the rewardless orchid *Orchis israelitica* has been shown to function as a mimic of the nectar-supplying *Bellevalia flexuosa*, even though the similarity between them is only at the very general level of inflorescence architecture, approximate floral shape, and white colour (Galizia *et al.* 2005). Another orchid species, *Dactylorhiza sambucina*, produces no reward and mimics a range of rewarding species. Within Europe there are two colour morphs of this species, yellow and magenta, and recent reports indicate that negative frequency-dependent selection maintains this colour polymorphism, as naïve bees preferentially visit the rarer morph, having not yet learned to associate it with a lack of reward (Gigord *et al.* 2001). This form of mimicry occurs where a large number of flowering plants of different species grow in a defined area and flower at similar times. The flowers mimicked may be pollinated by any of the common pollinating animals, so the mimic may develop all the features (except the reward) of any of the standard pollination systems. This sort of pollination by deceit saves the plant energy, but it can only be successful when the mimic is at low frequencies in the community; otherwise the pollinators will abandon both the mimic and the genuine flowers. It is most successful when the effects of direct competition are reduced. For instance, Internicola and Harder (2012) showed that the non-rewarding orchid *Calypso bulbosa* sets most seed early in the season, while the pollinators are still relatively naïve and before the majority of the nearby rewarding plants are in flower. This form of deceit pollination is thought to evolve primarily by loss of nectar from a rewarding flower with appropriate floral morphology, rather than by morphological evolution of an already nectarless flower to match nearby rewarding species.

A less diffuse system of deceit pollination is called *sapromyophily*. Sapromyophilous flowers attract carrion and dung flies and carrion and dung beetles, by mimicking the appearance and scent of a piece of rotting flesh or of carnivore or herbivore faeces. The insects are attracted to the flower to feed or lay their eggs, and transfer pollen in the process. The flower provides no reward to the insect, and may in fact reduce its fitness by causing eggs to be laid on a structure which will not provide sustenance to any larvae that hatch. A range of different colours and scents are used by sapromyophilous flowers, but they are commonly dark red, brown, or blue-tinged. They may have a surface covering of fine hairs, which is believed to enhance the visual mimicry of dead flesh. Most flowers that attract egg-laying insects also release a strong scent. A recent report of the components of the scents of a range of sapromyophilous flowers in the family Apocynaceae (Gentianales) observed that different species released different mixtures of hexanoic acid, carboxylic acids, pyrazines, heptanal, octanal, dimethyl oligosulphides, indole, and cresol, and that the mix of compounds released gave specific corpse, urine, or dung odours (Jürgens *et al.* 2006). Remarkably, the stinkhorn fungus, *Clathrus archeri*, has convergently evolved a similar set of volatiles, which attract flies for spore dispersal (Johnson and Jürgens 2010). To facilitate the dispersal of scent, some flowers are thermogenic, generating heat through a salicylic acid-activated signal transduction cascade. The fine hairs present on many such flowers might also enhance scent production and dispersal. There are many examples of sapromyophilous flowers, particularly in the orchid family, but the most famous example is the world's largest inflorescence, the titan arum (*Amorphophallus titanium*, Alismatales) (see Fig. 14.5a).

The most specialized system of deceit pollination is that used by flowers which mimic the female of a species of insect, and invite the male to attempt to mate with them. The colours and scents produced by these mimicking flowers are highly species specific, and may include the release of compounds that

Figure 14.5 Floral mimicry. (a) The titan arum (*Amorphophallus titanium*, Alismatales) attracts pollinators by releasing a strong scent reminiscent of rotting flesh. Image kindly provided by Cambridge University Botanic Garden. (b) *Ophrys episcopalis* (Asparagales), which mimics female insects to achieve pollination through pseudocopulation. Image kindly provided by Richard Bateman (Royal Botanic Garden, Kew). (c) The composite inflorescence of *Gorteria diffusa* (Asterales) mimics its pollinating flies. See also Plate 9.

mimic the insect's pheromones. Colours usually involve yellows, purples, blues, and browns, and may be distributed on the flower to mimic the appearance of the wing cases or thorax of the female. No reward is provided to the pollinating animal, but because this kind of mimicry releases instinctive behaviour, the animal cannot learn to avoid these flowers, and so they are not as frequency-limited as food deceivers. A classic example of a flower pollinated in this way is the fly orchid (*Ophrys insectifera*), which mimics the female of the scoliid wasp (*Campsoscolia ciliata*). Although the visual similarity between the flower and the wasp is very striking, it is the scents produced by the flower that are most astonishing. The orchid releases a range of very specific volatiles that closely mimic the sex pheromones of the female wasp. Some of these, such as (omega-1)-hydroxy acids and (omega-1)-oxo acids, are not found elsewhere in the plant kingdom (Ayasse *et al.* 2003). A number of related species produce similarly impressive mimics (reviewed by Schlüter and Schiestl 2008; see Fig. 14.5b).

A less extreme but nonetheless remarkable example is the South African 'beetle daisy' (*Gorteria diffusa*, Asterales). The composite inflorescence of these daisies contains bright orange ray florets, several of which develop large black spots (see Fig. 14.5c). The spots are composed of both intense pigmentation and specialized cells that add three-dimensionality and sheen. They attract male bee flies (*Megapalpus capensis*) by mimicking the female flies that often spend the night in the closed inflorescences (Johnson and Midgely 1997), and by inducing aggregating and mating behaviours (Ellis and Johnson 2010). These are perhaps some of the most extreme examples of coevolution between plant and pollinator, and certainly give much credit to the idea that floral features have evolved in response to pollinator preferences.

The different pollination syndromes discussed in this chapter vary from the extremely diffuse (fly pollination) to the very specialized (some of the deceit pollination examples). Although there is little doubt that some specialized flowers have evolved traits that make them more attractive to particular animals, there is a great deal of debate in the literature as to whether the majority of 'ordinary' flowers can really be said to show pollination syndromes. In the next few chapters we shall discuss the mechanisms by which the great diversity of shapes and colours is generated, before returning to consider the adaptive significance of these features in Section IIIB.

CHAPTER 15

Diverse floral shape and structure

This section of the book is focused on the changes that can occur to floral form, and on the effects of those changes on pollinator behaviour. In Chapter 10 we discussed the ABC model of flower development, and the molecular changes that occur to generate the correct arrangements of the four whorls of floral organs—sepals, petals, stamens, and carpels. In Chapter 14 we considered the traditional idea that different pollinators might have preferences for different shapes, structures, and sizes of flowers. In later chapters we shall investigate the evidence in support of this hypothesis, along with the evidence that such features are labile in evolutionary time, but before that we must consider the molecular and developmental processes which lead to diversity of floral form. In this chapter, then, we start with a basic flower, such as the flower of Arabidopsis, and consider the ways in which its form can be altered by known developmental programmes. We shall focus on the petals, or corolla, as the whorl of organs of primary importance in attracting potential pollinators, and consider changes to its size, its symmetry, the shape of its component petals, and its position within an inflorescence.

15.1 Controlling corolla size

As we saw in Chapter 10, a petal develops from a primordium of dividing cells, which arises on the floral meristem in response to the activity of organ identity genes. The final size of the petals, and thus the corolla they form, is a factor of the combined amount of cell division and cell expansion that occurs throughout the development of each organ. Only these processes can influence plant organ size. In animals, cell death and cell migration may also play roles, but plant cells do not migrate, and

programmed cell death in plants results only in dead structures (such as xylem vessels), as there is no mechanism for removing dead cells (Meyerowitz 1996). Cell division is the process by which a single cell replicates its DNA and then divides into two daughter cells, partitioning the DNA equally between the two daughters. Cell expansion is the process by which post-mitotic cells increase in size, largely through vacuolation and associated cell wall loosening. Earlier cell growth is driven by an increase in cytoplasmic mass, but the vast majority of increase in cell size is due to post-mitotic expansion through vacuolar expansion (Sugimoto-Shirasu and Roberts 2003). In theory, then, it would seem relatively easy to change final petal size. An increase in cell division rate should result in more cells, increasing petal size, while a decrease in cell division rate should have the opposite effect. Similarly, an increase in the extent to which cells expand should increase overall organ size, while a decrease should result in smaller cells and a smaller organ.

The extent to which the control of organ size is both robust and delicate can be seen from the astonishing degree to which petal shape and size are uniform within a species, but have the potential to be extraordinarily variable between species. Analysis of patterns of cell division and expansion within a single petal has shown that the processes are controlled in a coordinated way. Using markers for cell cycle and electron microscopical techniques to follow cell expansion, Reale et al. (2002) described the patterns of cell behaviour in a developing petal of Petunia hybrida (Solanales). They found that cell division was initially distributed uniformly throughout the tissue. However, the decline of cell division was not uniform, beginning first at the base of the petal and then spreading to form a gradient from base to tip. Cell expansion was coordinated with decline of cell division, beginning first in the basal part of the petal and then spreading throughout the whole tissue. The complexity of this developmental pattern confirms that control mechanisms must be at work to ensure appropriate final size of the organ.

An increasing body of evidence suggests that the control of organ size is indeed a rather complex process. In particular, it is becoming apparent that cell division and cell expansion are necessary to organ growth, but that the rates of these two processes alone do not control the final organ size. Instead, there appear to be mechanisms present that monitor organ growth and that balance the number and size of cells to achieve the correct end point (Shpak *et al.* 2003). A number of studies have now shown that altering either cell division or cell expansion through the use of transgenes does not necessarily lead to a change in organ size, because the other process adapts to compensate. For example, ectopic expression of the auxin-binding protein 1 (ABP1) from Arabidopsis in tobacco (Solanales) causes an increase in cell expansion. However, cell division is reduced and the organs therefore end up the same size as wild type, but are composed of fewer, larger cells (Jones *et al.* 1998). Similarly, ectopic expression of *CycD3;1*, a cyclin gene that induces progress through the cell cycle, causes an increase in cell number in Arabidopsis but no overall change in organ size. Instead, organs are found to be composed of numerous but very small and poorly differentiated cells (Dewitte *et al.* 2003). Studies such as these suggest that an intrinsic mechanism coordinates cell division and cell expansion, to control overall organ growth.

If the rate of cell division is not a factor in overall organ size, it is becoming clear that the duration of the phase of cell division is. Analysis of the *AINTEGUMENTA* (*ANT*) gene of Arabidopsis suggests that it might function as a checkpoint through which control of organ size is exerted. Unlike the examples described above, changes in ANT function do result in changes in overall organ size. Loss of ANT function results in leaves and petals smaller than wild type, while increased *ANT* expression in transgenic plants increases the sizes of all organs (Mizukami and Fischer 2000). These changes

result from changes in total cell number, which are not compensated for by changes in cell size. ANT appears to control cell number not by affecting the rate of cell division, but by regulating the duration of cell division in a developing organ. ANT sustains the expression of D-type cyclins, which are necessary for progress through the cell cycle, increasing the window of time during which cell division can occur. The ANT protein is a transcription factor with similar domains to APETALA2, the A function protein we met in Chapters 9 and 10, and it may act by directly maintaining transcription of these cyclin genes (Mizukami 2001). It has been shown that an auxin-inducible gene, *ARGOS*, prolongs the expression of *ANT*, and therefore acts upstream of it to maintain this competence for division (sometimes called meristematic competence) of the cells within developing organs (Hu *et al.* 2003). The involvement of auxin in this process suggests that larger-scale forces are at work, and that overall organ size is likely to represent a balance between a variety of environmental signals integrated at the point of controlling cell division and expansion. Indeed, *ANT* expression is inhibited by the auxin-responsive transcriptional repressor AUXIN RESPONSE FACTOR 2 (ARF2), which in turn can be inactivated by brassinosteroid signalling (Schruff *et al.* 2006; Vert *et al.* 2008). Interestingly, the *ARGOS-LIKE* (*ARL*) gene has been shown to respond to brassinosteroid application, and maintains not cell division but cell expansion in developing tissues. Transgenic plants expressing reduced levels of *ARL* through RNAi inactivation have smaller lateral organs than wild type, including petals (Hu *et al.* 2006). *ARL* has significant sequence similarity to *ARGOS*, both encoding ER-localized proteins with putative transmembrane domains, and it is likely that this protein family is essential to the control of organ size through the integration of a variety of signals.

An unrelated protein, BIG BROTHER (BB), has been shown to control maximum organ size by targeting the proteins necessary for cell division for degradation, and thus controlling the length of time during which cell division can occur. Plants mutant at the *BB* locus have much larger floral organs than wild type, and transgenic plants ectopically expressing *BB* have much smaller organs (see Fig. 15.1). The protein regulates the duration of the phase of

BB ox   WT   bb-1

Figure 15.1 The *bb* mutant (right) has larger floral organs than wild type, while plants overexpressing *BB* (left) have smaller floral organs. Images kindly provided by Michael Lenhard (University of Potsdam).

cell division and proliferation, not the rate or intensity of that proliferation, and affects final cell number, not cell size. *BB* encodes an E3 ubiquitin ligase, a protein that binds other proteins to ubiquitin conjugating enzymes, preparing them for degradation. These data strongly suggest that maximum organ size is regulated by preventing further cell division through degradation of key proteins necessary for the cell cycle (Disch *et al.* 2006).

The importance of the duration of the cell division phase is shown by a third example, the *kluh* mutant of Arabidopsis, which has small petals and leaves as a result of premature arrest of the period of cell division in the different organs. *KLUH* encodes a cytochrome P450 monooxygenase (CYP78A5) that induces a mobile signal to maintain cell division in developing organs (Anastasiou *et al.* 2007). This signal appears to integrate organ growth not only within a flower but also across a whole inflorescence, and therefore represents a mechanism by which flower size might be held constant within an individual plant (Eriksson *et al.* 2010).

Although a majority of the factors regulating petal size have been shown to work through control of the phase of cell division, it is also apparent that regulation of cell expansion can influence organ growth. One of two possible transcripts of the *BIG PETAL* gene, *BPEp*, has been shown to limit the extent of cell expansion in the Arabidopsis petal, limiting final organ size (Szecsi *et al.* 2006). *BPEp* encodes a bHLH transcription factor, presumably regulating genes involved in cell wall loosening and turgor. Its expression is in turn positively regulated by jasmonate signalling, implicating a third plant

growth regulator in the complex network of signals that regulate petal size (Brioudes *et al.* 2009).

Changing petal size may not be as simple a matter as we initially thought, but it has clearly happened many thousands of times throughout the evolutionary history of the angiosperms, to create the enormous range of petal sizes we see today. Understanding the control mechanism that coordinates cell division and cell expansion will be necessary for us to understand how petal size can be adapted in complex flowers.

15.2 Controlling corolla symmetry

The Arabidopsis flower is radially symmetrical, as are the flowers of many common species such as buttercup (Ranunculales), rose (Rosales), lily, and tulip (both Liliales) (see Fig. 15.2a). It is possible to draw many lines of symmetry on the face of these flowers. A radially symmetrical flower (also called an actinomorphic or polysymmetric flower) is the default form that develops through the activities of the ABC genes, as they direct identical development in all regions of each concentric whorl on the floral meristem. However, many plants produce flowers that are bilaterally symmetrical (also called zygomorphic or monosymmetric flowers). It is only possible to draw a single line of symmetry on the face of these flowers, as not all organs within a whorl are identical (see Fig. 15.2b). The changes in shape that generate zygomorphy may be to any of the four types of floral organ. Zygomorphy of the reproductive organs can influence pollen placement on or capture from pollinating animals, and can also

Figure 15.2 Zygomorphy and actinomorphy. (a) Many flowers are radially symmetrical, or actinomorphic. (b) The flowers of *Antirrhinum* species are bilaterally symmetrical, or zygomorphic. See also Plate 10.

affect the extent of within-flower self-pollination. Important as these forms of zygomorphy may be, the most striking zygomorphic flowers have alterations to either the petals or to petaloid sepals. Classic examples of zygomorphic flowers include Antirrhinum, sweet peas (Fabales), and many orchid species (Asparagales). A consequence of zygomorphy is that flowers come to have a clear top and bottom, and so are also said to have dorsoventral symmetry, or dorsoventrality. One advantage of dorsoventrality is that it confers position on each organ within the flower, allowing organs in different positions to adopt different morphologies. Organs can be identified and described as the dorsal (top) petals, lateral (side) sepals, ventral (bottom) stamens, and so on. When each developing organ has a specific position then it is possible to modify the development of individual organs so that the flower produced has

a clear structure, such as the ventral keel on sweet pea flowers.

The key to producing a zygomorphic flower is to supply it with a signal that marks a region of the developing floral meristem as top or bottom. It is not possible to study this type of process in a species like Arabidopsis, with actinomorphic flowers, but a considerable amount of work has been done on this problem in Antirrhinum. The Antirrhinum flower is zygomorphic with respect to both petals and stamens. The flowers have five petals—two large dorsal ones, two smaller lateral ones, and a single ventral petal which has a distinctive bend in it, creating the lip or hinge that acts as the landing platform for pollinating bees. There are also five stamen primordia, but the single dorsal stamen aborts, leaving two lateral stamens and two longer, hairier ventral stamens.

Analysis of the generation and maintenance of zygomorphy in Antirrhinum has been primarily concerned with four mutant lines—*cycloidea* (*cyc*), *dichotoma* (*dich*), *radialis* (*rad*), and *divaricata* (*div*). The *cyc* mutant has an incomplete conversion from zygomorphy to actinomorphy. The lateral petals of the mutant are converted into ventral petals, and the dorsal petals are morphologically somewhere between dorsal and lateral petals. The ventral petal is unaffected, giving a flower with three ventral petals and two dorsal/lateral petals. This incomplete loss of zygomorphy results in a flower described as semi-peloric, based on Linnaeus's description of fully actinomorphic mutants of *Linaria vulgaris* as peloric (or monstrous). The *dich* mutant produces flowers with a similar phenotype to those of *cyc*, but slightly less dramatic in their conversion. However, if the two mutant lines are crossed together to create a *cyc/dich* double mutant then the flowers produced are fully peloric or completely radially symmetrical, with all the petals converted to the ventral form (Luo *et al.* 1996).

The key to understanding these mutants is the ventralization of organs in the mutant form. This led to the conclusion that *CYC* and *DICH* provide dorsalizing signals to the developing floral meristem. In fact, both the *CYC* and *DICH* genes are expressed in overlapping domains in the dorsal portion of the floral bud, and remain so expressed throughout flower development. *CYC* is expressed in the dorsal petals and dorsal stamen primordium, while *DICH* is expressed in a slightly narrower range encompassing the dorsal stamen primordium and the dorsal part of the dorsal petals. *CYC* and *DICH* are paralogues, created by a gene duplication event within the Plantaginaceae (Gübitz *et al.* 2003), and both encode TCP family transcription factors. Both proteins seem to act early in development to limit growth in the dorsal region of the meristem, ensuring that the dorsal organ primordia stay within the region they define. The mutant floral meristem starts to differ from wild type at the time that sepal primordia become visible, and this restriction of dorsal primordium growth culminates in the abortion of the dorsal stamen. Later in flower development, CYC and DICH promote petal lobe growth so that the dorsal petals end up larger than the others. The result of this expression pattern is that every organ primordium in a developing Antirrhinum flower has a unique combination of expression of *CYC*, *DICH*, and the organ identity (ABC) genes. The only primordia that have the same patterns are those which are mirror images of each other, and this provides the bilateral symmetry to the flower (Luo *et al.* 1996).

DIVARICATA, by contrast, acts to provide a ventralizing signal to the floral meristem, and the *div* mutant has an abnormal ventral petal with little or no hinge and lip region. *DIV* encodes a transcription factor from the MYB family, and is expressed throughout the developing flower (Galego and Almeida 2002). The explanation for the ability of *DIV* to act only in the ventral and lateral petals while expressed throughout the flower comes from an analysis of the role of *RADIALIS* (*RAD*). *RAD* also encodes a MYB transcription factor, but one from a different subfamily to DIV. Most plant MYB proteins contain two repeats of the MYB domain, and are classed as R2R3 MYBs. The RAD protein contains a single MYB domain, and is therefore classed as a one-repeat MYB. There is some evidence to suggest that the MYB domain in one-repeat MYBs is not able to bind DNA. *RAD* is expressed in the dorsal regions of the developing flower, in a spatial and temporal pattern which suggests that it is activated by CYC and DICH (Corley *et al.* 2005). This activation is confirmed by the reduction of *RAD* expression in a *cyc* mutant, and its complete absence

from the *cyc/dich* double mutant. Corley *et al.* (2005) suggested that RAD acts as a repressor of transcription, and proposed that it competes with DIV in the dorsal part of the flower, inhibiting DIV binding to promoters of target genes and thus preventing any ventralizing activity. Zygomorphy in Antirrhinum can therefore be seen as a result of the activities of two TCP proteins and two MYB proteins. The TCP proteins and one MYB protein provide dorsalizing signals, while the other MYB protein provides antagonistic ventralizing signals to the floral meristem.

A rarer form of symmetry seen in the flowers of some species is left–right asymmetry, most commonly seen in the reproductive organs. Asymmetrical female reproductive organs result in a condition known as enantiostyly, which occurs in at least 12 families of angiosperms (Endress 2001; Jesson and Barrett 2005; see Fig. 15.3). This curving of the stigma and style to one side of the flower occurs mainly in flowers that are buzz-pollinated (pollinated through vibratile pollen release by pollen-gathering bees), and it may be that the curvature reduces damage to the style when large animals land on and grasp the stamens. Jesson and Barrett (2002) showed that enantiostyly increased outcrossing rate, even where both right- and left-styled flowers are present on the same individual plant, presumably by decreasing the frequency with

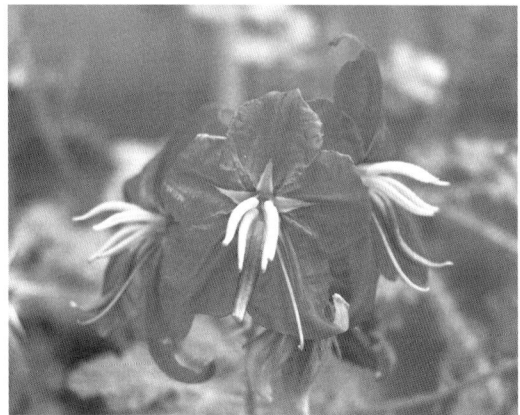

Figure 15.3 The enantiostylous flower of *Solanum heterodoxum* (Solanales). The style is the filamentous pink organ. Image kindly provided by Sandy Knapp (Natural History Museum, London). See also Plate 11.

which self pollen is passed to the stigma. They further showed that enantiostylous flowers set more seed than straight-styled flowers when pollen availability was a limiting factor on female reproductive ouput (Jesson and Barrett 2005). In most described cases, both left- and right-handed flowers are found on the same individual, suggesting an essentially random process. This situation is known as monomorphic enantiostyly. However, in three families, left- and right-handedness is separated on different individual plants (Barrett *et al.* 2000), a situation known as dimorphic enantiostyly. In one of these cases, *Heteranthera multiflora* (Commelinales), it has been shown that style direction segregates in a Mendelian fashion, with the right-styled form dominant over the left (Jesson and Barrett 2002). Little is yet known about the developmental control of this form of asymmetry.

A little understood form of floral symmetry is the development of petals contorted to the left or the right, such that each petal is asymmetric but the whole corolla has rotational symmetry. Such contortions are most prominent in the floral bud. It is possible for contort flowers to show fixed behaviour, where only the right- or left-handed form is present in a species. Alternatively, an unfixed pattern where left- and right-handedness is shown by an individual plant is also possible. Within the rosids, most contort species show an unfixed pattern. However, in the other subclass of eudicots, the asterids, a fixed pattern is usually shown at the species or higher taxonomic level. Currently nothing is known about the development of this form of asymmetry, and there is no explanation as to why the rosids and the asterids should show such different behaviour (Endress 2001).

15.3 Controlling petal shape

The shape of a petal is a result of the distribution of cell divisions and the distribution and direction of cell expansion which occur during its development. As such, the development of petal shape is an integral part of the development of petal size, controlled by the coordinating forces described earlier in this chapter. The development of petal shape has been most successfully studied using clonal analysis and modelling. Rolland-Lagan *et al.* (2003) used

a line of Antirrhinum with a temperature-sensitive transposon inserted in the *PALLIDA* gene. Flowers that are *pal* mutant are white. Transposon excision was induced by exposure of plants to a 15°C growth regime, resulting in clones of red epidermal cells descended from the single cell in which the transposon had excised from the *PAL* locus. Transposon excision was induced at a number of time points in petal development, and the size and shape of the resulting sectors were mapped on to models of petal development. This study revealed that petal growth rate was relatively constant, and that the asymmetric shape of the Antirrhinum petal results from changes in the direction of growth rather than from local differences in the rate of growth. However, petal growth was shown to be anisotropic (stronger in one direction than another), with an average of 15% more cell divisions along the principal axis of growth than along the axis perpendicular to it. The main growth direction was shown to rotate as the petal grew, giving the final asymmetric shape. Models such as this will be useful in describing the development of other petal shapes, and comparisons between species should allow us to dissect the factors responsible for the many different petal shapes seen in modern angiosperms.

Final petal shape is also dependent upon the generation of surface curvature. Curvature of the petal surface is produced by differential growth of the margins relative to the central region of the tissue. In fact, many petals are essentially flat, showing no such curvature. However, the chances of this occurring by chance are very low, implying the existence of genetic mechanisms that regulate growth rate to ensure no curvature is produced (Nath *et al.* 2003). In some species, petal curvature is an important part of corolla appearance, and in these cases it is likely that similar genetic mechanisms regulate relative growth differently to produce the final petal shape. In Antirrhinum, organ curvature is inhibited by the action of the *CINCINNATA* gene. Plants that are *cin* mutants have curved leaves as a result of excessive growth of the margins relative to the central region of the tissue. Further analysis indicated that this was a result of delayed cell expansion/division in the centre of the leaf relative to the margins. The *CIN* gene encodes a transcription factor from the TCP family, to which *CYC* and *DICH* also belong

(Nath *et al.* 2003). Subsequent studies in Arabidopsis have shown that CIN-like proteins interact with the ASYMMETRIC LEAVES 2 (AS2) transcription factor to downregulate expression of class 1 KNOX genes, which play an important role in inducing cell proliferation (Li *et al.* 2012). This suggests that the phenotype of *cin* mutants can be explained by a loss of CIN activity resulting in increased KNOX activity and thus increased cell division in the leaf margins.

Similar processes may be involved in the control of petal curvature, although CIN itself does not seem to operate in the same way in the petal. Plants that are *cin* mutants have smaller petals than wild type flowers, and fewer cells. These data indicate that CIN promotes petal growth by extending the cell division window in the petal lobes (Crawford *et al.* 2004). A different approach to petal curvature was taken by Liang and Mahadevan (2011), who analysed the development of the lily flower (Liliales) from bud stage to fully mature petal using a combination of imaging and mathematical modelling. They concluded that the bud burst and subsequent recurvature of the petals to generate the fully open flower were the result of differential growth across the petal, with excess growth at the petal margins driving the curvature of the mature organs and providing the strain necessary to open the flower.

The petals of some species show distinct architectural features. For example, the ventral petal of Antirrhinum has a distinctive hinge or lip region, which forms the landing platform for pollinating bees. Although all Antirrhinum petals show differences between the basal portions (which fuse to form the corolla tube) and the distal regions (which remain separate as lobes), the division between the two is most marked in the ventral petal. Elaboration of landing platforms is a common developmental process in many animal-pollinated flowers, and one which involves a very clear fold in a petal. A number of genes have been shown to play roles in landing platform development in Antirrhinum. The *div* mutant has a highly disturbed ventral petal, with no hinge formation. This may reflect the loss of the ventralizing signal and the lack of activity of downstream genes which contribute to hinge development. One such gene is the *Antirrhinum majus MYB MIXTA LIKE 1* (*AmMYBML1*) gene, expression of

which is significantly reduced in the *div* mutant flower. *AmMYBML1* encodes a MYB transcription factor expressed in the ventral petal, in the epidermis of both lobe and tube, but also in the adaxial-most mesophyll in the hinge region. Examination of this region indicated that the adaxial-most mesophyll cells expand more than the abaxial ones, causing the bend of the petal to make the hinge. Ectopic expression of *AmMYBML1* in tobacco petals causes the excess expansion of all petal mesophyll cells, supporting the idea that the gene is important in hinge development (Perez-Rodriguez *et al.* 2005). The *CIN* gene is also expressed in the cells of this region, and may act to reinforce other signals causing tissue bending, possibly through its repression of cell division (Crawford *et al.* 2004).

Elaboration of epidermal surfaces may also contribute to final petal appearance. The ventral petal of Antirrhinum contains a number of specialized epidermal cell types, including conical-papillate cells in the lobes, long trichomes on the lip, and twin lines of shorter trichomes leading down into the throat. These specialized cells may act as nectar guides or to trap pollen from visiting bees, and they may also be involved in regulating petal temperature and scent production (Perez-Rodriguez *et al.* 2005).

15.4 Generating a nectar spur

The presence of one or more nectar spurs is another way in which floral form can be varied. Nectar spurs can form on one particular petal (or sepal) of a zygomorphic flower, as is seen in *Linaria vulgaris* (toadflax, Lamiales) and many species of orchids (Asparagales), or on all petals or sepals, as is seen for instance in various species of *Aquilegia* (Ranunculales) (see Fig. 15.4). These spurs usually contain nectar, either secreted from a nectary within the spur or collected as it runs off a nectary elsewhere in the flower. This nectar is only accessible to particular pollinators, and so nectar spurs act to reproductively isolate different floral morphologies through pollinator choice. For example, *Linaria vulgaris* is primarily pollinated by long-tongued bumblebees that can access the base of the nectar spur, although shorter-tongued insect species may rob the flower of nectar by chewing through the nectar spur from the

Figure 15.4 The petals of *Aquilegia formosa* (Ranunculales) are heavily modified to produce nectar spurs. Image kindly provided by Scott Hodges (UCSB). See also Plate 12.

outside (Newman and Thomson 2005). Sometimes spurs do not contain nectar, and their presence acts as a dishonest signal, encouraging pollinator probing and pollen transfer without a reward.

Petal spurs require elaborate outgrowth of a small region of the petal, and their development is not well understood. An actinomorphic mutant of *Linaria* with all petals converted to the ventral form and making spurs was described by Linnaeus and has been reported many times since. This mutant has been shown to be defective in the function of a *CYC*-like gene (through an epigenetic effect), demonstrating that spur development acts downstream of genes that control floral symmetry (Cubas *et al.* 1999). More recently, two mutants of *Antirrhinum majus* which produced ectopic nectar spurs on the ventral petal were identified. Antirrhinum does not normally produce a nectar spur, although many other members of the Antirrhineae do. Both mutants were shown to contain transposon insertions in

genes (*HIRZINA* and *INVAGINATA*) encoding homeodomain transcription factors, similar to SHOOT MERISTEMLESS (STM) of Arabidopsis. STM is required for apical meristem maintenance, and causes cells to undergo division. Ectopic expression of *STM*-like genes in the developing petals of these Antirrhinum mutants resulted in nectar spur development (Golz *et al.* 2002). These data suggest that normal nectar spur development involves the activity of homeodomain transcription factors to cause plentiful cell division in appropriate petal tissue. In further support of this hypothesis, Box *et al.* (2011) isolated orthologues of *HIRZINA* and *INVAGINATA* from spurred *Linaria vulgaris*, and found *HIRZINA* in particular to be strongly expressed in developing petals. Ectopic expression of these *KNOX* genes in tobacco resulted in the formation of ectopic outgrowths of the petal, resembling short nectar spurs. However, the authors also noted that the majority of nectar spur growth in *Linaria vulgaris* was attributable to cell expansion, with only a short period of cell division early in development, suggesting that factors besides *KNOX* genes must also be involved in spur outgrowth. Similarly, Puzey *et al.* (2011) described the growth of nectar spurs in various *Aquilegia* species, and found that anisotropic cell expansion accounted for the measurable spur development, not additional cell division. This finding also supports a role for a regulator of cell expansion, rather than or in addition to *KNOX* genes, in nectar spur development.

15.5 Generating a composite inflorescence

Although modifications to corolla size and shape can have dramatic effects on the appearance of the whole flower, an even more dramatic effect can be achieved by modification of the entire inflorescence, so that it mimics the appearance of a single large flower. This form of floral development is found in at least 17 plant families, but is particularly characteristic of the daisy family (Asteraceae), where the entire inflorescence is condensed into a head-like structure (the capitulum), containing hundreds, or sometimes thousands, of flowers (often referred to as florets). The common daisy is a good example of

Figure 15.5 Composite inflorescences in the daisy family. (a) The capitulum of *Gerbera hybrida*, with zygomorphic outer florets (each with one large petal) and actinomorphic inner florets. (b) The capitulum of *Senecio vulgaris* usually contains only actinomorphic disc florets. See also Plate 13.

this growth form, but other familiar plants with this type of composite inflorescence include dandelions, the cultivated chrysanthemums and *Gerbera* (see Fig. 15.5a), and the groundsels. A composite inflorescence may be composed of a single type of flower, such as the inflorescence of common groundsel (*Senecio vulgaris*; see Fig. 15.5b), or of multiple different morphological forms. Where different floral forms are present they may vary in colour (for example, the yellow and white flowers of the daisy inflorescence), symmetry (daisies have both zygomorphic flowers, which are the outer white ones, and actinomorphic flowers, which are the central yellow ones), and sexuality. Where differences in symmetry occur in a capitulum, the zygomorphic flowers are known as ray florets and the actinomorphic flowers are known as disc florets. It is relatively common for the different flowers in a composite inflorescence to vary in sexuality. For example, *Gerbera hybrida* has hermaphrodite central flowers but female outer flowers.

The development of a composite inflorescence involves a reduction in internode length relative to an ordinary inflorescence. This causes all the flowers to develop closely together, and is associated with a widening of the inflorescence meristem to form the typical head-shaped structure. Flowers are usually produced in a spiral phyllotaxis in members of the Asteraceae, and the developing floral meristems produce some of the clearest examples of spiral patterning seen in plants. It was traditionally thought that the inflorescence meristem of the Asteraceae was indeterminate, with the potential to produce an infinite number of flowers. Most species in fact produce a relatively consistent number of flowers, but this was explained as a function of spatial constraint on the densely packed head. However, recent work has demonstrated that the inflorescence meristem of *Gerbera hybrida* is indeterminate, and requires the activity of a gene called *GRCD2*, encoding a MADS box transcription factor, to confer determinacy and ensure formation of the correct number of flowers. Reduction of *GRCD2* expression levels, using antisense technology, resulted in the continued development of the centre of the inflorescence, producing new flowers long after a wild type inflorescence had stopped and formed a single seed head (Uimari *et al.* 2004).

The flowers of members of the Asteraceae typically lack a whorl of sepals, which would be cumbersome when producing a composite inflorescence that mimics a single flower. Instead, most species have a whorl of fine hairs, known as pappus bristles, around the base of each corolla (seen developing in Fig. 11.5). These bristles are involved in dispersal of the resulting seed, using the wind as the agent of dispersal, and are highly modified sepals, produced through the activities of genes orthologous to those that produce sepals in Arabidopsis (see Chapter 11).

Analysis of the differentiation of the flowers of a composite inflorescence into multiple types has focused on floral symmetry and the activities of *CYC*-like genes. This work is discussed in Chapter 18, where we consider how floral symmetry and the various other features of diverse floral forms change in an evolutionary context.

CHAPTER 16

Colouring the flower

One of the most obvious ways in which flowers differ from one another, more immediately striking even than the differences in size, shape, and symmetry discussed in the previous chapter, is in their colour. This colour is usually the result of pigment deposition in the petals, but in some species brightly coloured bracts, sepals, or stamens can provide a similarly striking display. The colour of a flower has traditionally been viewed as one of the ways in which plants attract pollinating animals, and there is considerable literature on the preferences of different pollinators for different colours. In recent years these data have been reinterpreted, with the current emphasis more on contrast against vegetation and on search image formation, rather than on particular colours as 'favourites' of particular animals. In this chapter we shall consider the different biochemical pathways through which flowers can become coloured, paving the way for a discussion of the regulation, patterning, and enhancement of colour in the next chapter.

16.1 Colour as a signal

Colour is a signal used by plants for a wide variety of purposes. Colour is used to signal ripening of fruit, and to attract animals to eat that fruit and disperse the seeds it contains. Colour can also be used to signify the unpalatibility or toxicity of certain tissues, again often including fruits, particularly unripe fruits. Colour can also be an accident—the orange colour of autumn leaves is usually attributed to the breakdown and reabsorption of green chlorophyll by the plant, leaving behind the orange and brown waste products of the breakdown reactions and the orange and yellow carotenoids that were masked by the green chlorophyll. Similarly,

visible red colouration of leaves and stems is often a side effect of the production of UV-absorbing anthocyanins and flavonoids in response to a range of stresses, particularly high light levels.

The bright colours of flowers are primarily a signal to attract pollinating insects by making the floral tissue stand out against a green background. As we noted in Chapter 14, Darwin himself observed the importance of pigmentation in flowers, and argued that colour is simply a means of providing flowers with contrast against vegetation. This argument is supported by modern analysis of insect visual acuity, which indicates that vegetation is visually very similar to bark, soil, and stone from an insect's point of view (Kevan *et al.* 1996). All of these materials weakly reflect light across the whole range of an insect's visual spectrum. Leaves differ from the rest only in absorbing red light, but since red is at the very periphery of the visual spectrum for most insects, this makes little difference. All flowers, then, whatever their colour, stand out against this generally dull background (Kevan *et al.* 1996). However, it has also been argued that particular colours may act as signals to particular groups of animals. It is certainly the case that the visual ranges of different animals vary somewhat. Insects are able to see into the ultraviolet but usually, with the exception of the Lepidoptera, they are unable to easily detect light at the red end of the spectrum. Birds often have tetrachromatic vision, seeing from the ultraviolet right through to red, while bats have very poor colour vision (see Chapter 20 for a full description of bee and bird visual capability). In these circumstances there is little point in a bat-pollinated flower reflecting UV through specialized surface structures, or in a beetle-pollinated flower investing heavily in anthocyanins (which allow only red to be reflected), and the colours of flowers may indeed provide clues as to the identity of their pollinators. It is also the case that some rather dramatic changes in floral colour are associated with pollination and changing levels of nectar availability. For example, it has been shown that pollination of *Viola cornuta* (Malpighiales) is necessary to cause a change in pigmentation from white to purple, a change which presumably signals a change in nectar availability and floral receptivity to pollinating insects (Farzad *et al.* 2002). There are many reports of similar changes in floral colour, often associated with anthesis or pollination. As another example, Darwin forwarded a letter to *Nature* in 1877 describing the change in colour of a Brazilian species of *Lantana* (Lamiales), which progresses from yellow, through orange to purple, following pollination. Recent reports have confirmed that this change in flower colour serves to direct pollinating butterflies to flowers that are still receptive, while retaining a large coloured inflorescence which is more visible from a distance than a smaller inflorescence would be (Weiss 1991). Even more strikingly, a recent report by Willmer *et al.* (2009) demonstrated that this sort of colour change is reversible in the African legume *Desmodium setigerum* (Fabales). The flowers of this species are receptive for only one day, and change slowly from lilac to turquoise over a period of 24 hours. Mechanical tripping of the keel by a pollinator speeds up the colour change so that the flower becomes white/turquoise within 2–3 hours. However, if little pollen is deposited on the stigma during this pollination event, the flower can revert to lilac in the afternoon, becoming more visible to pollinators and gaining a second opportunity for pollination. When we consider these examples, in which flower colour is tightly controlled and changed by the plant, it is clear that flower colour must be acting as a more specific signal than simply contrast with vegetation. However, analysis of the attractiveness of particular colours to particular animals has resulted in less convincing evidence for the significance of colours as unique signals for individual animals. This issue will be considered in more detail in Chapters 20 and 21.

Biological colours can be produced in two different ways. Some animals, and a few plants, produce 'structural' colours, caused by the reflection of particular wavelengths of light from nanoscale physical structures. However, most plant colour is produced through the synthesis of pigments that absorb subsets of the visible spectrum, reflecting back only what they do not absorb, and causing the tissue to be perceived as the reflected colours. Chlorophyll absorbs light in both the red and blue parts of the spectrum, reflecting only green light, and causes leaves to appear green to us. Similarly, a flower that we perceive as red contains pigments that absorb yellow, green, and blue light, leaving red light as the only wavelength visible to us which is reflected. White tissues reflect all visible wavelengths equally, while black tissues absorb all visible wavelengths (light reflection by white, black, and red flowers is shown in Fig. 16.1). Where an animal has a different set of photoreceptors to us, similar black and white effects can be obtained by absorbing or reflecting light across all the wavelengths visible to that animal.

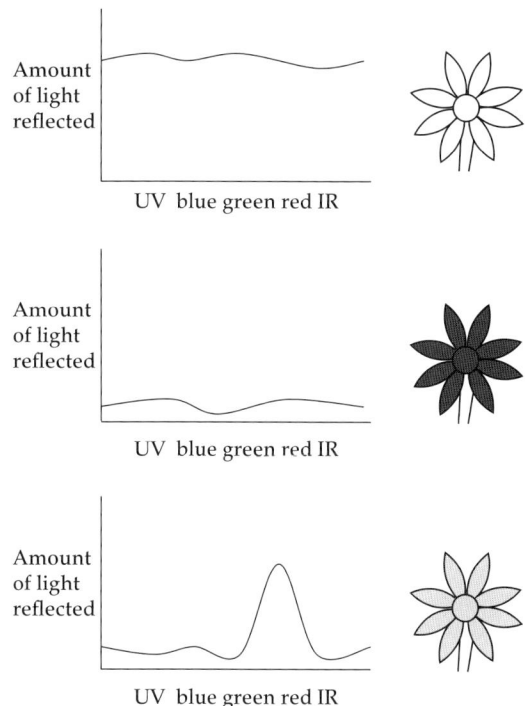

Figure 16.1 Flower colour depends on absorption and reflection of light by pigments. If a flower contains no pigment, and absorbs no light, the reflected light of all colours makes it appear white (top). If the flower absorbs light of all visible wavelengths it appears black. If the flower absorbs all light except red, and reflects the red light, it will appear red (bottom).

16.2 Plant pigments

Whichever organs they are found in, plant pigments can be divided into three chemical classes—the flavonoids, the betalains, and the carotenoids. The flavonoids are the major floral pigments in angiosperms, and different subsets of the flavonoids give rise to ivory and cream colours (flavonols and flavones), yellow and orange colours (aurones and chalcones), and the red-pink-purple-blue range (the anthocyanins). Flavonoids are water soluble and accumulate in cell vacuoles. They also play a range of other important roles in plants, including defence against pathogens and predators, protection against damaging UV light, and participation in pollen development and germination. The betalains are a group of pigments found exclusively in the angiosperm order Caryophyllales, and nowhere else in the plant kingdom (although they are also produced by Basidiomycete fungi). They give the red colour to beetroot, but are also responsible for flower colouration within this order (for example, in pink and red cactus flowers). The carotenoids are much more widespread, although less significant as floral pigments than the anthocyanins. They are lipid-soluble and are found in plastids throughout the plant. Carotenoids give yellow and orange colour to some flowers, although their more important role is as an essential component of the photosynthetic apparatus.

16.3 Carotenoid synthesis

Carotenoids are among the most widespread pigments in the natural world, and have the most varied functions of any group of pigments. In plants they play important roles in photosynthesis, where they act as accessory light-harvesting pigments and as photo-protectants (Hirschberg 1999), as well as acting as floral and fruit pigments. In mammals they are the precursors for vitamin A synthesis, and in fish they are essential for phototropic responses (Goodwin 1980). Animals cannot usually synthesize carotenoids directly, but modify plant-derived carotenoids present in their diets for diverse purposes. For example, the red plumage of flamingos is produced using modified carotenoids (Grotewold

2006). However, a recent report showed that the pea aphid (*Acyrthosiphon pisum*) can, uniquely, produce its own carotenoids because its genome contains genes that encode carotenoid cyclase–carotenoid synthase enzymes and carotenoid desaturases, derived by lateral gene transfer from fungal genomes (Moran and Jarvik 2010).

The carotenoids are a family of isoprenoid derivatives. Isoprenoids themselves are lipid molecules, with an estimated 22,000 different types known. They have essential roles as membrane sterols, components of chlorophyll, cytokinins, abscisic acid, and strigolactones, and a variety of roles in plant secondary metabolism. They are usually classified by the number of carbon atoms they contain. The carotenoid family contains conjugated polyene molecules, composed of 40 carbon atoms. The molecule is produced by the joining of eight 5-carbon isoprenoid units such that the arrangement of the units is reversed at the centre. Carotenoid hydrocarbons are known as carotenes, and include pigments such as zeaxanthin. The addition of oxygen to these molecules creates the oxygenated carotenoids, or xanthophylls, which are also important pigments.

The absorption spectrum of a particular carotenoid molecule is determined by the conjugated polyene system and by additional structural features. Each conjugated double bond increases the wavelength of maximum absorption by 7–35 nm (Goodwin 1980). Thus deep orange flowers may contain lycopene, orange flowers contain β-carotene, and yellow flowers often contain highly oxidized xanthophylls (see Fig. 16.2).

Carotenoids are synthesized within plastids, although the genes encoding the enzymes necessary for their synthesis are located within the nucleus. The enzymes must therefore be transported after translation across the plastid membrane. All isoprenoids, including carotenoids, are produced from a 5-carbon isoprenoid called isopentenyl pyrophosphate (IPP). The major source of plastidic IPP is from pyruvate and glyceraldehyde-3-phosphate (Hirschberg 1999). IPP is converted into dimethylallyl pyrophosphate (DMAPP) by an isomerization reaction catalysed by isopentenyl pyrophosphate isomerase (IPPI). This enzyme has been shown to be the rate-limiting step in carotenoid synthesis in bacteria and yeast (Kajiwara *et al.* 1997).

Figure 16.2 Carotenoids give the yellow and orange colours to (a) *Freesia* (Asparagales), (b) *Gerbera hybrida* (Asterales), and (c) lilies (Liliales). The photo in (b) is a modified version of a photo by Mauro Girotto (Wikimedia Commons). See also Plate 14.

Assuming that IPP and DMAPP are both present, the synthesis of carotenoids proceeds along a regular pathway, which can be followed in Fig. 16.3. Prenyl transferases with different substrate specificities build up a 20-carbon molecule in a succession of steps. One IPP molecule is condensed with one DMAPP molecule to produce a 10-carbon compound known as geranyl pyrophosphate (GPP). The addition of another IPP molecule produces

the 15-carbon farnesyl pyrophosphate (FPP), and a final addition creates the 20-carbon molecule geranylgeranyl pyrophosphate (GGPP). The prenyl transferases involved in these different steps have not yet been well characterized. One reason for this is that carotenoid synthesis occurs in all organs of all plants, but different genes may encode the enzymes that are active in different locations and at different times. It is therefore quite difficult to define the precise enzymes involved in the production of a particular carotenoid in the flowers of a particular species. However, it is believed that different prenyl transferases interact with the different substrates to produce each of GPP, FPP, and GGPP. In Arabidopsis, ten nuclear genes encode GGPP synthase. Of the proteins encoded by these genes, at least two are located in the plastids, at least two in the endoplasmic reticulum, and at least one in the mitochondria. Some of the genes are expressed throughout the plant, one is specific to flowers and root tissue, one is entirely flower-specific, and one is entirely root-specific (Okada *et al.* 2000). When a simple plant with a small genome, such as Arabidopsis, shows such complexity in the activity of only one of these enzymes, it is not surprising that we cannot yet say precisely which genes are active in the petal carotenoid pathway of many species.

Up to this point in the pathway, intermediates may be used to make cytokinins, monoterpenes, sterols, triterpenes and sesquiterpenes, phytol, and gibberellins, as well as carotenoids. The first committed step of carotenoid synthesis is the condensation of two molecules of GGPP to form the 40-carbon molecule phytoene, which is a colourless molecule formed by the head-to-head condensation of the two GGPP molecules. This generates the reversal of the molecular arrangement at the centre characteristic of carotenoids. This coupling reaction, an electrophilic attack on a double bond, is a two-step process and is catalysed by the enzyme phytoene synthase (Chappell 1995). It is considered the bottleneck in carotenoid synthesis, and is the target of attempts to increase carotenoid production through genetic engineering (Farre *et al.* 2011; Ruiz-Sola and Rodriguez-Concepcion 2012). Genes encoding phytoene synthase have been isolated from a number of species, including a single gene copy from Arabidopsis, and share domains encoding prenyl

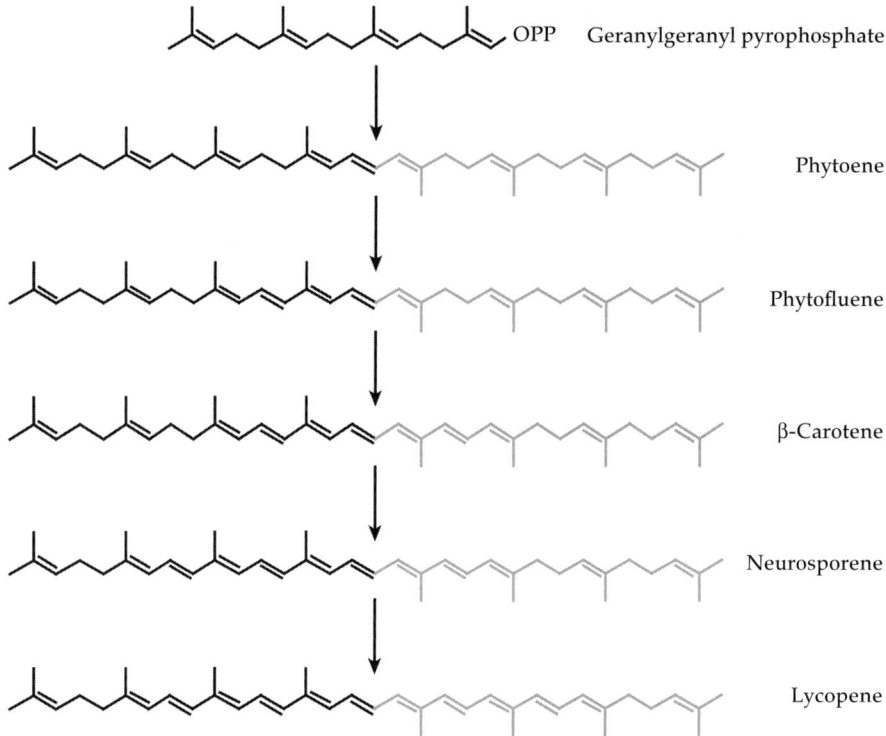

Figure 16.3 The carotenoid synthesis pathway produces lycopene, from which all other carotenoids can be derived.

transferase activity with other enzymes, notably squalene synthases (Chappell 1995).

The conversion of phytoene to lycopene involves two steps, although in bacteria and fungi the conversion uses only one bifunctional enzyme (Moise *et al.* 2005). In plants, phytoene is converted first into a ζ-carotene by phytoene desaturase. Phytoene desaturase uses plastoquinone and then oxygen as electron acceptors during this reaction, and is encoded by genes expressed at constant levels throughout flower development (Al-Babili *et al.* 1996). The enzyme ζ-carotene desaturase (ZDS) converts the ζ-carotene to 7,9,9',7'-tetra-*cis*-lycopene, and then lycopene itself is produced using the enzyme carotenoid isomerase. Mutation in genes encoding carotenoid isomerase results in the accumulation of tetra-*cis*-lycopene and a subsequent shift in the absorption spectrum of the tissue (all reviewed by Farre *et al.* 2011, and in detail by Ruiz-Sola and Rodriguez-Concepcion 2012). The pathway to this point is shown in Fig. 16.3.

Lycopene itself is a carotenoid, and all other carotenoids can be derived from it. The different pathways that give rise to the different carotenoid end products are complex. However, all carotenoids can be formally derived from lycopene by reactions involving hydrogenation, dehydrogenation, cyclization, insertion of oxygen, double-bond migration, methyl migration, chain elongation, and chain shortening (Goodwin 1980). Differential cyclization of one or both ends of lycopene is the branch point that determines which carotenoids are produced by further reactions.

Final flower colour can be attributed as much to the amounts of carotenoid present as to the types of carotenoid involved. Moehs *et al.* (2001) showed that variation in marigold (Asterales) flower colour was correlated with the amount of xanthophylls present, and also with the expression levels of all the carotenoid biosynthetic genes. These data suggest that regulation of carotenoid synthesis occurs primarily through the transcriptional control of the genes encoding the biosynthetic enzymes, perhaps in ways similar to the regulation of anthocyanin synthesis (discussed in Chapter 17).

16.4 Flavonoid synthesis

The flavonoids are a group of phenolic compounds in which two six-carbon rings are linked by a three-carbon unit. All flavonoids absorb in the wavelength ranges 250–270 nm and 330–350 nm, and some in the range 520–550 nm, so that they appear as visible pigmentation to most animals, although the colour that they appear will vary with the animal's ability to perceive reflected ultraviolet light. Flavonoids are water soluble and accumulate in the vacuoles of plant cells. Flavonoids play a number of roles in plant physiology and development. They may be involved in defence against pathogens and predators, they are a component of the legume–*Rhizobium* signalling pathway, they are required for correct pollen development and pollen tube growth, they protect sensitive tissues from ultraviolet radiation, and they act as antioxidants and metal chelators. Within the flower their main role is as pigments and co-pigments, with different classes of flavonoids contributing different colours.

The first committed step of flavonoid synthesis is the condensation of three molecules of malonyl CoA (derived from acetyl CoA) with one molecule of p-coumaroyl CoA (derived from the shikimate pathway and general phenylpropanoid metabolism) to produce chalcone, the key intermediate in the synthesis of all flavonoids (Martin and Gerats 1993). Chalcone itself is usually yellow or orange, and may accumulate as a pigment in its own right (as for example in safflower, *Carthamus tinctorius*, Asterales; Tanaka *et al.* 2008). In a few species it is converted into a yellow aurone, the pigment which provides the yellow colour to some other daisy flowers, such as *Dahlia* and *Cosmos* species (Nakayama 2002), and the yellow throat to the Antirrhinum flower. Aurones are derived from chalcones by hydroxylation of the B-ring moiety and/or oxidative cyclization. In Antirrhinum, both of these reactions are catalysed by aureusidin synthase, which is encoded by a gene with strong similarity to more general plant polyphenol oxidases (Nakayama *et al.* 2000; Sato *et al.* 2001).

Usually, however, chalcone is modified to a colourless flavanone, and flavanone then feeds into one of three further pathways. First, flavanone may be directly converted into flavones by enzymes called flavone synthases. Flavones vary in colour from very pale to bright yellow, depending on their degree of hydroxylation. Flavones may provide some yellow colouration to flowers, but are more usually thought of as co-pigments (see Chapter 17). Alternatively, flavanone can be converted into dihydroflavonol by the enzyme flavanone 3-hydroxylase, a dioxygenase. Dihydroflavonols can be modified by another dioxygenase, flavonol synthase, to form various flavonols. The flavonols are usually colourless, but act as co-pigments, stabilizing and modifying the colour of other pigment molecules (see Chapter 17). They also absorb light strongly in the UV, and so may appear coloured to insects that can see in that range. Thirdly, dihydroflavonols may be converted to anthocyanidins by the activity of two enzymes, dihydroflavonol 4-reductase and anthocyanidin synthase. The unstable products of anthocyanidin synthase may be further modified by glycosylation, methylation, and acetylation to form anthocyanins. Anthocyanins provide a number of different colours, depending on a number of factors (see Fig. 16.4). One factor is the degree of hydroxylation of the B ring. There are thought to be up to 1000 different forms of anthocyanin in nature, although many of them probably occur only rarely (Andersen and Jordheim 2006). The colours that they produce range from orange/brick red (pelargonidins; one hydroxyl on the B ring), to red/magenta (cyanidins; two hydroxyls on the B ring), to purple/blue (delphinidins; 3 hydroxyls on the B ring), with increased blueness determined by an increase in hydroxyl groups. Methylation of an anthocyanin tends to shift its colour towards red, compared with bluer unmethylated molecules. Acylation with aromatic groups shifts the pigments towards the blue range, whereas aliphatic acylation increases the stability of the anthocyanins in solution.

The amount of pigment present in a tissue can also determine the perceived colour. Flavonoid concentrations vary dramatically both between and within species. For example, wild cornflower petals (Asterales) are reported to comprise 0.7% dry weight anthocyanin, while more intensely coloured

Figure 16.4 Anthocyanins give the purple, magenta, and pink colours to (a) *Petunia hybrida* (Solanales, delphinidin and petunidin), (b) *Antirrhinum majus* (Lamiales, cyanidin), and (c) *Pelargonium* (Geraniales, pelargonidin). The photo in (c) is adapted from a photo by Rameshng (Wikimedia Commons). See also Plate 15.

garden varieties accumulate up to 15% dry weight anthocyanin. Similarly, the concentration of anthocyanin in deep purple pansy varieties may be as high as 30% dry weight.

Anthocyanin synthesis has been characterized both from a biochemical perspective and from a molecular one, and the pathway is summarized in Fig. 16.5. The most detailed studies have used the petals of Antirrhinum and petunia (Solanales) as models. Antirrhinum flowers normally produce a magenta anthocyanin, called cyanidin. Petunia flowers produce a purple anthocyanin, delphinidin, as well as petunidin. Since the blueness of the anthocyanin is determined by the degree of hydroxylation of the B ring, in theory a plant can always hydroxylate the molecule less, and thus make less blue anthocyanins, but it will not necessarily have the enzymes to hydroxylate it more, and so cannot make bluer anthocyanins. In fact, investigation of this hypothesis in petunia revealed that the enzymes of the biosynthetic pathway have a greater efficiency for conversion of their usual substrates, and cannot therefore easily produce orange pelargonidins (Mol *et al.* 1998).

The p-coumaroyl CoA for flavonoid biosynthesis comes from general phenylpropanoid metabolism, and is shown in Fig. 16.5. The first enzyme of phenylpropanoid metabolism is phenylalanine ammonia lyase (PAL), which catalyses the conversion of phenylalanine to cinnamate (Martin and Gerats 1993). Cinnamate is converted to a hydroxycinnamic CoA ester (4-coumaroyl CoA) through two steps, and this intermediate feeds into a number of pathways, including those responsible for lignin production and the synthesis of cyanogenic compounds (Martin and Gerats 1993).

The first committed enzyme of flavonoid synthesis, which condenses malonyl CoA with this 4-coumaroyl CoA to produce a yellow chalcone, is chalcone synthase (CHS). The *NIVEA (NIV)* gene encodes CHS in Antirrhinum, and loss of gene function through mutation results in pure white flowers in which no flavonoids are produced at all (Wienand *et al.* 1982). A large number of *nivea* mutant alleles have been described, and it is clear that *NIVEA* function is determined partly by accurate regulation of its expression. Although *NIV* appears to be the only locus encoding chalcone synthase in Antirrhinum, multiple CHS encoding genes are present in many species. Eight CHS loci have been described in petunia (Koes *et al.* 1986), and the potential for redundancy between so many genes may explain the absence of a CHS mutant in this species. Maize has also been shown to contain two loci encoding CHS (Coe *et al.* 1981; Dooner 1983), although the expression patterns of the two transcripts do not appear to overlap.

The next step in the pathway, the isomerization of chalcone into a colourless flavanone by the closure of the C ring, occurs spontaneously at a low rate but is accelerated by the enzyme chalcone isomerase (CHI) (Mol *et al.* 1985). CHI mutants usually accumulate chalcone as well as reduced quantities of anthocyanin, indicating that although CHI may not be strictly necessary for anthocyanin synthesis, it is required to prevent this step of the pathway limiting the rate of pigment production. Petunia has

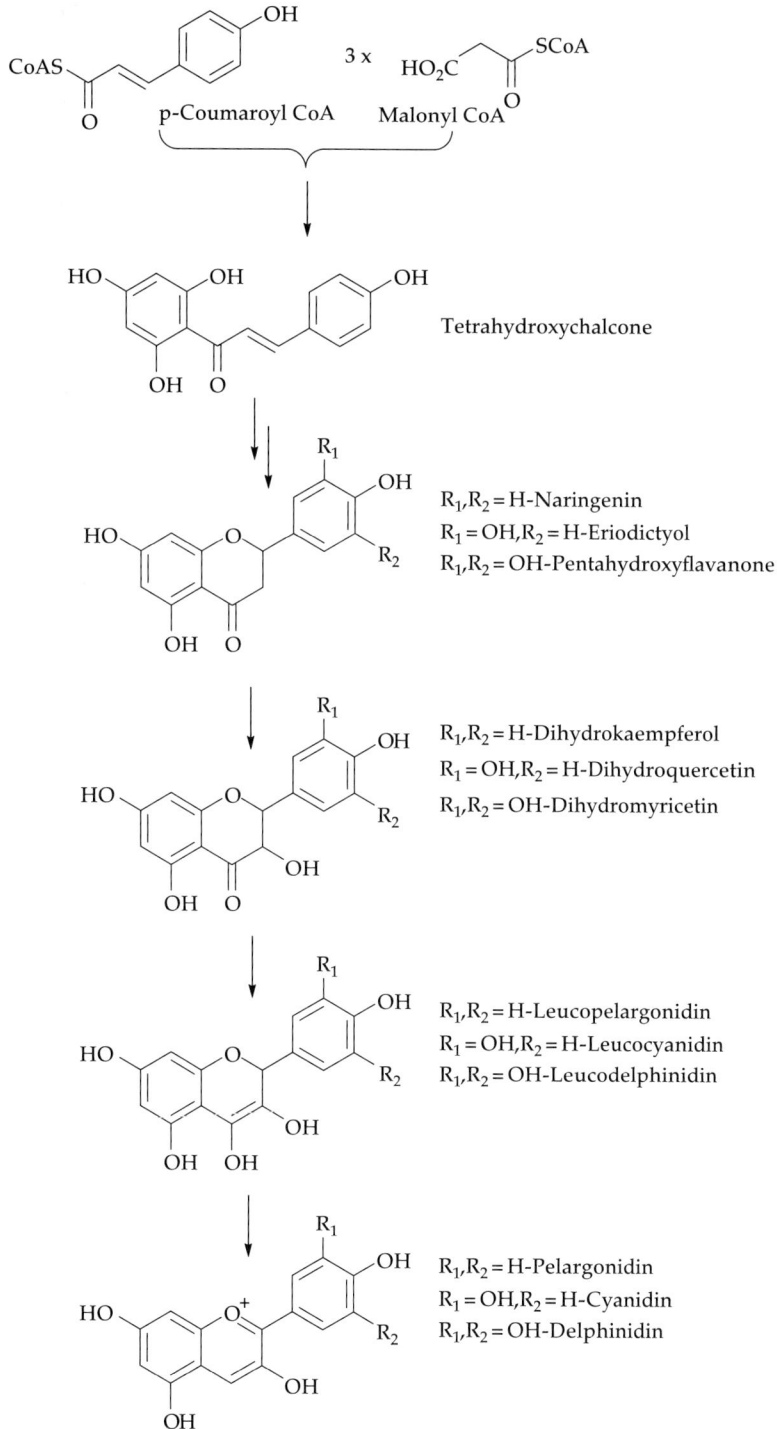

Figure 16.5 The anthocyanin synthetic pathway. The different products depend on the groups found at positions R_1 and R_2.

been shown to contain two transcripts of *CHI*, with different expression patterns (van Tunen *et al.* 1988).

The flavanone produced in the second step is one of the molecules which can be hydroxylated on the B ring to generate bluer end products. The dihydroflavonol produced in the next step (below) is the other intermediate that can act as substrate for extra hydroxylations. In Antirrhinum, flavonoid 3′-hydroxylase (F3′H) activity adds a single hydroxyl group to the B ring, leading to the final accumulation of cyanidin (magenta) as the pathway's end product, instead of pelargonidin (orange/brick red). The *EOSINA (EOS)* gene of Antirrhinum encodes F3′H, and *eos* mutants accumulate pelargonidin as a consequence (Martin and Gerats 1993). Two loci encode F3′H in petunia, each active in different parts of the flower (Brugliera *et al.* 1999). In some plants, including petunia, a related enzyme, flavonoid 3′5′-hydroxylase (F3′5′H), is able to add two hydroxyl groups to the B ring, leading to the final production of the blue anthocyanin, delphinidin. Two genes encode F3′5′H in petunia, and their expression patterns differ, ensuring that both the petal tube and limbs contain the enzyme (Stotz *et al.* 1985; Holton *et al.* 1993). Interestingly, sequence analysis of F3′H and F3′5′H from various species indicates that F3′5′H was derived from F3′H before the divergence of angiosperms and gymnosperms. However, within the daisies (Asteraceae) a separate, and quite recent, origin of F3′5′H has occurred, also by duplication and modification of F3′H (Seitz *et al.* 2006).

In the third step of the anthocyanin pathway, flavanone-3-hydroxylase (F3H) catalyses the hydroxylation of flavanone to produce very palely pigmented dihydroflavonols. These dihydroflavonols may be used as substrates for flavonol production by flavonol synthase or stay within the anthocyanin synthetic pathway. Within the anthocyanin pathway they may also be subject to one or more extra hydroxylations by F3′H and F3′5′H. The *INCOLORATA (INC)* locus of Antirrhinum encodes F3H, and mutations at this locus result in an ivory phenotype. Unusually for a locus encoding an enzyme, the wild type *INC* allele is only semi-dominant. Therefore the heterozygote *INC/inc* produces a reduced amount of anthocyanin relative to the wild type homozygote. This observation suggests that F3H may be a rate-limiting step in anthocyanin production in Antirrhinum (Martin and

Gerats 1993). In petunia the *AN3 (ANTHOCYANIN3)* locus encodes F3H.

The ivory dihydroflavonol produced in step three is reduced by dihydroflavonol 4-reductase (DFR) to give an ivory-coloured leucoanthocyanidin. The *PALLIDA (PAL)* gene of Antirrhinum is responsible for DFR production, and mutation results in ivory-coloured flowers (Martin *et al.* 1985; Coen *et al.* 1986), while the *ANTHOCYANIN6 (AN6)* locus encodes DFR in petunia (Beld *et al.* 1989). It is DFR in petunia that exhibits substrate specificity and that is therefore responsible for the failure of the petal to produce pelargonidin if F3′H and F3′5′H activities are not present. Petunia DFR shows strong substrate specificity for dihydroquercetin (one extra hydroxylation) and dihydromyricetin (two hydroxylations), as opposed to dihydrokaempferol (no extra hydroxylations) (Martin and Gerats 1993).

Leucoanthocyanidin is converted into a coloured anthocyanidin by the action of a leucoanthocyanidin dioxygenase (also called anthocyanidin synthase). The Antirrhinum locus *CANDICA (CANDI)* is required for these conversions, and encodes a protein with strong sequence similarity to other dioxygenases, including INC (Martin *et al.* 1991). Mutations at this locus again result in ivory-coloured flowers. The petunia orthologue of *CANDI, ANT17,* is highly expressed in petals and its transcript is absent from regulatory mutants which do not produce anthocyanin (Weiss *et al.* 1993).

Anthocyanidin is glycosylated to form an anthocyanin by UDP glucose-flavonoid 3-O-glucosyltransferase (UF3GT). The addition of this glucose molecule stabilizes the pigment and enables it to be transported across the vacuolar membrane from the cytosol, where all previous steps in the pathway have occurred. In maize the *BRONZE1 (BZ1)* locus encodes UF3GT, and in *bz1* mutants anthocyanidin accumulates in the cytosol and polymerizes with proteins to form a bronze-coloured complex (Larson and Coe 1977). Recent reports suggest that there might be some species-specific diversity in the order in which glucose molecules are added and in the activities and even origins of the enzymes concerned (reviewed by Tanaka *et al.* 2008).

Anthocyanin is thought to be transported across the vacuolar membrane by a number of methods. The best studied system requires glutathione

S-transferase (GST) as a molecular chaperone for the transporter. The *BRONZE2* (*BZ2*) locus of maize (Poales) encodes GST, and *bz2* mutants accumulate anthocyanin in the cytoplasm, where it is oxidized to a bronze colour (Marrs *et al.* 1995). GST transports anthocyanin to the vacuole via a multidrug resistance-like protein (a type of transporter), encoded by a pair of orthologous genes in maize. The *ZmMRP3* locus encodes the transporter active in vegetative tissue, while *ZmMRP3* encodes a nearly identical protein active in the developing kernels (Goodman *et al.* 2004). In petunia the *ANTHOCYANINLESS 9* (*AN9*) locus also encodes a GST responsible for the final step of floral anthocyanin synthesis. However, plants contain many GSTs, and phylogenetic analysis suggests that *BZ2* and *AN9* evolved independently from one another and from different types of GST. This conclusion is particularly striking in light of the ability of *BZ2* to rescue an *an9* mutant, and vice versa. However, since both genes are transcriptionally regulated by the same proteins, which control the anthocyanin biosynthetic pathway in both species, it is likely that their recent evolutionary history has followed very similar patterns (Alfenito *et al.* 1998).

An alternative route for vacuolar transport of anthocyanins might involve the use of MATE (multidrug and toxin extrusion) transporters. The Arabidopsis MATE protein TT12 (TRANSPARENT TESTA 12) is necessary for the vacuolar transport of the proanthocyanidins that colour the seed coat, and has also been shown to transport anthocyanins *in vitro* (Marinova *et al.* 2007).

In Antirrhinum, petunia, and many other species the anthocyanin undergoes a second glycosylation, adding an additional sugar. In both these cases the sugar is rhamnose, and the glycosylation is catalysed by the enzyme rhamnosyl transferase (RT). This second glycosylation may serve to stabilize and enhance colour—petunia mutants lacking RT activity are slightly less blue than wild type (Martin and Gerats 1993). The activities of acyl transferases and methyl transferases can also influence anthocyanin colour at this stage. These enzymes can generally modify any anthocyanin, irrespective of the degree of hydroxylation of the B ring, but the final colour produced will depend on the combination of core anthocyanin structure and the various modifications (sometimes known as decorations) resulting from the activites of these late-acting enzymes.

Within the vacuole, anthocyanin is sometimes stored in membraneless matrixes, known as anthocyanic vacuolar inclusions, which appear to act as traps for anthocyanin (Markham *et al.* 2000). The packing and storage of anthocyanins also influences final petal colour (Grotewold 2006), and will be discussed in Chapter 17.

16.5 Betalains

The betalains have been less well studied than the carotenoids and anthocyanins. They replace anthocyanins as the colouration in flowers and fruit of the order Caryophyllales (see Fig. 16.6). However, within that order the families Caryophyllaceae (the pinks) and Molluginaceae (the carpetweeds) do not

Figure 16.6 Betalains give the yellow, purple, and pink colours to (a) *Portulaca oleracea*, (b) *Mirabilis jalapa*, and (c) *Sesuvium portulacastrum* (all Caryophyllales). All images kindly provided by Sam Brockington (Cambridge). See also Plate 16.

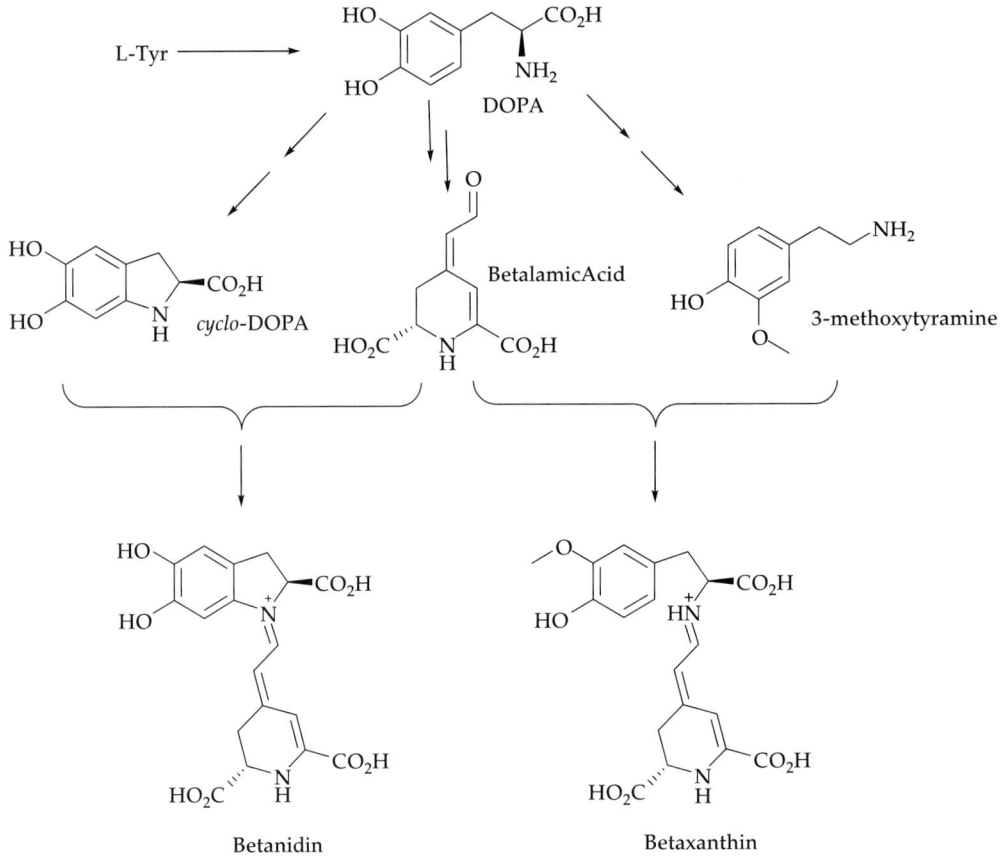

Figure 16.7 The final product of betalain synthesis depends on what is conjugated to betalamic acid.

make betalains and do make anthocyanins. It has been proposed that an ancestor of the Caryophyllales made both betalains and anthocyanins, but that within each family one pathway was lost (Strack et al. 2003). It is not unusual for a plant species to produce both carotenoids and betalains, or both flavonoids and betalains, but there is no known species that produces both anthocyanins and betalains. This mutual exclusivity may reflect the overlapping absorption spectra of anthocyanins and betalains, and therefore the lack of advantage in producing two similarly coloured pigments (Grotewold 2006). Common betalain-pigmented plants include beetroot, the petals of the Christmas cactus, and the brightly coloured bracts of *Bougainvillea*.

The betalains are water-soluble nitrogen-containing pigments synthesized from tyrosine by the condensation of betalamic acid, and come in two main colour groups. The betacyanins, formed when betalamic acid interacts with a derivative of dihydroxyphenylalanine, are red to purple, while the betaxanthins, formed from the condensation of betalamic acid with an amino acid, are yellow to orange (see Fig. 16.7). The betacyanins can also be modified by the addition of glycosyl or acyl groups, and over 50 betacyanins are currently known (Tanaka et al. 2008).

The early and late steps of betalain synthesis are thought to be catalysed by enzymes, but many of the intermediate reactions in the synthetic pathway may occur spontaneously (Strack et al. 2003). The absence of molecular and genetic model systems within the Caryophyllales has, to date, precluded any detailed analysis of betalain synthesis. Feeding experiments and other biochemical approaches have suggested a variety of potential routes to betalain synthesis, but there is as yet no consensus on whether the same pathway is used by all betalain-producing plants or on which pathway that might be. Tanaka *et al.* (2008) provide a detailed review of the current biochemical hypotheses, and only a brief summary is given here. Betalain synthesis begins with the hydroxylation of tyrosine to give two precursor molecules of L-5,6-dihydroxy-phenylalanine (L-DOPA) (Piatelli 1981). This reaction is catalysed by a tyrosinase or phenol-oxidase complex (Delgado-Vargas and Paredes-Lopez 2002), and genes encoding such enzymes have been isolated from *Portulaca grandiflora*, *Beta vulgaris*, and *Phytolacca americana* (Joy *et al.* 1995, Steiner *et al.* 1999). L-DOPA is then converted to betalamic acid by DOPA-4,5-dioxygenase (DOD). A sequence encoding DOD has also been identified from *Portulaca grandiflora,* where it was shown to induce betalain production on transient expression in a white-flowered variety (Christinet *et al.* 2004). Betalamic acid then condenses either with *cyclo*-DOPA to produce betanidin and later betacyanins, or with amino acids to generate betaxanthins (see Fig. 16.6).

Like flavonoids, betalains are stored in the vacuole of plant cells in a glycosylated form. However, unlike anthocyanins they are not pH sensitive and so are considerably more stable. Phylogenetic analyses indicate that the genes encoding the betalain-specific glucosyltransferases are very closely related to those encoding flavonoid-specific glucosyltransferases (reviewed by Strack *et al.* 2003).

The observation that betalain glucosyltransferases are related to flavonoid glucosyltransferases is one of several pieces of data pointing to a complex and labile evolutionary history for betalains. Comparative genetic studies suggest that many of the genes encoding enzymes of anthocyanin synthesis are expressed in betalain-producing species (Shimada *et al.* 2005). Similarly, Christenet *et al.* (2004) found that homologues of the sequence encoding DOD could be isolated from a number of anthocyanin-producing taxa. Recent character mapping studies by Brockington *et al.* (2011), based on updated phylogenies of the Caryophyllales, suggested that multiple transitions between betalain synthesis and anthocyanin synthesis have occurred within the order, in both directions. Taken together these various lines of evidence suggest that the switch between synthesis of anthocyanins and synthesis of betalains might involve only a few mutations to genes encoding key enzymes, allowing for a very labile evolutionary history.

CHAPTER 17

Enhancing flower colour

The production of coloured tissues, particularly insect-attracting petals, depends upon the synthesis of the pigments described in Chapter 16. However, very few flowers are coloured simply by the synthesis of a single pigment, and equally few petals are composed of a single block of unchanging colour. In the same way as an artist's palette can be used to produce an almost infinite variety of colours, plants are able to mix, modify, and enhance pigments to produce a vast array of final petal colours. These colours are usually distributed across the flower in patterns, which vary in their degree of regularity and complexity between different species. While colour contrast is much more important than pattern for attracting pollinators from a distance, pattern becomes important at close range, and allows animals to distinguish between flowers of different species and to learn to 'handle' flowers (extract the reward) with the minimum possible expenditure of energy (Kevan et al. 1996). In this chapter we shall consider the effects of mixing pigments together, the regulation of pigment distribution in the flower, and a number of other ways in which plants are able to modify the final colour of the flower. Although much of the work described in this section has been conducted using human-visible pigments, it should be borne in mind throughout that the patterning we perceive on a flower may be very different from that perceived by an insect, and therefore that the UV-absorbing pigments visible to the insect eye are also likely to be subject to the same regulatory mechanisms.

17.1 Mixing pigments

The colour that a flower appears is very rarely the colour of a single pure pigment, either carotenoid or flavonoid. In almost all species investigated, multiple pigments are present within petals and it is their combination that gives the final colour. This combination will vary according to whether the pigments are mixed within the vacuole (i.e. multiple flavonoids), or are produced in different plastids, or a combination of these factors. In particular, plastid distribution may vary across the flower, giving very different colour outcomes. Mature red rose petals (Rosales) contain no pigment-rich plastids, so the red colour is purely a result of anthocyanin deposition in vacuoles. In contrast, yellow and sometimes red carotenoid-containing chromoplasts are present throughout the petals of wallflower, *Erysimum cheiri* (Brassicales), blending visually with the magenta anthocyanin in the vacuole to produce the orangey-brown colour of the mature flowers (Weston and Pyke 1999). The importance of plastid distribution in determining pattern has been shown in cultivated primroses (Ericales), where the central eye of the flower is composed of carotenoid-containing pigments, while the rest of the petal is pigmented by flavonoids. The carotenoids in this case give a much stronger yellow colour than the paler flavonoids (Anon. 1921).

Some combinations of pigments have more dramatic effects than others. In particular, a combination of a yellow flavonoid (aurone) with a yellow carotenoid may appear little different from the carotenoid alone, because aurones are rarely produced in significant quantities. In general, carotenoids impart much stronger and heavier colours to petals than do apparently similarly coloured flavonoids. For example, the bright yellow of garden *Forsythia* flowers (Lamiales) is due to the yellow carotenoids present, while the yellow flavonoids also produced in the petals apparently have little effect on final flower colour (Rosati *et al.* 1998).

17.2 Co-pigmentation

A co-pigment is a compound that modifies the apparent colour of a pigment despite having little or no colour itself. Co-pigments are believed to have the ability to modify all varieties of anthocyanin, and this modification may involve a shift in absorption spectrum or simply an enhancement of the apparent brightness of the pigment. This enhancement is usually as a result of stabilization of the structural form of the anthocyanin molecule (Markovic *et al.* 2005).

Many flavonols and flavones act as co-pigments, and are thought to form stacked complexes with anthocyanin, causing a shift in the absorption spectrum of the anthocyanin molecule (Mol *et al.* 1998). These stacked complexes can also contain metal ions, further modifying final colour. The pigment/co-pigment/metal structure that generates the blue colour of cornflowers (Asterales) is described in Section 17.4. Flavonols are usually colourless or pale yellow, and are derived from dihydroflavonols through enzymatic conversion by flavonol synthase. There are therefore four ways in which flavonols may affect final flower colour: by being yellow, and mixing with reddish anthocyanins to give a final orange colour; by absorbing UV and thus changing final flower colour to an insect's eye; by competing with the anthocyanin synthetic pathway for dihydroflavonol (see Fig. 16.5 in Chapter 16) and thus reducing the final concentration of anthocyanin produced; and by acting as a co-pigment, where they enhance or modify anthocyanin colour. Analysis of petunia (Solanales) mutants with lesions in flavonol synthase activity, along with investigation of the effects of ectopic expression of the gene encoding flavonol synthase, have shown that flavonols play roles in both substrate competition and co-pigmentation in this species (Holton *et al.* 1993).

Cinnamic acids have also been shown to play an important role in co-pigmentation. They often constitute the main acyl groups in acylated anthocyanins, and their stabilizing effects on the pigments enhance the final petal colour, which is otherwise subject to bleaching when exposed to acidic pH (Figueiredo *et al.* 1996). Aliphatic acyl groups such as malanyl are particularly effective in stabilizing anthocyanins, while aromatic acyl groups such as sinapayl shift the pigment towards blue. Recent evidence indicates that cinnamic acids also play a role in co-pigmentation of non-acylated anthocyanins, changing the absorption spectrum of common pigments such as malvin (Markovic *et al.* 2005).

17.3 Regulation of pigment distribution

Since the type of pigments and co-pigments produced has such a range of effects on flower colour, it is hardly surprising that the pigmentation pathways should be under tight regulatory control. This regulation is both spatial and temporal, generating colour patterns on almost all flowers, and in some species changing the colours of petals during their development. For example, the flowers of *Viola cornuta* cultivar 'Yesterday, Today and Tomorrow' (a garden variety of pansy, Malpighiales) change from white through pale lilac to purple over the space of 5–8 days (see Fig. 17.1a). This change in colour has been shown to be the result of a steady increase in anthocyanin production over the time period, which is very likely to be the result of transcriptional regulation of the genes encoding the enzymes of anthocyanin biosynthesis. Although such a colour change may seem extraordinary, its dependence upon pollination (without which the petals remain white) suggests that plant growth regulator-mediated signals from germinating pollen tubes trigger the change in anthocyanin regulation, perhaps to change the attractiveness or visibility of the flower to potential pollinators (Farzad *et al.* 2002). A similar, but reversible, colour shift in the petals of *Desmodium setigerum* (Fabales) was described in Chapter 16 (Willmer *et al.* 2009), although its biochemical and molecular basis is not yet known.

It has been shown that the activities of the genes encoding the enzymes of anthocyanin biosynthesis are predominantly regulated at the transcriptional level, rather than through post-transcriptional processes. It can be inferred from this that the majority of observed petal pigmentation patterns are specified by the expression patterns of regulatory genes that control the activity of the biosynthetic genes (Mol *et al.* 1998). The analysis of pigment regulation

Figure 17.1 Pigment regulation. (a) *Viola cornuta* 'Yesterday, Today and Tomorrow' is fully purple 5–8 days after pollination (left), but opens as a white flower (middle) in which pigmentation steadily increases (right). Image kindly provided by Martha Weiss (Georgetown University, Washington, DC). (b) The *delila* mutant of Antirrhinum lacks pigmentation in the tube as a result of loss of activity of a bHLH transcription factor. (c) The *Venosa* locus produces pigmentation over the petal veins in a pale Antirrhinum flower. *VENOSA* encodes a MYB transcription factor. (d) The *an11* mutant of petunia lacks pigmentation as a result of loss of activity of a WD40 protein. The transposon in the *AN11* locus excises somatically, generating patches of wild type red tissue. Image kindly provided by Ronald Koes (Vrije Universiteit, Amsterdam). See also Plate 17.

has focused on anthocyanins rather than carotenoids, since carotenoids are essential to many other aspects of plant growth and so cannot be manipulated as easily. By convention, anthocyanin regulatory loci are divided into those that specifically control anthocyanin deposition (by regulating 'late' biosynthetic genes, some way down the pathway), and those that control the synthesis of anthocyanins and other flavonoids (by regulating 'early' biosynthetic genes). A number of genes have been shown to be important in both pathways, and three key groups of proteins have been shown to be involved. The most important of these are transcriptional activators of the MYB and basic helix-loop-helix (bHLH) families. In maize (Poales), the MYB protein C1 and the bHLH protein R have been shown to activate expression of the structural genes encoding all enzymes of anthocyanin synthesis (early and late), from chalcone synthase onwards, in kernel aleurone cells (reviewed in Martin *et al.* 1991). However, it appears that the evolution of complex floral pigmentation patterns in eudicots has involved subfunctionalization of the roles of multiple duplicated members of the MYB and bHLH families, so

that genes encoding enzymes required early in the anthocyanin synthetic pathway can be activated independently of genes encoding enzymes that act late in the pathway.

17.3.1 bHLH proteins

The best characterized example of a floral pigment regulatory protein is the bHLH transcription factor DELILA (DEL) of *Antirrhinum majus*. Mutations at the *DEL* locus result in wild type red lobes, but the corolla tube of the flower is white (Goodrich *et al.* 1992; see Fig. 17.1b). The DELILA protein is very similar in sequence to R of maize. The bHLH domain that defines members of this family is also common to the MYC family of transcription factors in animals, and functions as a DNA-binding domain. Extensive analysis of DEL function has determined that the protein activates the late anthocyanin biosynthetic genes, specifically those encoding F3H, DFR, and UFGT (Martin *et al.* 1991). The regulation of DFR by DEL has been analysed in detail. In the wild type Antirrhinum flower, pigmentation is first deposited in a ring at the base of the tube, then in the lobes, and finally in the body of the tube. Mutation of the promoter region of *PALLIDA* (encoding DFR see Chapter 16) releases it from the control of DEL. Analysis of a number of promoter mutations has established that DEL is responsible for PALLIDA activity in each of the regions of the flower at the correct developmental stage (Almeida *et al.* 1989). DEL has also been shown to be functional in heterologous hosts, activating anthocyanin synthesis in flowers of tobacco and leaves of tomato (both Solanales). Interestingly, although these experiments were conducted using a constitutive promoter, and *DEL* transcript was detected in all tissues, DFR was only activated in a subset of plant organs. This implies that DEL functions in conjunction with other, developmentally regulated factors (Mooney *et al.* 1995).

The DEL orthologue in petunia is JAF13, a bHLH protein very similar in sequence to DEL, which has also been shown to activate DFR expression (Quattrocchio *et al.* 1998). However, recent reports indicate that another bHLH protein, less similar to DEL and R, is also required for anthocyanin regulation in petunia. ANTHOCYANIN1 (AN1) encodes a bHLH protein with only limited sequence similarity to JAF13 (and DEL and R), but a closer relationship to another maize protein, INTENSIFIER (IN). The IN locus is an inhibitor of anthocyanin synthesis in maize kernels, but AN1 has been shown to activate DFR expression, in the same way that JAF13 does, and also to interact with an MYB protein, AN2 (Spelt *et al.* 2000, 2002). It also appears that AN1 may play a role in cell pH, as the *an1* mutant is allelic to the *ph6* mutant and the AN1 protein interacts physically with PH4, a MYB transcription factor with a role in regulating vacuolar pH (Quattrocchio *et al.* 2006; see Section 17.5).

A similar situation seems to exist in Arabidopsis, where two of the 139 bHLH transcription factors show particular sequence similarity to IN and AN1. Both GLABRA3 (GL3) and MYC-146 (also known as AtMYC2) were shown to activate anthocyanin synthesis when transiently expressed in white varieties of *Matthiola incana* (Brassicales) (Ramsay *et al.* 2003). While *GL3* was initially identified as a locus involved in trichome development, it has been shown to act semi-redundantly with EN-HANCER OF GL3 (EGL3) and to play a critical role

in leaf anthocyanin production in response to stress (Zhang *et al.* 2003; Feyissa *et al.* 2009). Later studies suggest that AtMYC2 plays a number of important roles in photomorphogenesis and hormone signalling, and may have only a limited endogenous role in anthocyanin production under stress conditions (Yadav *et al.* 2005).

17.3.2 MYB proteins

The joint function of the bHLH factor R and the MYB protein C1 in anthocyanin activation in maize led to a general expectation that MYB proteins would also be important in the regulation of anthocyanin synthesis in eudicot flowers. A number of studies have demonstrated that MYB proteins do play important roles in regulating anthocyanin synthesis, and that they have particularly key roles in controlling the patterning of pigmentation. All MYB proteins share 1, 2, or 3 repeats of a helix-helix-turn-helix domain, which binds DNA. The protein family is ancient, with MYB transcription factors also known to play important roles in animal cell proliferation.

In petunia, an MYB protein, ANTHOCYANIN2 (AN2), is necessary for anthocyanin synthesis. *AN2* is not particularly similar in sequence to *C1* outside of the conserved MYB domain, and it is thought that pigment regulation in monocot and eudicot lineages might be under the control of different groups of MYB transcription factors. Nonetheless, expression of petunia *AN2* is able to complement a maize *c1* mutant (Quattrocchio *et al.* 1999). AN2 activates late biosynthetic genes, particularly *DFR*, and is hypothesized to function in concert with AN1. Hoballah *et al.* (2007) explored the function of AN2 in *Petunia axillaris*, a white-flowered species, relative to pigmented *Petunia integrifolia*. They found that wild-collected and botanic garden accessions of *P. axillaris* contained alleles of *AN2* with a variety of frameshift and nonsense mutations at different positions. They further showed that introduction of a functional *AN2* allele into *P. axillaris* using a transgenic approach resulted in pigmented flowers but decreased attractiveness to the usual hawkmoth pollinator, a clear demonstration that regulation of anthocyanin synthesis is critical to pollination success.

Additional MYB transcription factors regulate vegetative anthocyanin production in petunia, and also specify pattern-specific pigmentation in the flower. *DEEP PURPLE (DPL)* and *PURPLE HAZE (PHZ)* encode proteins that cluster with AN2 and the pH-regulating AN4 in phylogenetic analyses, suggesting that they are the products of recent duplication events (Albert *et al.* 2011). Ectopic expression of either gene results in pigmented leaves and stems, but their endogenous expression in leaves occurs in response to high light levels. In the flowers, DPL was shown to regulate the production of anthocyanins over the veins in the corolla tube, while PHZ is responsible for pigment production in petal surfaces exposed to light at the bud stage. Both of these phenotypes are masked by the full pigment in petunia lines with a functional *AN2* allele, but can be detected in *an2* mutants (Albert *et al.* 2011).

Three lines of Antirrhinum, *venosa*, *rosea^colorata*, and *rosea^dorsea*, show variation in pigment patterning through the activities of genes encoding MYB transcription factors (Schwinn *et al.* 2006). The *ROSEA* locus has been shown to comprise two genes, *ROSEA1* and *ROSEA2*, and a number of mutations in each coding sequence contribute to the different patterning and pigment intensity seen in the *rosea^colorata* and *rosea^dorsea* mutant lines. The *ROSEA* genes were isolated using sequence similarity to *AN2* and maize *C1*, but are themselves likely to be the product of a recent intrachromosomal duplication event (Schwinn *et al.* 2006). ROSEA1 and ROSEA2 activate the expression of overlapping subsets of the anthocyanin synthetic genes. ROSEA1 was shown to activate transcription of all of the late biosynthetic genes, with particularly strong effects on F3'H, F3H, DFR, and UFGT. In contrast, ROSEA2 seems to only significantly enhance expression of F3'H. Both proteins contain the appropriate amino acid motif for interaction with a bHLH protein, and double mutant analysis confirms that there is an interaction between the *ROSEA* loci and the bHLH-encoding *DELILA*.

The dominant *Venosa* allele is only visible in an Antirrhinum background which is not fully pigmented, but in a paler flower it results in the production of magenta anthocyanin in the epidermal tissue overlying the petal veins (see Fig. 17.1c). The VENOSA protein is very similar in sequence to ROSEA1 and 2, and it has been suggested that ROSEA and VENOSA arose by duplication either within

Antirrhinum or within a recent ancestral lineage, before further duplication of ROSEA to give ROSEA1 and ROSEA2. The VENOSA protein also contains the necessary amino acid motifs for interaction with a bHLH protein. Analysis of transcription of biosynthetic genes in different genetic backgrounds revealed that VENOSA induces expression of *CHI, F3H, F3'H,* and *UFGT* (Schwinn *et al.* 2006). The expression domain of *VENOSA* itself is restricted to the cells surrounding the veins (Shang *et al.* 2011). However, anthocyanin is only produced in the epidermis. Shang *et al.* (2011) proposed that the specificity of anthocyanin to the epidermis over the veins is provided by the interaction between vein-associated VENOSA and epidermal-specific DELILA.

The genes encoding two more MYB transcription factors, MYB305 and MYB340, are also known to be expressed in Antirrhinum flowers, although mutant phenotypes have not been described for either locus (Moyano *et al.* 1996). These genes are worth considering because they provide a unique insight into the precision with which flavonoid production can be regulated. The primary function of these genes appears to be in the regulation of flavonols in the carpel and stamens. The MYB305 and MYB340 proteins are virtually identical across their DNA-binding domains, differing in only four amino acids, and they both bind to the promoters of the genes encoding PAL, CHI, and F3H, and activate transcription of the genes. MYB340 is the stronger activator of the two proteins, producing more transcript of all three of the structural genes than MYB305. However, MYB305 has a much stronger binding affinity for biosynthetic gene promoters than does MYB340. Since the weaker activator binds more strongly, and the stronger activator binds much more weakly, coexpression of the two genes results in 50% less biosynthetic gene transcript than does expression of *MYB340* alone, because the proteins are in competition with each other (Moyano *et al.* 1996). These data show that flavonoids are subject to very precise regulation by gearing mechanisms, and that similar proteins may interact to regulate petal pigmentation. A gene with very high sequence similarity to *MYB305* and *MYB340* has been described from pea (*Pisum sativum*, Fabales). *Myb26* is expressed in developing pea flowers, with transcript levels increasing as anthocyanin deposition occurs. However, the Myb26 protein is not able to complement the petunia *an2* mutant, suggesting that Myb26, like MYB340 and MYB350, activates a set of earlier biosynthetic genes (Uimari and Strommer 1997).

17.3.3 WD repeat proteins

The third group of proteins with an important role in pigment regulation are the WD repeat (WDR) proteins. These proteins produce a seven-bladed propeller motif and act as surfaces for the interactions of other proteins, such as the transcription factors previously discussed (Ramsay and Glover 2005). The first WDR protein shown to play a role in anthocyanin regulation was ANTHOCYANIN11 (AN11) of petunia. The *an11* mutant has reduced activity of the late biosynthetic genes (from DFR onwards), and produces flowers with white petals as a result (Fig. 17.1d). WDR proteins are so conserved across enormous evolutionary distances that the human sequence with most similarity to *AN11* was able to complement an *an11* mutant phenotype in a transient expression experiment (de Vetten *et al.* 1997). Although no WDR protein has yet been shown to be involved in pigment regulation in Antirrhinum, the characterization of the Arabidopsis protein TTG1 as a WDR protein required for both anthocyanin synthesis and trichome development suggested that the role of a WDR protein in this regulatory pathway might be ubiquitous to the eudicots (Walker *et al.* 1999). More recently, this hypothesis has been strengthened by the identification of WDR proteins involved in anthocyanin species in a wide range of species, particularly fruit crops such as apple (Rosales), grape (Vitales), and pomegranate (Myrtales), and ornamental flowers such as morning glory (Solanales).

17.4 The effects of metal ions

Interactions between floral pigments and metal ions can also alter the final colour of the petals. For example, the bright blue colour of cornflowers is not simply due to the presence of a blue pigment. Instead it stems from an interaction between six molecules of the magenta anthocyanin, cyanidin, six molecules of a flavone co-pigment, one ion each of the metals iron and magnesium, and two calcium ions.

The combination of the pigment and co-pigment with the metals results in a molecule that gives the flower a very bright blue colour, even though the anthocyanin alone is red/magenta (Shiono *et al.* 2005). The packaging of multiple pigment and ionic components in this way leads to the production of structures known as supermolecular pigments, or superpigments. In the case of cornflower this structure is called protocyanin (Shiono *et al.* 2005).

Equally remarkably, the bright blue colour of Himalayan poppy, *Meconopsis grandis* (Ranunculales), is due to the interaction between iron and the magenta anthocyanin cyanidin (see Fig. 17.2a). Yoshida *et al.* (2006) were able to reproduce the blue colour of Himalayan blue poppy petals *in vitro* by mixing cyanidin with flavonol co-pigments, iron, and magnesium under low pH conditions. The iron was the most crucial factor apart from the cyanidin. Unusually for a blue flower, vacuolar pH in the Himalayan poppy is very acidic (see Section 17.5).

Another good example of this type of interaction is the variable colour of the flowers of *Hydrangea* (Cornales), which is commonly grown in gardens (see Fig. 17.2b). *Hydrangea* flowers are blue if there is aluminium in the soil, as aluminium and the pigment delphinidin form a very stable, very blue complex. The aluminium ions affect pigment colour not only by complexing directly with the delphinidin, but also by encouraging its stacking into a superpigment molecule (Schreiber *et al.* 2010). If there is less aluminium available in the soil, and more molybdenum, the same pigment interacts with the molybdenum ions, causing the flowers to appear light pink instead. The presence of this changeable anthocyanin pigment in *Hydrangea* is exploited by gardeners, who may water the soil around their plants with a solution containing the appropriate ion to generate the final flower colour of their choice. The aluminium–delphinidin complex is known to be acid-stable, giving the very blue colour even in an acidic pH which would normally make the delphinidin appear red (Kondo *et al.* 2005).

Since anthocyanins are water soluble and retained in the vacuole, complexing with metal ions requires a mechanism to transport these ions across the vacuolar membrane. In 2009, Momonoi *et al.* described the isolation of a vacuolar iron transporter from the purple and blue tulip, *Tulipa gesneriana*

Figure 17.2 Metals and pH both affect flower colour. (a) The Himalayan blue poppy owes its blue colour to an interaction between anthocyanin and iron. Photograph kindly supplied by Cambridge University Botanic Garden. (b) Hydrangea flowers can be blue or pink, depending on the metal ions present in the soil. (c) Morning glory flowers have a high vacuolar pH. Image kindly provided by Felix Jaffe. (d) An unstable *pH4* mutant of petunia, with revertant wild type red (acidic) sectors on a mutant bluish-pink (more alkaline) background. Image kindly provided by Ronald Koes (Vrije Universiteit, Amsterdam). See also Plate 18.

(Liliales). The tepals of this flower are purple, but with a much bluer patch at the base, and previous studies had shown that cells from the blue patch had a higher iron content than cells from the rest of the tepal (Shoji *et al.* 2007). Monomoi *et al.* (2009) isolated the iron transporter on the basis of its sequence similarity to Arabidopsis and rice iron transporters. They showed that the gene is most strongly expressed in the blue regions of the petal, and that its transient expression in purple cells causes them to blue. It is likely that similar transporters are necessary to move aluminium and magnesium into the vacuole in other species.

17.5 The importance of pH

The pH of petal cells can also affect the final colour of the flower, as pH determines the conformation of the anthocyanin molecule and its resulting absorption spectrum. Acidic conditions stabilize red forms of anthocyanin, whereas alkaline conditions cause a colour shift towards blue. For example, the light blue petals of *Ipomoea tricolor* (morning glory, Solanales) owe their colour to the effect of a high petal pH on their anthocyanin (see Fig. 17.2c). The closed buds of these flowers are purplish red and their cells have a pH of 6.6. However, when the flowers open, the petal cell pH increases to 7.7, and the pigment changes colour to sky blue (Yoshida *et al.* 1995). To confirm that this change in colour was solely due to a change in pH, Yoshida *et al.* (1995) also showed that aluminium and iron had no role in morning glory colouration, and that the anthocyanin was stabilized by the same caffeic acid co-pigment in both the bud and mature flower stages. Treatment of open flowers with an acidifying environment (carbon dioxide) resulted in a reduction in petal cell pH and a reversion of the colour to purple. When investigating the mechanism of this *in vivo* pH change, Yoshida *et al.* (2005) observed that an Na^+/H^+ exchanger was present only in petal vacuolar membranes, and was present at particularly high levels in mature flowers. This suggests that the increased pH is due to active transport of Na^+ and/or K^+ from the cytosol to the vacuoles. Further support for this mechanism comes from the characterization of a mutant of *Ipomoea nil* that fails to undergo colour change on maturity and remains purple. Fukada-Tanaka *et al.* (2000) confirmed that the difference between purple mutant and blue wild type tissue was entirely due to a difference in pH, with the mutant tissue retaining a pH about 0.7 lower than wild type petals. Isolation of the gene perturbed in the purple mutant, *InNHX1*, confirmed that it encoded a protein with strong sequence similarity to Arabidopsis and rice vacuolar Na^+/H^+ exchangers. The *InNHX1* transcript is most abundant in petals 12 hours prior to bud opening, an expression pattern that is strongly correlated with the timing of change in colour (Yamaguchi *et al.* 2001). A similar gene, *InNHX2*, was also isolated from floral cDNA, and was also found to increase in expression levels in the 12 hours before bud opening. Using mutant strains of yeast sensitive to cation presence, Ohnishi *et al.* (2005) demonstrated that both InNHX1 and InNHX2 catalyse the transport of both Na^+ and K^+

into vacuoles, confirming their importance in regulating vacuolar pH and thus flower colour.

The ability of vacuolar pH, controlled by membrane transporters, to alter flower colour without any change in the types of pigment produced gives plants the flexibility to alter petal colour temporally, after pigments have been made in the mature flower. Stewart *et al.* (1975) observed that the colours of several wild flower species became bluer as they aged, and that this blueing was correlated with an increase in petal pH.

Petal pH has been shown to be under the control of genetic factors in a number of species, including *Primula sinensis* (Ericales), *Papaver rhoeas* (Ranunculales), *Tropaeolum majus* (Brassicales), *Lathyrus odoratus*, and the clover *Trifolium pratense* (both Fabales), and this is likely to be the case in many other species, too. Petunia has proved to be the best model system in which to investigate effects less dramatic than those of morning glory, and seven loci that determine petal cell pH have been identified (de Vlaming *et al.* 1983; Mol *et al.* 1998). In petunia, these loci work to maintain the acid status of the vacuole and thus the red colour of the anthocyanin. Wild type petunia flowers have a vacuolar pH of 5.5, whereas it rises to as much as 6 in some of the *ph1–7* mutants (de Vlaming *et al.* 1983). *PH6* was isolated first, and shown to correspond to the bHLH-encoding locus *AN1* (Spelt *et al.* 2002), while *PH4* encodes a MYB-related transcription factor (Quattrocchio *et al.* 2006). The *pH4* mutant shows an increased vacuolar pH, from the wild type 5.5 to around 6.0, and as a result the flower appears a slightly bluer shade (see Fig. 17.2d). The *PH4* gene is expressed in the petal epidermis, and the MYB transcription factor that it encodes can interact with the bHLH proteins AN1 and JAF13. These data suggest that AN1 interacts with different MYB proteins to activate different pathways—with AN2 to activate anthocyanin synthesis, but with PH4 to regulate cell pH. Mutation of *PH4* had little effect on the expression of anthocyanin synthetic genes (Quattrocchio *et al.* 2006).

The mechanism by which two transcription factors regulate vacuolar pH was explained when the *PH5* gene was isolated using transposon-tagged lines (Verweij *et al.* 2008). *PH5* is expressed in the petunia petal, and encodes an H^+-ATPase (from a subfamily known as type P_{3A}) that is localized to the vacuolar membrane (and possibly also to the plasma membrane, the more usual localization of this type of transporter). The PH5 protein was able to complement a yeast mutant with defects in two genes encoding H⁺-ATPase. The transcription factors AN1 and PH4 interact with the WDR protein AN11 to activate transcription of *PH5*. However, ectopic expression of *PH5* in an *an1* or *ph4* mutant background could not fully restore wild type pH, indicating that the transcription factors also regulate the expression of other genes with a role in vacuolar acidification (Verweij *et al.* 2008).

Yamaguchi *et al.* (2001) isolated the petunia orthologue of the *InNHX1* gene, responsible for increased pH of the *Ipomoea nil* flower, but found no particularly strong expression of the gene in petals. This is unsurprising, considering that petunia employs an opposite mechanism to that of *Ipomoea*, namely an active reduction in petal pH.

17.6 The role of petal cell shape

A very subtle way in which the colour of a petal can be enhanced is by distribution of specialized cell types in the epidermal layers. In a study that examined the surface structures of 201 species of flower from 60 families, 79% were found to have some form of cone-shaped cells on the epidermis oriented towards potential pollinators, described as conical-papillate cells (Kay *et al.* 1981). The frequency of this specialized cell morphology within the flowering plants, and its almost universal restriction to the petal, argue for an adaptive explanation involving the function of the petal in pollinator attraction. It was proposed that conical-papillate cells increased the amount of light absorbed by the pigments in flowers, enhancing the perceived colour of the petal (Kay *et al.* 1981). Conical-papillate cells might also scatter light reflected back from the inner layers of the petal more evenly than would flat cells, resulting in a sparkling effect, or a velvety texture to the petal.

In Antirrhinum, the conical-papillate cells (see Fig. 17.3a) are found only on the adaxial epidermis of the petal lobes (where they will be seen by potential pollinators as they approach the flower). The *mixta* mutant of Antirrhinum fails to develop conical-papillate petal cells and instead has flat

petal cells, more similar to those of the leaf epidermis (see Fig. 17.3b). The significance of the conical-papillate cells in enhancing colour is shown by the fact that the *mixta* mutant was originally identified in a screen of mutagenized plants because it was paler in colour than wild type flowers (see Fig. 17.3c). The mutant petal has a matt appearance, unlike the velvety sparkle of the wild type petal (Noda *et al.* 1994).

By comparing the ability of epidermal cells to focus light in the wild type and *mixta* mutant lines, conical-papillate cells have been shown to enhance visible pigmentation. Conical-papillate cells focus light approximately twice as well as the mutant flat cells, and they focus it into the region of the epidermis where the pigment is contained (Gorton and Vogelmann 1996). It has also been shown that the wild type Antirrhinum petals reflected significantly less light away from the flower than *mixta* mutant petals did, and absorbed significantly more light. These differences can be attributed to the focusing of the light on to the pigments in the epidermal cells, and to the reduction in reflection of light at low angles of incidence, resulting in the greater depth of colour of wild type conical-celled flowers (Gorton and Vogelmann 1996). This trick is used by many plant species to enhance petal colour.

Isolation of the *MIXTA* gene revealed that it encodes a MYB transcription factor, presumably responsible for the activation of structural genes involved in cytoskeletal arrangement and cell wall deposition (Noda *et al.* 1994). Transgenic experiments have demonstrated that the MIXTA protein is sufficient to activate the conical-papillate cell form, even when expressed in other tissues, such as the leaf (Glover *et al.* 1998).

The *mixta* mutant has been used to test the idea that conical-papillate petal cells, and the brighter colour that they generate, enhance pollination success. Plants of the wild type and *mixta* mutant lines were grown in a field plot and the stamens were removed from the flowers to prevent self-pollination. The subsequent development of fruit was used as an indicator of a pollinator visit. The shape of the petal cells had a highly significant effect on the likelihood of fruit set (and thus a pollinator visit). Flat-celled flowers set significantly fewer fruits than conical-papillate celled flowers (Glover and Martin

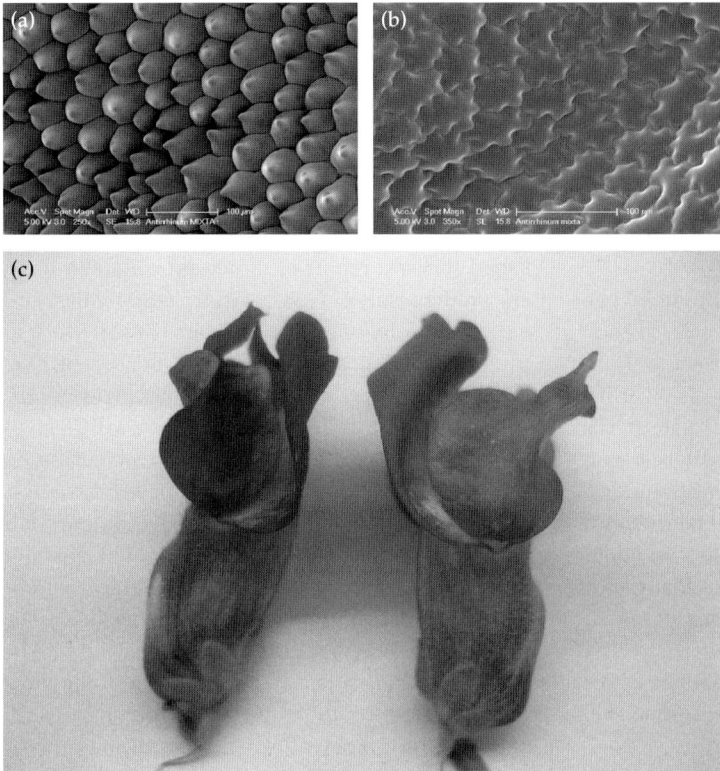

Figure 17.3 Petal cell shape affects flower colour. (a) Wild type Antirrhinum petal epidermis, composed of conical cells. (b) *mixta* mutant petal epidermis, composed of flat cells. (c) Wild type (left) and *mixta* mutant (right) flowers, showing the difference in colour attributable to the cell shape. See also Plate 19.

1998). However, a similar result was observed when conical-papillate celled white Antirrhinum flowers were compared with flat-celled/white flowers (both generated by crossing wild type and *mixta* mutant lines to the *nivea* mutant, which lacks chalcone synthase and thus all of flavonoid synthesis). Although conical-papillate and flat-celled surfaces were identical to the human eye in the absence of pigmentation, the flat-celled flowers experienced reduced fruit set and a reduction in pollinator visits (Glover and Martin 1998). These data provided the first indication that enhanced colour was unlikely to be the main explanation for the evolutionary success of conical-papillate cells, and more recent experiments have confirmed that bumblebees have no intrinsic preference for the wild type Antirrhinum colour compared with the *mixta* mutant (Dyer *et al.* 2007).

An alternative explanation for the specialized shape of petal epidermal cells is that the texture of a flower may provide tactile benefits in the form of increased grip, or cues to pollinator position on the flower. It has long been known that bees can be trained to recognize the petals of different species by the shape of the cells. Bees were provided with a food reward when they touched epidermal layers composed of certain cell types, but no reward when they touched cells of other shapes. They learned very quickly to associate the reward with the texture of the petal, and would search for food only when presented with the epidermal tissue that usually accompanied the reward (Kevan and Lane 1985). Recent experiments with resin casts of wild type and *mixta* mutant Antirrhinum petals confirmed that bees can discriminate between them using touch alone. More significantly, bee preference for conical epidermal surfaces was stronger when artificial flowers were presented vertically rather than horizontally (Whitney *et al.* 2009a). When the same experiment was repeated with Antirrhinum flowers, it was found that the preference for the wild type line was lost when flowers were presented horizontally so that bees could land on them easily. Similarly, Alcorn *et al.* (2012) showed that bumblebee preference for the wild type conical-celled form of petunia over the flat-celled *ph1*mutant was increased when the flowers were gently shaken (on a lab shaking platform). These recent studies suggest that, although conical cells do visibly enhance flower colour, their primary adaptive function might be to enhance pollinator grip on the flower.

A third way in which conical cells might be perceived as adaptations to enhance the attractiveness of the petal is through a role in regulating intrafloral microclimate. The temperature within a flower has been shown to influence nectar secretion rate, nectar evaporation rate, and nectar concentration. It may also influence scent diffusion. There is also some evidence to suggest that a high intrafloral temperature is itself attractive to insects, particularly in cooler climates and at dawn (Dyer *et al.* 2006). Analyses of intrafloral temperature have demonstrated small, variable, but sometimes significant differences between the wild type and *mixta* mutant lines of Antirrhinum under different conditions (Comba *et al.* 2000; Whitney *et al.* 2011).

Finally, conical petal epidermal cells can also influence the degree of petal reflexing and thus the apparent surface area of the petal. This change in apparent surface area may play a role in enhancing flower visibility at a distance.

It is likely that conical petal cells influence pollinator behaviour through each of these different methods in different species. However, in the context of flower colour, it is certainly clear that conical petal epidermal cells do enhance the appearance of anthocyanin by focusing light into the cell vacuoles.

17.7 Structural colour and structural enhancement of colour

A final way in which plants can enhance the colour of their flowers is through the use of nanoscale structures. 'Structural colour' is the production of colour by a transparent material that interferes with incident light as a result of its nanoscale characteristics. The most familiar examples are those of petrol on a puddle, or a soap bubble. In both cases, thin film interference occurring at the surface where the petrol and water (or detergent and air) meet causes reflection of rainbow colours, which vary with the angle of observation—they are iridescent. Structural colours are well known in the animal kingdom, where butterflies, beetles, and birds commonly produce bright and iridescent effects using nanoscale structures. Whitney *et al.* (2009b) published the first report of structural colour in flowers, on the petals of *Hibiscus trionum* (Malvales) and various tulip species and cultivars. Recent work in my laboratory has confirmed that the same structures are widespread throughout the angiosperms. In these flowers, colour is produced by a diffraction grating—a series of regular folds of the cuticle on top of flat petal epidermal cells. These folds act in the same way as the data grooves on a compact disc to reflect different wavelengths of light at different angles. The rainbow effect is most visible on a dark-pigmented background, such as that of the 'Queen of the Night' tulip variety (see Fig. 17.4a and b). The developmental control of these cuticular folds is not yet well understood, but is likely to involve coordination of the timing and amount of cuticle synthesis relative to the timing and anisotropy of cell growth. Whitney *et al.* (2009b) demonstrated that the iridescence arising from a diffraction grating could be learned by bumblebees as a cue to identify rewarding flowers.

Structural effects can act more subtly to enhance pigment colour, as we have seen with conical cells. Highly glossy surfaces can be produced by the presence of a very smooth, flat cuticle layer, and effective light scattering can be achieved by larger-scale disordered structures, such as conical cells. One particularly impressive example of structural enhancement of pigment colour is shown by the buttercup, *Ranunculus repens* (Ranunculales) (see Fig. 17.4c and d). In the common children's game, reflection of yellow light from the buttercup flower on to the chin signals that the child likes butter. In fact, Vignolini *et al.* (2012) demonstrated that the strong yellow reflection stems from a double mirror system within the buttercup petal. The smooth cuticle layer acts as a mirror, reflecting a significant portion of light away before it enters the petal. Such light as does enter passes through the

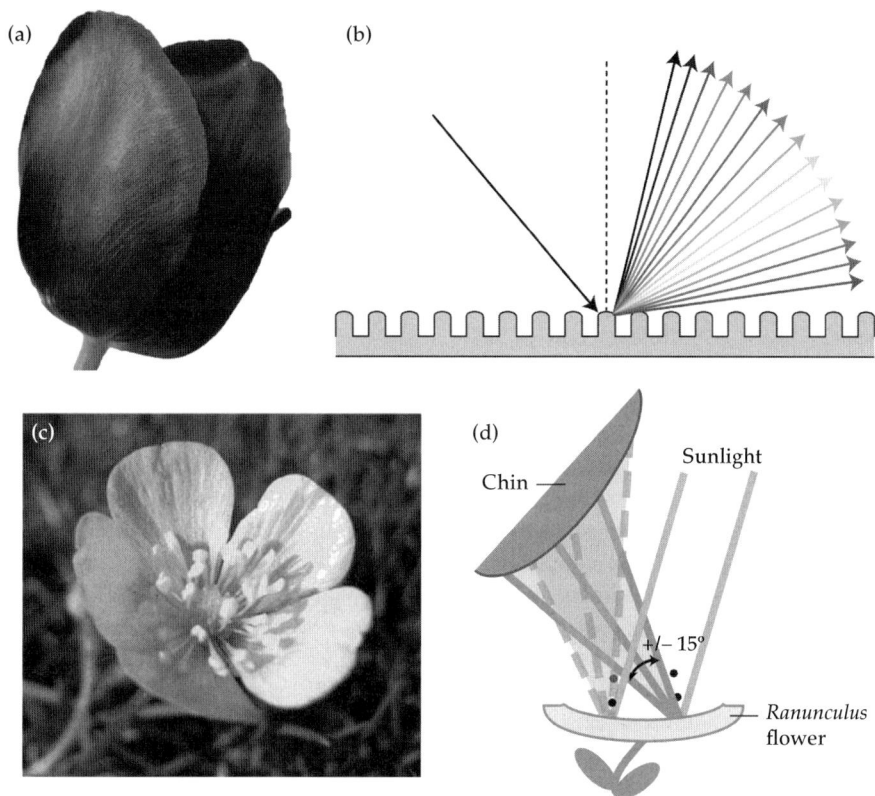

Figure 17.4 Structural colour. (a) Tulip 'Queen of the Night' has iridescent rainbow colours on top of purple pigmentation. (b) This iridescent effect is caused by a diffraction grating. (c) The bright yellow buttercup reflects yellow light very strongly. (d) The buttercup acts as a double mirror, reflecting yellow and white light together on to nearby surfaces such as a child's chin. See also Plate 20.

carotenoid-containing epidermal layer, which absorbs wavelengths other than yellow. This yellow light is then reflected back from a second mirror, a thin air interface that separates the epidermis from a diffusive starch-containing layer of cells below. The yellow light joins with the reflected white light to give a very strong beam of colour.

The complexity of the many ways of altering final flower colour has added support to the idea that colour is intricately linked to the attraction of specific pollinators. In the next section of this book we shall consider the lability of floral form, and whether pollinators and plants are really linked in an evolving series of specific relationships.

CHAPTER 18

Lability of floral form

This section of the book is focused on the changes that can occur to floral form, and on the effects of those changes on pollinator behaviour. In Chapters 15, 16, and 17 we discussed the molecular genetic and developmental processes that lead to a diverse range of floral forms, considering size, shape, symmetry, and colour. In each of these contexts we considered the generation of a particular floral morphology in the species in which it has been best studied. However, to understand whether floral form is sufficiently labile in an evolutionary sense to respond to shifts in pollinator type or preference, generating the sorts of pollination syndromes that were discussed in Chapter 14, we need to look at change in floral form between closely related species and at repeated evolution of traits across the angiosperm phylogeny. In this chapter, then, we shall take key aspects of floral form and discuss the evidence that they are evolutionarily labile. Where traits are shown to change repeatedly we shall also discuss what is known about the molecular genetic basis of morphological evolution. In many cases these molecular genetic data are limited by the lack of genomic resources available for the comparator species, or by their genetically intractable nature. However, despite these difficulties, great advances in understanding floral lability have been made in recent years.

18.1 Lability of floral size

Flower size is enormously variable across the angiosperms, spanning a range varying from less than 1 millimetre in some aquatic plants to the 1 metre diameter of the giant flowers of *Rafflesia* (Malpighiales). Changes in flower size are often generally correlated with changes in plant size, changes in ploidy level, and changes in habitat. They may also be involved in trade-offs with flower number. Sargent *et al.* (2007) explored flower size and flower number in 251 species from 63 families, and found a significant negative correlation between the size and number of flowers. However, specific changes just to flower size can indicate a shift in breeding system or pollinator type. For example, selfing flowers are usually smaller than outcrossers, and bat-pollinated flowers are often larger than insect-pollinated ones (although it should be noted that the very largest flowers of all are often pollinated by beetles or by carrion flies, possibly as a result of unusually specific selective pressures; Davis *et al.* 2008). A recent analysis of flower size across a plant group was conducted in several genera of the coffee family, Rubiaceae (in the Gentianales), where flower size varies between approximately 2 mm in length and 40–50 mm in length (Razafimandimbison *et al.* 2012). The authors mapped flower size on to a phylogeny of the group, finding that large flowers were ancestral and that small flowers had probably evolved only once. Small flowers in this system are associated with a shift to pollinators with a shorter proboscis, particularly small bees and flies.

In Chapter 15 we discussed the control of flower size, noting that cell division and cell expansion were closely co-regulated to ensure reproducibility of organ size. However, a number of examples of proteins that could decouple these processes were discussed, and these represent candidate loci for molecular changes that might underlie the lability of flower size. Analysis of the developmental basis of smaller flowers in a number of selfers has concluded either that the duration of flower growth has been reduced, or that the rate of flower growth has decreased. For example, Sicard *et al.* (2011)

demonstrated that the reduction in petal size that accompanied the evolution of selfing in *Capsella rubella* (Brassicales) can be attributed to a shorter period of cell division than is seen in outcrossing *Capsella grandiflora*. The duration of the cell division window has previously been shown to be important in regulating organ size in Arabidopsis, and is controlled by the AINTEGUMENTA transcription factor (Mizukami and Fischer 2000). In *Capsella*, quantitative trait loci (QTL) analysis suggests that multiple loci interact to determine flower size, but it is likely that at least some of them encode the regulators of organ size, such as AINTEGUMENTA (discussed in Chapter 15).

18.2 Lability of floral symmetry

Floral symmetry is generally classified in a binary fashion, as either actinomorphic (radially symmetrical) or zygomorphic (bilaterally symmetrical), although there is considerable variation between flowers in the extent of zygomorphy and the numbers of organ whorls it encompasses. The developmental genetic basis of zygomorphy in Antirrhinum was discussed in Chapter 15. Dorsal identity is attributed to the action of a recently duplicated pair of TCP family transcription factors (CYC and DICH) and a MYB transcription factor (RAD), while another MYB protein (DIV) provides ventralizing signals (see Section 15.2 for a full description; Luo *et al.* 1996; Galego and Almeida 2002; Corley *et al.* 2005). Within the same tribe as Antirrhinum (the Antirrhineae, all part of the Lamiales), many species show much reduced or expanded zygomorphy compared with that of Antirrhinum, reflecting the evolutionary lability of the degree of zygomorphy (see Fig. 18.1). For example, the corolla of *Mabrya acerifolia* has almost complete actinomorphy. The flower of *Mohavea confertiflora* aborts the top three stamens (i.e. the dorsal and both lateral stamens), not just the dorsal one, but its corolla shows little deviation from radial symmetry. These differences have been correlated with differences in the expression patterns of the *CYC* and *DICH* orthologues. Unlike the situation in Antirrhinum, both genes in *Mohavea* are expressed in the lateral stamen primordia as well as in the dorsal one, resulting in the abortion of all three stamens. However, in the petals the

Figure 18.1 Variation in zygomorphy in the Antirrhineae. (a) Highly zygomorphic *Antirrhinum majus*. (b) Moderately zygomorphic *Maurandya scandens*. (c) Slightly zygomorphic *Mabrya acerifolia*. (d) Almost actinomorphic *Rhodochiton atrosanguineum*. All scale bars 1 cm. See also Plate 21.

expression of the *CYC* and *DICH* orthologues terminates much earlier than it does in Antirrhinum, possibly contributing to the reduced zygomorphy of the corolla (Hileman *et al.* 2003). A more extreme example of loss of zygomorphy in the Lamiales is seen in wind-pollinated *Plantago* species, where Preston *et al.* (2011) were unable to identify orthologues of *RAD* and *DIV*, and also found loss of a dorsally expressed *CYC*-like gene. They concluded that the entire floral symmetry gene network had disintegrated in this genus, associated with the shift from insect to wind pollination.

Zygomorphy has evolved approximately 38 times from actinomorphy across the angiosperm phylogeny, and is a hallmark of some of the most species-rich plant groups (Endress 2001). Work on the evolutionary lability of floral symmetry has focused on assessing whether evolution of zygomorphy involves repeated recruitment of the same molecular components to direct similar developmental programmes, or whether different developmental pathways apply in different taxa (Cubas 2003). Within the Fabales, reports from *Lotus japonicus*, *Lupinus nanus*, and *Pisum sativum* indicate that *CYC*-like genes have been recruited to the development of zygomorphy in this order and are also involved in shifts of floral symmetry within it. The *LjCYC2* gene was isolated from *Lotus japonicus* by Feng *et al.* (2006) and shown to be expressed in the dorsal part of developing flowers, just as *CYC* itself is in Antirrhinum. Loss of *LjCYC2* expression resulted in ventralization of the dorsal petal, while ectopic expression of *LjCYC2* resulted in dorsalization of the lateral and ventral petals. Wang *et al.* (2008) took a neutral, non-candidate approach and isolated genes from mutants of pea (*Pisum sativum*) with perturbations in floral symmetry. They discovered that mutations in *PsCYC2* and *PsCYC3* were responsible for the observed phenotypes, and further showed that *PsCYC1* also has a dorsal-specific expression pattern. Similarly, Citerne *et al.* (2006) showed that a *CYC*-like gene was expressed in the dorsal part of the developing flower of *Lupinus nanus*. Surprisingly, however, the radially symmetrical phenotype of a close relative, *Cadia purpurea*, was not caused by loss of expression of this *CYC*-like gene, but by its ectopic expression throughout the developing flower. Therefore *CYC*-like genes have not only been recruited to specify floral zygomorphy in the Fabales (a separate evolutionary origin of zygomorphy to that of Antirrhinum), but also their ectopic expression has resulted in a novel phenotype—dorsalization of the ventral organs and therefore actinomorphy of whole flowers within this group.

A number of similar studies have reported that *CYC*-like genes are involved in various independent transitions to zygomorphy across the eudicots. In the Brassicales, Busch and Zachgo (2007) showed that zygomorphic *Iberis amara* had strongly dorsal-specific expression of *IaTCP1*, a *CYC*-like gene, late in petal development when differential growth was most apparent. In the Dipsacales, Howarth *et al.* (2011) showed that *CYC2* was expressed throughout the corolla in radially symmetrical *Viburnum plicatum*, but restricted to the dorsal, or dorsal and lateral, petals in related zygomorphic species. Similarly, Zhang *et al.* (2010, 2012) showed that *CYC*-like genes were dorsally expressed in several zygomorphic species of the Malpighiales, but either absent, barely expressed, or expressed throughout the floral meristem in related actinomorphic species. These studies all provide evidence both for independent recruitment of *CYC*-like genes to the development of zygomorphy and for the lability of floral symmetry, occurring through changes to the function or expression of *CYC*-like genes.

The evolution of zygomorphy has been studied in less depth in the monocots. *CYC* itself is not present in monocots, as it is derived from a gene duplication event at the base of the core eudicots (Howarth and Donoghue 2006). However, related TCP family members have been shown to have asymmetric expression patterns in monocot flowers. Bartlett and Specht (2011) demonstrated that zygomorphic flowers from the Zingiberales had ventral-specific expression of genes encoding TCP transcription factors, while Preston and Hileman (2012) showed that similar genes were ventrally expressed in zygomorphic flowers of *Commelina* but expressed throughout the floral meristem of actinomorphic *Tradescantia* (both Commelinales). Preston and Hileman (2012) also showed asymmetric expression of B function genes in *Commelina*, supporting the hypothesis that differential expression of floral organ identity genes might be partially responsible for zygomorphy in monocot flowers, including orchids (Mondragon-Palomino and Theissen 2008).

Variation in floral symmetry is a hallmark of the composite inflorescences produced by members of the Asteraceae (Asterales). For example, garden daisies have both zygomorphic flowers (the outer white ones) and actinomorphic flowers (the central yellow ones). More complex capitula show a variety of intermediate floral forms, ranging from extremely zygomorphic through mildly zygomorphic to actinomorphic. The zygomorphic flowers in a daisy capitulum are known as ray florets, and the actinomorphic flowers as disc florets.

Zygomorphy in daisies has been studied in *Senecio, Gerbera,* and *Helianthus,* and differential expression of duplicated *CYC*-like genes appears to be responsible for the different floral symmetries. *Senecio vulgaris* has been a useful system in which to study daisy floret symmetry because in the UK it is found in two forms—a radiate morph, which has both zygomorphic ray florets and actinomorphic disc florets, and a non-radiate morph, which has only actinomorphic disc florets. The non-radiate form is believed to be ancestral, while the radiate form is thought to have arisen by introgression of genetic material from *Senecio squalidus* (the Oxford ragwort, a non-native but naturalized species with both actinomorphic and zygomorphic flowers; Abbott *et al.* 1992). The ray floret trait segregates as a single locus (the *RAY* locus), and Kim *et al.* (2008) showed that different alleles of two *CYC*-like genes segregated with presence or absence of ray florets, and that these *CYC*-like genes were linked to the

RAY locus. Surprisingly, constitutive expression of one of these genes, *RAY1*, in radiate *Senecio vulgaris* resulted in shortening and reduced zygomorphy of the ray florets. Constitutive expression of *RAY2* increased the ventralization of the ray florets, making them tubular and more actinomorphic. A similar system was reported in *Gerbera hybrida*, where there is a gradual transition in floret zygomorphy from the outer ray florets to the central disc florets. Broholm *et al.* (2008) showed that *GhCYC2* is expressed in a gradient, being most strongly expressed in the outer regions of the developing capitulum, and barely detectable in the central regions. Constitutive expression of *GhCYC2* induced zygomorphy in the central disc florets. In *Helianthus annuus* (sunflower), *double-flowered* mutants, in which some or all of the disc florets are converted to a zygomorphic form, have been linked to capitulum-wide expression of *HaCYC2c*, which is usually expressed only in the outer whorl of ray florets (see Fig. 18.2;

Figure 18.2 Variation in zygomorphic and actinomorphic florets in sunflower. (a) Wild type sunflower has zygomorphic ray florets surrounding actinomorphic disc florets. (b) A *double-flowered* mutant has zygomorphic florets throughout the capitulum. (c) In the wild type sunflower (left), *HaCYC2c* is expressed only in the outer florets (grey colour), but in the *double-flowered* mutant (right) it is expressed throughout the capitulum. Figures (a) and (b) kindly provided by Mark Chapman (University of Southampton).

Chapman *et al.* 2012). The same authors conducted a phylogenetic analysis of *CYC*-like genes in the Asteraceae, concluding that sunflower contained at least 10 members of the family, and that different members of the gene family had been shown to influence zygomorphy in *Gerbera*, *Senecio*, and *Helianthus*. The size of the *CYC*-like gene family in the Asteraceae is most probably attributable to a whole genome duplication at the base of the family, and subsequent subfunctionalization and neofunctionalization of these genes in different lineages may be partially responsible for the variety of floret symmetry and organization produced by different daisies.

18.3 Lability of nectar spur length

Nectar spurs have been described as a key innovation, facilitating pollinator specificity and subsequent plant reproductive isolation and speciation. In many different plant groups the presence or length of nectar spurs is apparently highly labile. For example, in the Antirrhineae, nectar spurs are variously long, short, or absent, and even within a single genus of the tribe, *Linaria*, nectar spur length varies from a few millimetres to several centimetres. These differences in spur length, and those in many other groups, have often been assumed to correlate with shifts between pollinators with different proboscis lengths. The North American species in the genus *Aquilegia* (Ranunculales) display a great variety of spur lengths. The shorter-spurred species are generally bee pollinated, the mid-length-spurred species are hummingbird pollinated, and

the flowers with the longest spurs are pollinated by hawkmoths. By mapping spur length and pollinator type on to a phylogeny of the genus, Whittall and Hodges (2007) showed that spur lengths in *Aquilegia* had increased over evolutionary time, and that the majority of spur length increase was associated with shifts from one pollinator type to another. The authors concluded that spur length evolution proceeded in a series of steps, with rapid evolutionary change associated with a change of range bringing a new combination of plant and pollinator into contact. Lability of nectar spur length has also been studied in a number of different orchid (Asparagales) groups. Box *et al.* (2008) mapped spur length on to a phylogeny of the Orchidinae, and established that the short spurs of *Gymnadenia odoratissima* and *G. austriaca* were derived by an earlier termination of spur growth in comparison with the ancestral long-spurred species *G. conopsea* (see Fig. 18.3). They also showed that the short spur of *Dactylorhiza viridis* was derived in a similar way relative to the long-spurred *D. fuchsii*. Spur length is highly variable in another orchid, *Disa draconis*, with populations in different habitats of South Africa showing significantly different spur lengths associated with different pollinator types. Johnson and Steiner (1997) showed that plants on a lowland sand plain had spurs on average 48 mm in length, and were pollinated by a tanglewing fly with a proboscis 57 mm long. Artificial shortening of these spurs reduced pollen receipt and fruit set in this population. However, mountain populations of the same plant had spurs around 35 mm in length,

Figure 18.3 Variation in nectar spur length. (a) *Gymnadenia conopsea* flowers have very long nectar spurs. (b) *Gymnadenia rhellicani* flowers have almost no nectar spur. Images kindly provided by Matt Box (Sainsbury Laboratory, Cambridge University). See also Plate 22.

pollinated by a horsefly with a proboscis 22–35 mm long. This divergent response to pollinator-induced selection for different spur lengths is a likely driver of speciation in many different plant groups.

Analysis of the molecular events underlying evolutionary change in nectar spur development has been hampered by the lack of clear information on how the spur is built in any model system. In Section 15.4 we discussed the data from the Antirrhineae (*Linaria* and two spurred mutants of Antirrhinum), which suggest that *KNOX* genes might be important in regulating ectopic cell division to initiate nectar spur formation. However, these data are circumstantial and not yet well supported by matching data from other systems. Box *et al.* (2012) followed up their analysis of spur ontogeny in *Dactylorhiza fuchsii* by identifying *KNOX* genes expressed in the developing flower and demonstrating that ectopic expression of one of these genes in tobacco (a very distantly related model, in the Solanales) could induce petal tube outgrowth. However, the orchids are a genetically intractable system, and it is unlikely that more definitive evidence will emerge from these studies. In contrast, the variation in spur length within *Linaria* provides an opportunity to explore correlations between *KNOX* gene expression and spur development within an evolutionary context (B. J. Glover, unpublished data). Although *KNOX* genes have been implicated in the initiation of nectar spur growth, it is clear that the majority of spur length in both *Linaria* and *Aquilegia* is attributable to cell expansion, not cell division, which suggests that other genes might be responsible for evolutionary changes in nectar spurs. In *Aquilegia*, Puzey *et al.* (2012) found that differences in anisotropic cell expansion accounted for the measurable differences in spur morphology between species, suggesting that the molecular drivers of spur lability might be genes involved in determining cell shape.

18.4 Lability of flower colour

The synthesis of floral pigments was described in Chapter 16, and a variety of ways in which flower colour can be modified were discussed in Chapter 17. The colour of flowers is an extremely labile trait, as can be seen from the great diversity of hues observed in any garden. Breeders have used the variation in flower colour present in many species to select a range of horticultural varieties in different colours. Across the angiosperms, flower colour varies enormously. If we look more closely within families or genera it is also possible to see a great variety of flower colour, with evolutionary transitions between colour types. Within *Antirrhinum* the most common colours are ivory, yellow, and magenta, but various shades of pink, pink and ivory patterns, and even purple (in New World species) are observed. Similarly, within *Linaria*, species are found with white, yellow, pink, and purple flowers, and in *Aquilegia*, flowers can be red, yellow, blue, or white. The direction of change in flower colour in *Aquilegia* is associated with shifts in pollinator type. The work by Whittall and Hodges (2007) on nectar spur evolution in the genus indicated that there have been multiple shifts from bee to hummingbird pollination, and from hummingbird to moth pollination. Many of these shifts in pollinator type are associated with shifts in flower colour. Seven independent losses of anthocyanin production have been inferred in the genus (producing white or yellow flowers, associated with moth pollinators), and four transitions between red and blue anthocyanin (two in each direction) (Whittall *et al.* 2006; Hodges and Derieg 2009). In one of these independent losses of anthocyanin production, Whittall *et al.* (2006) reported that genes encoding enzymes acting late in anthocyanin biosynthesis, such as *DFR* and *ANS*, were downregulated, suggesting that the phenotypic shift and consequent pollinator shift was attributable to changes in a transcriptional regulator of anthocyanin pathway genes.

Perhaps the best studied case of lability of flower colour within a genus, associated with a shift in pollinator type, is that of *Mimulus* (Lamiales). In a landmark paper (discussed in more detail in Chapter 20), Bradshaw and Schemske reciprocally introgressed the carotenoid-inducing *YUP* allele from *M. cardinalis* into *M. lewisii*, and the carotenoid-repressing *YUP* allele from *M. lewisii* into *M. cardinalis* (see Figure 20.2; see also Chapter 20). The parental species are pollinated primarily by hummingbirds (red/orange *M. cardinalis*) and bumblebees (pink *M. lewisii*), respectively, but the novel-coloured lines experienced significantly increased visitation from the 'wrong' pollinator. The authors concluded that this particular flower colour polymorphism was attributable to mutation at a single locus, and that it had significant effects on the identity of pollinators attracted to the flowers.

The vast majority of studied instances of an evolutionary transition in flower colour are the result of loss of function of either a regulator of the anthocyanin pathway or of part of the pathway itself. Wessinger and Rausher (2012) reported that, although many white-flowered mutants have been found to have lesions in genes encoding anthocyanin biosynthetic enzymes, of the recorded cases of fixation of a white allele in a population or species all examples involved changes to the activity of a regulatory transcription factor. Those authors also investigated colour shifts between blue and red, noting that the great majority of such transitions involved the loss of blue pigmentation, usually through degradation of the gene encoding flavanone 3'5' hydroxylase (F3'5'H; see Chapter 16). From these data it is apparent that flower colour shifts usually involve loss of part of an established pigmentation pathway, and occur relatively frequently. For example, Smith and Rausher (2011) observed that there have been two shifts from blue flowers to red flowers in the genus *Iochroma* (Solanales), one of which (in *I. gesnerioides*) they attributed to deletion of the gene encoding F3'5'H along with downregulation of the gene encoding F3'H. However, in *Phlox drummondii* (Ericales) a unique gain of function allele has been reported (Hopkins and Rausher 2011). A dark red flowered form of *P. drummondii*, which usually has light blue flowers, is found in populations in the southern USA growing sympatrically with blue-flowered *P. cuspidata*. The novel colour morph of *P. drummondii*, which acts to reinforce reproductive isolation between *P. cuspidata* and *P. drummondii*, is the result of two molecular changes. Loss of the gene encoding F3'5'H is responsible for the loss of blue tone to the anthocyanin, as in previous studies. However, the intense dark red of the novel morph was attributed to gain of an allele that conferred stronger expression of a MYB regulator of anthocyanin synthesis.

Given that much of our understanding of the basic anthocyanin synthetic pathway is derived from work in *Antirrhinum majus*, it is not surprising that the evolution of flower colour and pattern has also been studied in this genus. The European species of *Antirrhinum* do not produce blue-toned anthocyanins, but a North American species of the Antirrhineae known variously as *A. kelloggii*, *Neogaerrhinum strictum*, and *Asarina stricta* does. It is not clear that this plant does fall within the genus *Antirrhinum*, but various phylogenies place it as sister to the European *Antirrhinum* group. Ishiguro *et al.* (2012) found that *A. kelloggii* contains two functional copies of a gene encoding F3'5'H, which have apparently been lost from the European section of *Antirrhinum*. These genes were shown to act in petunia (Solanales) to increase delphinidin synthesis, confirming their role in hydroxylation of the anthocyanin molecule. Within the European *Antirrhinum* complex, the variation in flower colour between yellow, magenta, and ivory is thought to be primarily controlled by three loci—*ROSEA* and *ELUTA*, encoding MYB transcription factors (see Section 17.3.2 in Chapter 17), and *SULFUREA*, which has not yet been isolated (Schwinn *et al.* 2006). It is difficult to assess the direction of evolutionary change within the European section of *Antirrhinum*, as the radiation within the group is relatively recent. However, Whibley *et al.* (2006) explored the genetic basis of flower colour variation between sympatric *A. majus* subsp. *pseudomajus* (magenta) and *A. majus* subsp. *striatum* (yellow), and found that they expressed different alleles of *ROS*, *EL*, and *SULF*. Crosses between these subspecies resulted in some orange flowers in the F2, but Whibley *et al.* (2006) hypothesized that this colour morph was not found in wild *Antirrhinum* species because it was unattractive to pollinating bees, suggesting that this mechanism might maintain reproductive isolation between *A. majus* subsp. *pseudomajus* and *A. majus* subsp. *striatum*. Many of the ivory-flowered European species of *Antirrhinum* have anthocyanic veins. This venation pattern is regulated by another MYB transcription factor, *VENOSA* (see Section 17.3.2 in Chapter 17; Shang *et al.* 2011), and is only visible in a background that is not fully anthocyanic. Shang *et al.* (2011) proposed that the veined form of *Antirrhinum* was ancestral to the fully pigmented form, with at least two shifts to full red flowers within the genus. Their pollinator studies, along with those of Whitney *et al.* (2013), suggested that veined flowers were visited by bumblebees as often as, or more often than, red flowers, with behavioural studies indicating that the cyanic veins might act as learned nectar guides.

Venation patterning is just one of a number of patterns of flower colour that are labile within plant groups. Another very variable trait is the presence of petal spots—that is, regions of the corolla with dense accumulations of pigment. These spots often contain a different pigment to that found throughout the rest of the petal tissue. Martins *et al.* (2013) investigated the molecular basis of petal spot variation in *Clarkia gracilis* (Myrtales), in which different subspecies produce spots at the base of the petal, in the centre of the petal, or not at all. They found that this variation in spot morphology could be attributed to the interactions of two genetic loci that regulate expression of a duplicate copy of the gene encoding dihydoflavonol reductase (*DFR2*). Early expression of *DFR2* in the spot region, in the presence of F3'H but not F3'5'H, resulted in an intense spot of red cyanidin pigment. Later expression of *DFR1* throughout the petal, in the presence of F3'5'H, resulted in malvidin production and a background mauve colouration (see Fig. 18.4). Petal spots appear to be ancestral in *Clarkia*, but this study shed useful light on how they might have evolved

in the ancestor of the genus, and on the mechanisms that might underlie their loss and positional change in various species and subspecies.

18.5 Lability of epidermal morphology

In Section 17.6 of Chapter 17 we discussed the idea that epidermal cell shape could influence flower colour, temperature, and the grip offered to foraging insects. The presence of conical-papillate petal epidermal cells is a trait that is apparently ancient within the angiosperm lineage, being found in *Amborella* (Amborellales) and among species such as *Cabomba* in the Nymphaeales. The MYB transcription factor subfamily, the *MIXTA*-like genes (also known as MYB subgroup 9), which regulates the production of conical petal cells, is also ancient. Brockington *et al.* (2013) showed that a duplication event generated two subgroup 9 lineages before the divergence of the gymnosperms and the angiosperms, and that within the eudicots additional duplication events have occurred in both branches, generating four clades of genes. Analysis of the evolutionary lability of petal epidermal morphology has focused on three transitions: the loss of conical

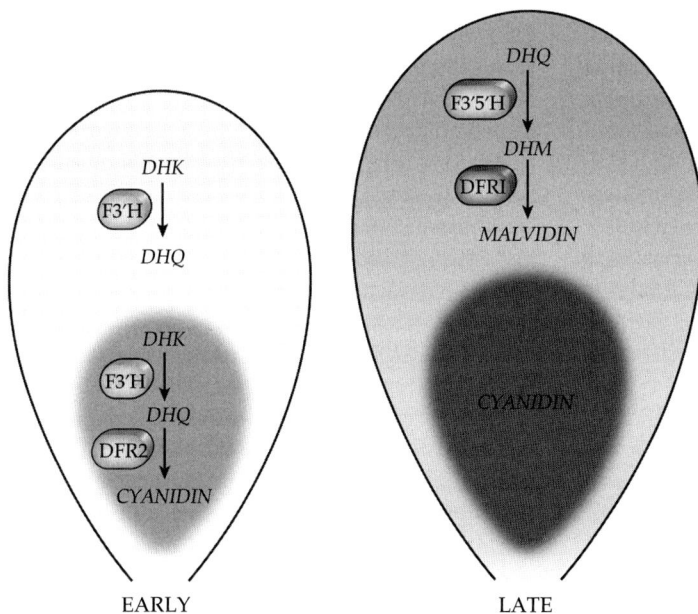

Figure 18.4 Development of the *Clarkia gracilis* petal spot. Early expression of *DFR2* in the spot region, in the presence of F3'H but not F3'5'H, results in red cyanidin pigment. Later expression of *DFR1* throughout the petal, in the presence of F3'5'H, results in mauve malvidin production. See also Plate 23.

cells in response to pollinator shifts, the variation in conical cell distribution on petals with different functions, and the gain of conical cells on petaloid bracts and stamens.

The loss of conical cells has been reported in many lineages. Around 20% of petaloid angiosperms have petal epidermal cells that are not conical, and many of these are found in groups where the ancestral state is very likely to be the presence of conical cells. For instance, a number of species of *Solanum* (Solanales) have been found to have flat petal epidermal cells, possibly in response to a shift to buzz pollination which requires little or no petal handling by the bee. Similarly, flat-celled species are found in *Nicotiana* (Solanales), often associated with a switch to hovering moth pollinators. In the Antirrhineae, flat petal epidermal cells are associated with a shift to hovering hummingbird pollination (all examples from B. J. Glover, unpublished data). These transitions suggest that conical cells might be most relevant to pollinators that land on and manipulate the petals, such as bees. It has also been suggested that the presence of conical cells might correlate with flowers which are presented at angles that are difficult to handle, but Rands *et al.* (2011) found no evidence to support this idea. Ojeda *et al.* (2009) recorded enormous variety in the presence of conical cells on the different petal types (dorsal, lateral, and ventral) of the zygomorphic legume flowers (Fabales). The functional basis of this variation is not known in most cases, but in a more detailed study, Ojeda *et al.* (2012) showed that four species of *Lotus* native to the Canary Islands had lost conical epidermal cells from the dorsal petal, in conjunction with a shift from bee to bird pollination. Phylogenetic character mapping indicated that this loss had occurred only once, before the radiation of the four species. It is therefore likely that this evolutionary transition occurred when the ancestor of the four *Lotus* species colonized the Canary Islands and acquired a new major pollinator.

Petal epidermal cells are not just labile in a binary presence/absence way between lineages, but also vary in their distribution on the different petals of zygomorphic flowers, and in their positions within individual petals. For example, Ojeda *et al.* (2009) reported that most legume flowers had conical cells on the dorsal petal (the standard, or flag),

but in the commonly grown broad bean (*Vicia faba*) they are found only on the lateral (or wing) petals. These differences may reflect variation in the way that pollinators interact with the flower. Bradshaw *et al.* (2010) examined the epidermal surfaces of the labellum (the complex inner tepal in the ventral position) of orchids from the sexually deceptive genus *Ophrys*. They found considerable variation in the position, extent, and cuticular patterning of both conical cells and trichomes, and noted that, in those species with highly reflective patches at the centre of the labellum, the presence of this glossy speculum correlated with the presence of flattened epidermal cells. Similarly, Mudalige *et al.* (2003) showed that different species of *Dendrobium* orchids had different distributions of conical and flat cells on the labellum and other tepals. Analysis of the relationships between these morphological transitions and the pollinator interactions of the different orchid species might provide insight into the functional significance of conical cells in these highly specialized flowers.

Conical epidermal cells might be a defining feature of a petal to a developmental biologist, but they are not limited solely to petals. Across the angiosperms a number of groups have experienced a loss of animal pollination and an associated loss of petals, followed by a reversion to animal pollination in some derived lineages. In these systems, sepals and/or stamens have been recruited to a role in pollinator attraction, and in some cases have become modified to be more petal-like. Di Stilio *et al.* (2009) showed that the reversion to insect pollination was accompanied by a gain of conical cells on the sepals of *Thalictrum thalictroides* (Ranunculales), and by a gain of conical cells on the stamens of *T. filamentosum*, whereas wind-pollinated *T. dioicum* had only flat floral epidermal cells. The authors also showed that this gain of conical epidermal cells on novel organs probably occurred through the activity of *MIXTA*-like genes, as the expression of these genes coincided with the presence of conical cells on the sepals and stamens on the insect-pollinated species, but was restricted to the carpels of the wind-pollinated one. It is clear that epidermal morphology is a highly labile floral trait, with numerous gains, losses, and changes of position reported across the angiosperm lineage.

18.6 Lability of floral scent

The production of floral scent has not been discussed much in this book, but it is a trait that is believed to be highly labile, and which clearly has important associations with particular pollinators. The biosynthetic pathway of floral scent production has been well studied in a number of species, particularly in petunia (Dudareva and Pichersky 2006). Verdonk *et al.* (2005) showed that scent production in petunia was regulated by an MYB transcription factor called ODORANT 1 (ODO1). In a landmark paper published in 2011, Klahre *et al.* showed that the pollinator shift between moth-pollinated *Petunia axillaris* and bird-pollinated *P. exserta* could be partly attributed to ODO1. In a cross between the two species, scent production (a *P. axillaris* trait) mapped to two QTLs, one of which was identified as *ODO1.* As with flower colour, it appears that the production of floral scent can evolve by mutation of the gene encoding the MYB transcription factor that regulates the pathway.

Various studies have shown that floral scents can attract specific sets of floral visitors, contributing to reproductive isolation and speciation. The specificity of the scent can be attributed either to a specific compound produced by an individual species, or to a particular mix of more commonly occurring scent compounds in particular ratios. Both the production of specific compounds and the constitution of floral scent mixes have been shown to be evolutionarily labile across groups of plants. For example, Waelti *et al.* (2007) showed that moth-pollinated *Silene latifolia* (Caryophyllales) produced floral scent comprising 44% monoterpenoids and 23% benzenoids, whereas bee- and butterfly-pollinated *S. dioica* produced 33% monoterpenoids and 46% benzenoids. This variation in scent composition

was shown to reduce interspecific hybridization—applying the same scent to both flowers increased cross-pollination. The great majority of research into species-level differences in floral scent and the consequences for pollinator attraction has been conducted in the orchids (Asparagales). Many orchids achieve pollination by food mimicry (resembling a rewarding species despite producing no nectar) or sexual mimicry (resembling a female insect to attract mate-searching males). In both cases the floral scent has been shown to be a key component of the mimicry. Various studies, particularly in the sexually deceptive genus *Ophrys*, have shown that floral scent mixes are highly labile and specific to particular species, and in turn attract specific insect species as pollinators (reviewed by Ayasse *et al.* 2011).

Studies of evolutionary lability are necessarily constrained by the need for a comparative approach within a solid phylogenetic context. It is rare to find a system in which the direction of floral trait change can be readily mapped on to a well-resolved phylogeny, identifying genetically amenable sister species for comparison of the molecular, as well as ecological, correlates of the morphological change. However, as the examples discussed in this chapter show, it is certainly possible to explore the question of floral lability for a number of traits, if the system is chosen carefully. The data discussed in this chapter demonstrate that floral traits can and do change, with mutations to the regulation or function of the transcription factors that control particular developmental pathways often being the source of the observed variation. In the next section of the book we shall discuss the evidence that this floral lability results in frequent occurrence of the pollination syndromes described in Chapter 14.

The Influence of Pollinators on Floral Form

CHAPTER 19

Are flowers under selective pressure to increase pollinator attention?

In the previous section of this book we discussed a variety of ways in which flower visibility and appearance can be enhanced, through changes to size, shape, structure, and colour. The underlying assumption of much of the work on flower development and morphology is that these features serve to increase the attractiveness of the flower to pollinating animals, thus maximizing pollinator attention, and consequently seed set and fitness. As we discussed in Chapter 14, it has long been believed that these elaborations are the consequence of adapting to attract particular types of pollinating animals, resulting in pollination syndromes. In Chapter 21 we will analyse in detail the evidence for the existence of these syndromes. However, before we consider whether pollination syndromes do exist, and indeed whether pollinating animals do exert selective pressure on floral form at all, we need to consider whether there is evidence that two underlying assumptions are met. In Chapter 20 we will look at the evidence that pollinating animals actually discriminate between the traits described in the previous section—if they cannot or do not, then those traits cannot be under selective pressure unless they enhance pollen transfer indirectly. In this chapter we shall address an even more fundamental issue—do plants actually benefit from increased pollinator attention and should floral attractiveness therefore be expected to increase across generations?

19.1 Competition for pollinator attention

The pollination syndrome concept rests on the idea that individual plants are competing for pollinator attention. If there is no pollen limitation in a given habitat, then the limits to the reproductive success of an individual plant are not imposed by the pollination system. If this is the case, increasing the showiness of the flower, or specializing in some way to attract a subset of pollinators, will have no positive effect on the plant's fitness. Indeed, since pigments and elaborate petal structures are costly in terms of energy and key nutrients such as nitrogen, in the absence of a beneficial increase in seed set or pollen export such structures will actually have a negative effect on plant fitness. Elaboration of floral attractants only enhances the reproductive success of a plant if by doing so more of that plant's pollen reaches an appropriate stigma, or more of its egg cells are fertilized by sperm from incoming pollen grains. If the plant is already dispersing pollen and being pollinated quite successfully, then there is no pressure to change.

Pollination competition falls into two different categories, which have different evolutionary consequences. The variety that we are most concerned with in this chapter is *exploitation competition*. Exploitation competition occurs when a plant receives insufficient visits by pollinators because one or more other plants nearby are more attractive. This situation results in the first plant setting less seed, or contributing fewer male gametes to the gene pool. The fitness of the plant is therefore compromised. Exploitation competition can be intraspecific or interspecific. When it is intraspecific it may be predicted to drive elaboration of floral structures and increase showiness of flowers. When it is interspecific it is more likely to result in pressure to alter habitat or flowering time, or else to specialize in attracting a smaller subset

Understanding Flowers and Flowering. Second Edition. Beverley Glover.
© Beverley Glover 2014. Published 2014 by Oxford University Press.

of pollinators. Interspecific exploitation competition is anecdotally of commercial importance, with many fruit farmers reporting that flowering weed crops decrease fruit set of their orchard trees. Free (1968) observed honeybee behaviour in fruit orchards containing dandelions, and confirmed that most bees foraged preferentially on dandelions, both for nectar and for pollen, at the expense of the fruit trees.

The second form of pollination competition, interference competition, is only ever interspecific. It occurs when pollen of the wrong species lands on a stigma, where it can begin to germinate and clog up the surface, reducing the chances of the right pollen grains germinating successfully and thereby decreasing female fitness. At the same time, placement of pollen on the stigmas of plants of another species effectively wastes that pollen, resulting in reduced male fitness. Interference competition can have more significant effects than exploitation competition, particularly in situations where density of a particular plant species is low relative to the densities of other species (Kunin 1993). Interference competition may be expected to drive changes that minimize the number of pollinating animal species visiting a plant or the number of plant species visited by each pollinator, preventing interspecific pollen transfer. Such changes might include alterations in habitat or flowering time, but might also include increased specialization towards the foraging preferences of a limited subset of pollinators. This latter set of changes is only possible if a sufficient number of specialist pollinators are available, and is heavily dependent on the density of both plant and pollinator.

19.2 Facilitation of pollination

Attempts to identify and quantify the importance of pollination competition are made more difficult by the complicating factor of facilitation of pollination. In some cases, the presence of more flowers nearby, whether of the same or different species, may actually increase the pollination success of a particular plant. Such an increase in pollination success is likely to be frequency dependent, and when a certain number of nearby plants is reached the balance will tip towards a competitive situation (illustrated in Fig. 19.1). For example, a single individual plant flowering in an otherwise barren environment may simply not be spotted by passing pollinators, or may be judged insufficiently rewarding to be worth stopping for. The addition of other plants in bloom to the same locality will increase the chances of a pollinator visiting the area and thus increase the chances of the first plant receiving pollination services. However, when the number of nearby plants increases sufficiently, the original plant will find itself competing with its neighbours, and its fitness may decline accordingly. Some of the methods used to identify pollination competition take into account the importance of facilitation, but other methods may not fully register its contribution.

19.3 Techniques for investigating the role of pollinator attention in limiting fitness

There are three main methods used to investigate the role of pollination competition in limiting plant fitness. The aim of each method is primarily to

Plot

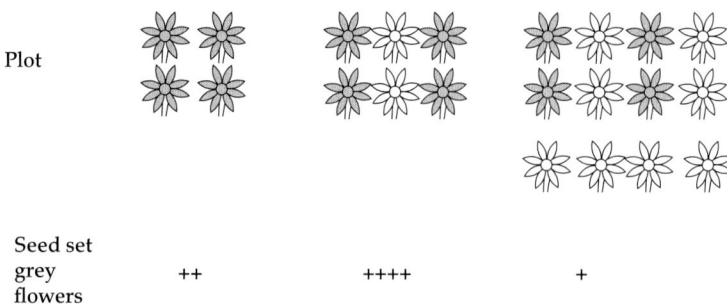

Seed set
grey
flowers

++   ++++   +

Figure 19.1 Facilitation and competition in mixed species plots. Addition of flowers of species B (white) to a plot of flowers of species A (grey) initially facilitates pollination of species A by attracting more pollinators to the plot. However, when relatively high proportions of species B are present, competition for pollination occurs and fitness of species A is compromised.

establish whether such competition occurs or has occurred in a particular situation, and then, if possible, to quantify the extent or importance of that competition. These methods only provide data on particular plants growing in particular habitats, and it has been difficult to extrapolate from these data, as they themselves are highly variable. Each method has advantages and disadvantages, but each provides useful information and therefore all three methods are used where appropriate in current studies.

19.3.1 Fruit and seed set following hand pollination

Perhaps the simplest method of identifying pollination competition is to hand-pollinate flowers and observe the effects on fruit and seed set, relative to control flowers not hand-pollinated but open to natural pollinators. In theory, such an experiment will reveal what proportion of ovules in the control plants is unfertilized because of an insufficient supply of suitable pollen. This may be due to a shortage of pollinator attention, or else to an excess of heterospecific pollen clogging the stigma. In practice, this method is open to several criticisms. First, it takes no account of the male component of fitness. It is not sufficient to assume that male fitness is directly related to female fitness, even though in many cases it might be. If the reduced seed set of the control flowers is attributable to insufficient pollinator attention, then male fitness might well be similarly affected. However, if the limitation is due to heterospecific pollen clogging the stigma, then variations in pollinator foraging habits may, for example, mean that the pollinator has moved on only to plants of the same species, and has actually caused the experimental plant's male fitness to be extremely high. Since this experimental method makes no observation of pollinator behaviour, and male fitness is notoriously variable, it is likely that erroneous conclusions about the degree of pollination competition, if not about its absolute occurrence, are often drawn from its use. Secondly, this method must be used with great caution to ensure that female fitness is accurately recorded. For example, simply scoring fruit set in hand-pollinated and control plants is insufficient, as many species can produce fruit even in

the absence of fertilization, a situation which obviously does not contribute to female fitness and will often actually decrease it through waste of resources. Similarly, seed number within fruit is also variable within species, and so it is necessary to quantify seed set, not fruit set, even if the plant is known not to produce fruit entirely free from seed. Ideally, the set of viable seed should be scored (by allowing the seed to germinate), to ensure that non-viable hybrid seed is not being counted. An even greater difficulty in correctly scoring female fitness is that increased seed set in one part of the plant, or one flowering season, may be offset by reduced seed set in other parts or future seasons. This reflects the costly nature of seed and fruit development, and the fact that seed set in many plants is limited not by pollination competition but by resource availability. It is therefore not sufficient to hand-pollinate all flowers on a single branch of a tree, and compare them with control branches—the control branches might be setting less seed because their resources have been allocated to the hand-pollinated branch. It is similarly inappropriate to hand-pollinate all the flowers on a tree in a single growing season, without reference to the subsequent seasons—in subsequent seasons seed set might be reduced to compensate for the increased resource use in the experimental season. Little is known about the effects of nutrient acquisition and use in one year on the reproductive success of a plant in subsequent years (Horvitz and Schemske 1988). However, it has been clearly shown in several tropical orchid species (Asparagales) that increased fruit set in one season (as a result of hand pollination) can result in significant negative effects on vegetative growth, inflorescence production, and flower production in subsequent years (Montalvo and Ackermann 1987; Ackerman and Montalvo 1990). Ideally, then, these experiments should compare seed set of a plant on which every single flower has been hand-pollinated for its entire life with seed set of a control plant. Such experiments are most easily performed with short-lived annuals, so many of the data generated using this method can only be extrapolated to long-lived plants with some caution. Several authors have pointed out that assuming that seed set is limited by either pollen limitation or female resource availability is an oversimplification, with many plants likely to be operating at an equilibrium between these two limiting factors (for a detailed discussion of the causes and consequences of pollen limitation, see Ashman *et al.* 2004). Wesselingh (2007) proposed an alternative approach to counter the problem of separating pollen limitation from resource availability. She argued that viewing the plant as made up of independent physiological units, consisting of individual inflorescences or branches or other units, according to plant growth habit, might yield more useful insight. Within one of these units resource reallocation could be expected to occur, but between them it might be less likely. Wesselingh suggested that studies focused on comparing seed set between hand-pollinated and open-pollinated units might therefore be revealing.

A variation on the method described above is to compare seed set of open-pollinated flowers under different natural pollinator regimes—for instance, at different times of the season or before and after events (such as disease or natural disasters) that dramatically alter pollinator availability. However, the same difficulties of interpretation apply to this form of analysis.

19.3.2 Mixed species plots

A direct way to measure the effect of pollination competition on fitness is to compare artificial plots containing a single plant species with those containing a mix of species. If the plots are in the same environment, the only variables should be the number and species composition of plants in them. It is then possible to measure a number of factors relating to fitness, including pollinator visitation rates, fruit set, and seed set. As before, seed set is a much more useful measure of female fitness than fruit set. Recording pollinator visitation will give a measure of the extent to which any negative consequences of mixed plots are due to competition for pollinator attention as opposed to the inhibitory effects of heterospecific pollination, and from that it may be possible to estimate the male component of fitness. It is also possible, using such plots, to assess where facilitation of pollination turns into competition. If a range of plots containing a decreasing proportion of the plant of interest is used, then it becomes possible to chart the increase in seed set associated with

facilitation and the point at which seed set begins to fall, marking the onset of competition.

The major drawback of this method of identifying pollination competition is the extent to which it is labour intensive. The difficulties associated with producing a range of plots and recording pollinator visitation as well as scoring seed set are such that only small plots are used, containing only two or three species, and data are only available from a very small number of habitats. Furthermore, it is usually only possible to use small fast-growing plants, or those that can be easily transplanted, in these artificial plots. It is accordingly difficult to extrapolate from these data to the likely importance of pollination competition for larger, long-lived plants such as trees, and for plants growing in extremely species-diverse habitats, such as tropical forests.

19.3.3 Character displacement

The third method used to identify pollination competition is the least direct one. In this method, plant species within a phylogenetic group growing in a particular environment are analysed to see whether characters appear to have been displaced by the need to avoid pollination competition (illustrated in Fig. 19.2). For example, a study might consider all the species of a particular genus in one locality, predicting that, if pollination competition had been a problem in the past, present-day floral features would vary between species to attract different pollinators. Alternatively, all the species in a habitat which attract a single type of pollinator might be studied, to assess whether the timing of their flowering is staggered to minimize competition. The power of these studies is that they can look at a large number of species in a wide range of habitats. Plant life habit does not matter in these studies, so trees can readily be compared with shrubs and annuals. Geographically distant localities can be compared, often using previously published data. An obvious null hypothesis in these experiments is that flower form or flowering time (or whatever feature is studied) is constrained by phylogeny, and thus that closely related species should have similar characters. Deviation from this null hypothesis is easy to spot and can be readily analysed using statistical methods. However, the major drawback of

(a)

Spring   Summer

(b)

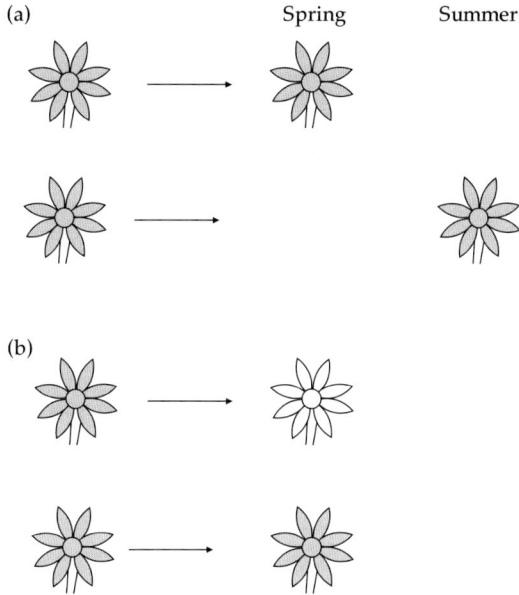

Figure 19.2 Character displacement to minimize pollination competition. (a) Two plants which would otherwise compete for pollinator attention can stagger their flowering time to prevent overlap. (b) Alternatively, other floral characters, such as colour, can be displaced to attract different pollinators.

this approach is that it can only ever be correlative. No direct proof of current pollination competition is produced, and its previous presence can only ever be indirectly inferred with caution. Without knowing whether the species under analysis evolved sympatrically or allopatrically it can be very difficult to draw conclusions from simple observations of current characters. Indeed, some authors have argued that plants seldom coexist under stable conditions for long enough for pollination competition to drive character displacement, suggesting that even where such displacement is observed it may be too simplistic to attribute it to previous competition for pollinator attention (Feinsinger 1987).

19.4 Evidence based on fruit and seed set following hand pollination

In the last 30 years many studies have been published describing the consequences of hand pollination of fruit and seed set. These studies cover many different species, although predominantly short-lived and annual plants, including the insect-pollinated relative of *Arabidopsis thaliana, A. lyrata* (Sandring and Agren 2009). One of the best studies of pollination limitation using hand pollination was published in 1988 by Horvitz and Schemske, who concluded that their system under investigation was not statistically significantly limited for pollinator attention. Plants of *Calathea ovandensis* (Zingiberales) growing in a Mexican forest were examined. This species is herbaceous, flowering in multiple years, but only usually producing a single inflorescence per plant per year. Pairs of plants at similar developmental stages and within a few metres of one another were chosen for analysis, and all flowers in the inflorescence of one plant were hand-pollinated, while no flowers on the control plant were pollinated. Since all flowers within an inflorescence were treated, and most plants produce only a single inflorescence per year, these authors effectively hand-pollinated entire plants for comparison with controls. They found that hand-pollinated inflorescences initiated 24% more fruits than controls, but since not all initiated fruit developed fully, they only set 6.8% more mature fruit. Total seed set did increase by 12.6% in the hand-pollinated inflorescences, but this difference was not statistically significant. The authors could not therefore conclude that there was direct evidence that this species was suffering pollination limitation in this locality. Instead, they concluded that seed set was likely to be resource limited, a conclusion strongly supported by the abortion of fruit early in development in the hand-pollinated plants. It should be noted here that pollination limitation and resource limitation are not mutually exclusive conditions for a plant, and that it is possible that the interaction between these limiting factors actually determines reproductive success. Campbell and Halama (1993) showed that pollen availability limited seed number per flower of *Ipomopsis aggregata* (Ericales), but that nutrient availability limited the number of flowers produced.

In a more recent example of this approach, Ghazoul (2006) hand-pollinated four flowers on each of 50 *Raphanus raphanistrum* (wild radish, Brassicales) plants in experimental plots, and merely tagged four flowers at the same developmental stage on 50 control plants. Pollen limitation was analysed by comparing fruit and seed set of the hand-pollinated plants versus the controls. Although this experiment lacks the thoroughness of the previously described study, it was conducted with plants in defined plots and could control for environmental differences. The author found that hand-pollinated flowers set 50% more seed than control flowers, and that this difference was statistically significant.

A recent review of pollen limitation analysed all of the published literature describing hand pollination as an attempt to supplement pollen supply. Knight *et al.* (2005) found that most studies recorded fruit set, rather than seed set. Of the 482 published reports on the effect of hand pollination on fruit set, 304 (63%) showed significant pollen limitation in the control plants.

Studies observing fruit and seed set in wild situations with variable pollinator profiles have also found evidence for pollination limitation. Bumblebee-pollinated *Antirrhinum majus* plants in an experimental field plot in Norfolk flowered all summer, but the bees were most active in June and early July. Early in the summer as many as 88% of the flowers were pollinated (measured by emasculating flowers to prevent self-pollination, tagging, and then scoring fruit set). Later in the summer when the bees were less numerous only 41% of flowers were pollinated (B. J. Glover, unpublished data).

A more striking example of this approach was provided by Rathcke (2000), who analysed fruit set of a bird-pollinated shrub in the Bahamas in response to hurricanes. *Pavonia bahamensis* (Bahama swamp-bush, a member of the Malvales and endemic to the Bahamas) is pollinated by two bird species, the Bahama Honeycreeper (or Bananaquit) and the Bahama Woodstar. During the winter flowering season of 1994–95, both bird species were frequently observed on Bahama swamp-bush flowers, 98% of stigmas that were monitored received pollen, and fruit set was close to 100%. Following a severe hurricane in October 1996 (Hurricane Lili, a Category 2 storm), both bird species were virtually absent in the flowering season of 1996–97. At the same time only 49% of stigmas received pollen, and fruit set dropped to only 11%. Since almost all fruit in this species contains the same number of seeds, fruit set is a reliable indicator of seed set and thus of total female reproductive fitness. It is clear from these data that this particular species was limited for pollinator attention when its pollinator population declined following severe weather conditions. Other circumstances that might be expected to have an impact on pollinator populations include changes in predator populations, disease epidemics, and alteration to habitats, including habitat loss. It is likely that many different sets of circumstances frequently result in pollination limitation for a range of different plant species growing in variable habitats all over the world.

19.5 Evidence from mixed species plots

Analysis of pollination competition in mixed species plots has an advantage over other methods in that it allows facilitation of pollination to be monitored at the same time. A study conducted in 2006 by Ghazoul provides a textbook example of how relative population densities determine whether mixing flowering species together results in competition or facilitation of pollination. This result is consistent with the proposition that facilitation and competition are simply opposite ends of a continuum of interactions. For our purposes, the existence of such a continuum means that, even where facilitation seems to be occurring in a particular habitat, it is possible that at other times in the flowering season or in other localities with variable population densities, competition for pollination is occurring between the same groups of plant species. Such competition may be expected to result in adaptations to increase floral attractiveness or to make flowers more distinctive than those of other members of the community. At the same time, facilitation may actually increase intraspecific competition for pollinator attention, by ensuring that there is maximum seed set by those flowers of each species that are most attractive.

Ghazoul (2006) established seven field plots containing 50 *Raphanus raphanistrum* (wild radish, Brassicales) plants and 50 *Cirsium arvense* (creeping thistle, Asterales) plants. The number of inflorescences on the *C. arvense* plants varied naturally, and was recorded twice daily over 8 days. At the same time the number of pollinating insects visiting the *R. raphanistrum* flowers within a 2-minute time window was also recorded. The total number of pollinating insects within the plot was then re-corded in a subsequent 2-minute window, and the proportion of insects visiting the *R. raphanistrum* flowers was calculated from these two observa-tions. These data thus allowed a comparison of the proportion of pollinators visiting species 1 with the number of inflorescences of species 2 present. At relatively low numbers of *C. arvense* inflorescences (fewer than 32), the presence of extra inflorescences increased pollinator visitation to *R. raphanistrum*. Therefore small numbers of flowers of a second species facilitated pollination of one plant species. However, as the number of *C. arvense* inflorescences increased above 32, a decreasing proportion of pol-linators visited the *R. raphanistrum* flowers. When over 300 *C. arvense* inflorescences were present in the plot, pollinator visits to *R. raphanistrum* ceased almost entirely. Competition for pollinator attention thus became extreme when *R. raphanistrum* was at a lower relative density within the mixed plot. This study suffers from a number of problems, notably that the presence of insect visitors does not neces-sarily equate to subsequent seed set. However, pre-vious experiments reported within the same paper had shown a correlation between flower visitation and seed set, so it is likely that the same correla-tion held true in this particular study. It is certainly the case that an absence of insect visitors, as seen when large numbers of competing inflorescences were present, will result in a significant reduction in seed set.

In a study that used similar methods, Brown *et al.* (2002) investigated the consequences of pollina-tion competition from an invasive Eurasian species in the Myrtales, *Lythrum salicaria* (purple looses-trife), for seed set in the American *Lythrum alatum* (winged loosestrife). Over 2 years they measured pollinator visits at 15-minute intervals and seed set per fruit for 15 fruits per plant, comparing plants in single species plots and mixed species plots. They found that increasing the number of *L. alatum* plants in a single species plot had no significant effect on seed set per fruit. It did cause a reduction in visi-tor numbers per flower per 15-minute interval, sig-nificantly so when the number of plants within a plot was tripled. However, since there was no effect on seed set it seems likely that this increased com-petition for pollination attention was of no fitness

consequence to the plants. In contrast, when *L. sali-caria* was added to the plots, *L. alatum* plants expe-rienced reductions in both seed set and number of visitors per flower. The reduction in pollinator vis-its was more significant than that observed in the single species plots, suggesting that the reduction in seed set might reflect insufficient pollinator visits, as well as the transfer of heterospecific pollen. This is another example of a specific situation in which competition for pollinator attention may occur. However, the mobility of most plant species, often aided by the activity of humans, makes it very like-ly that large numbers of plants regularly come into contact with invasive species and suffer pollination competition as a result. A similar finding was re-ported by Chittka and Schürkens (2001), who com-pared bumblebee visits to flowers, and seed set per inflorescence, of plots of *Stachys palustris* (Lamiales) growing on German riverbanks with and without the Asian invasive *Impatiens glandulifera* (Ericales). This species has been shown to have a higher sugar production rate than any native European flower in which this has been measured. Not surprisingly, bumblebee visits to *S. palustris* and seed set of the inflorescences were both significantly reduced in those plots that also contained *I. glandulifera* plants. Again, in this situation it is likely that selection will favour any variation in the *S. palustris* flowers that makes them more attractive to pollinators generally, or more attractive to a specific subset of pollinators. Another recent example of this type of analysis was provided by Kandori *et al.* (2009). These authors demonstrated that the Japanese native dandelion (*Taraxacum japonicum*, Asterales), growing in both artificial and wild plots, received fewer pollinator visits and set less seed when the invasive *Taraxacum officinale* was also present, even though *T. officinale* is apomictic. The relatively higher nectar production of *T. officinale* is probably responsible for its greater attractiveness, and also probably explains its long history of being blamed for the depletion of fruit tree pollinators (Free 1968).

Using genetic markers to assess outcrossing in the progeny of plants in mixed species plots led Bell *et al.* (2005b) to conclude that pollination competi-tion affected seed quality as well as seed quantity. In plots containing *Lobelia siphilitica* (Asterales) as well as *Mimulus ringens* (Lamiales)—plants that both have similar blue flowers and coexist in American meadows—the *Mimulus* plants set significantly fewer seeds per fruit than in single species plots. This reduction in seed set was also associated with a reduction in outcrossing rate, from 0.63 to 0.43, indicating that pollination limitation had resulted in an increased rate of self-pollination and thereby reduced seed quality and likely fitness of progeny.

19.6 Analysis of character traits potentially displaced by pollination competition

Studies that search for evidence of pollination competition through displacement of character traits have primarily focused on flowering time (commonly referred to as flowering phenology in the ecological literature, although not in the literature concerned with the molecular regulation of flowering-time control). The suggestion that competition for pollinator attention might lead to displacement of flowering time in plants growing together is based on two processes. First, it is possible that evolutionary displacement of flowering times may occur, with those individuals flowering at times slightly offset from their competitors being most successful and setting most seed. Alternatively, an ecological sorting process may serve to eliminate species from a locality if their flowering competes too directly with other species, resulting in assemblages of plants that do not flower together. However, several factors will also limit the extent to which flowering time can be displaced. To start with, flowering time is under strict genetic control, as discussed in Section IIA of this book. The many different environmental and endogenous factors that interact to regulate flowering time are not easily perturbed, as there are so many checkpoints and fail-safes within the complex network of interactions. For displacement of flowering time to occur, then, it is necessary for mutations to occur in one or more key genes that are sufficient to influence the end point of the entire network. Such mutations do occur in crop plants, where they have allowed breeders to manipulate flowering time to commercial advantage, and so it is likely that they will also occur from time to time in wild species. It is also necessary for competition for pollination to

be sufficiently strong for these mutants to be at a selective advantage strong enough to enable them to persist and eventually dominate a population. At the same time, the environment in which a plant finds itself will have a significant effect on flowering time. No amount of pollination competition should be sufficient to drive a species to flower under extremely cold conditions, or with insufficient time to set seed before frost hinders metabolic and developmental activity. In tropical environments, variations in rainfall pattern will also be important, and flowering may be constrained to those seasons with adequate water availability, unless the plant has adaptations that allow water storage. However, even with these limitations, a number of studies have found apparent displacement of flowering time in a manner that strongly suggests the existence of competition for pollinator attention.

Aizen and Vázquez (2006) considered the flowering time of a group of unrelated bird-pollinated species in South America. They used data for 13 different species, each from a different genus, but found growing together at three different sites. Each plant produced ornithophilous-appearing flowers, usually red and tubular with large quantities of nectar. For each species, the hummingbird *Sephanoides sephanoides* had been recorded as the main flower visitor, and had also been shown to transport pollen of each species. These hummingbird-pollinated species were shown to have a much broader range of flowering times than was usual for plants flowering in these three localities. Other, non-bird-pollinated plants tended to flower in late spring, whereas the hummingbird-pollinated species flowered throughout spring and summer. Flowering times were compared with a number of models, and the authors concluded that flowering times of these 13 species were overdispersed, flowering over a broader range of times than would be predicted based on the behaviour of other plants in the same environment. They therefore concluded that this system provided evidence that competition for the attention of the hummingbird pollinator had occurred, resulting in displacement of flowering times by either evolutionary or ecological processes.

In a similar study, using a group of more closely related plant species and their bat pollinators, Lobo *et al.* (2003) also showed that flowering time appeared to be displaced to minimize pollination competition. Flowering patterns of trees from the family Bombacaceae (Malvales) were compared in three different geographic locations that differed in rainfall patterns. All of the trees studied had been reported to be pollinated by nectarivorous bats, and also by moths. Within each locality, the flowering time of the trees was staggered so that maximum flower production of each species occurred sequentially, rather than together. This did not seem to be related to rainfall, but did correlate with phylogeny (the same species tended to flower at the same time in different habitats) and with the presence of other bat-pollinated species and the need to avoid concurrent flowering. Interestingly, analysis of the pollen content of bat faeces suggested that the pollinating bats specialized on whichever tree was at the peak of its flowering in a single location, essentially feeding on only one pollen type at a time, for periods of up to a month. This specialization by a major group of pollinators on the most abundantly flowering species will act to reinforce selective pressures inhibiting concurrent flowering, by strongly punishing trees that flower earlier or later than the mean for their species. In this way clear displacement of flowering time can be maintained and further pollination competition prevented.

A recent study confirms that concurrent flowering of species that usually flower sequentially can have detrimental effects on fitness. The organ pipe cactus (*Stenocereus thurberi*, Caryophyllales), has a peak flowering time approximately 1 month later than two other species of cactus that grow alongside it in Mexico. All three species are bat-pollinated. Fleming (2006) showed that the fruit set by those individual organ pipe cacti that flower early, and therefore concurrently with the two competing species, contained sterile seeds. He further showed that pollination with heterospecific pollen caused the development of fruit, but that the seed within lacked embryos. Selective pressure will therefore act to maintain displacement of flowering time in these species, as concurrent flowering reduces viable seed set. A similar study by Waser (1978) drew similar conclusions for the interaction between *Delphinium nelsonii* (Ranunculales) and *Ipomopsis aggregata* (Ericales) in Colorado. These species flower sequentially, and are both pollinated by broad-tailed

hummingbirds. Waser reported that plants which flowered at the margins of the flowering period for their species, and that therefore overlapped with the other species, set 30–45% fewer seeds per flower than plants which flowered at displaced times. Similarly, plants grown in mixed species plots set 25–50% fewer seeds than those in single species plots, and plants hand-pollinated with pollen from the other species set significantly fewer seeds than control plants. These studies provide evidence for the importance of displaced flowering time in maintaining reproductive success in these species.

An extremely detailed analysis by Stone *et al.* (1998) demonstrated that assemblages of closely related species can show displacement of a number of characters to minimize competition for pollination. Working with ten species of *Acacia* (Fabales) growing together in mixed habitats in Tanzania, they first showed that, although most species flowered during the rainy season, two species flowered in the dry season. This restriction of flowering to a less favourable time of year can be interpreted as a mechanism to avoid competition. They also showed that there was some partitioning of pollinator species between tree species, but that this partitioning was only ever partial, and substantial overlap in pollinating species existed. However, they were able to show that a novel character trait had been displaced to minimize competition for pollination and heterospecific pollen transfer. In habitats where multiple *Acacia* trees flowered together after the autumn rains, the timing of pollen release within a single day had been partitioned. Thus all trees of each *Acacia* species released their pollen together, but the timing of pollen release was different for each species. Some species had peak pollen availability as early as 6 a.m., while others released maximum pollen as late as 4 p.m. The window of peak pollen availability was usually little more than an hour, allowing ample time for all co-flowering species to independently release pollen each day. The local pollinator populations had learned to follow this daily synchronous release of pollen, foraging from each species at the most rewarding time. Megachilid bees, honeybees, and flies showed temporal patterns of flower visitation that ensured they visited each tree species at the time of maximum pollen availability. This partitioning of pollen release ensured that the pollinators served only one tree species at a time, minimizing both competition for pollinator attention and the interfering effects of heterospecific pollen transfer.

A number of studies have considered displacement of characters other than timing of flowering and pollen release. For example, Armbruster *et al.* (1994) considered length of nectar tube and position of pollen placement on the pollinators of a group of species of *Stylidium* (Asterales) flowering together in Western Australia. They reported that overlap in pollination system (determined by nectar tube length and the position and length of the pollen-depositing column) was virtually absent from groups of co-flowering species, suggesting that character displacement had occurred. Similarly, Smith and Rausher (2008) found that the presence of *Ipomoea purpurea* (Solanales) flowers resulted in increased selection favouring clustering of anthers around the stigma of *I. hederacea* flowers, possibly to maximize self-pollination.

Not all studies of character displacement have found evidence to support the idea that pollinator attention is limiting. Ollerton *et al.* (2007) reported that the parasitic plant, *Orobanche elatior* (Lamiales), and its host, *Centaurea scabiosa* (Asterales), flower concurrently in England and share a common pollinator, the bumblebee *Bombus pascuorum*. The host species is also pollinated by a number of other insects, but the authors postulated that the use of a common pollinator species might drive character displacement. However, they were unable to find any differences in flowering time, placement of pollen on the bee's body, or daily timing of pollinator visits, suggesting that competition between these two plant species was insufficient to drive such character displacement.

In this chapter we have considered evidence, collected through a number of different methods, that shows the presence of competition for pollinator attention, either in the present or in the evolutionary past of a group of plants. At least one of these methods, namely the analysis of displaced characters such as flowering time, in fact shows an absence of competition for pollination in the current environment. The data considered show that the interaction between flowering plants of the same or different species can range from positive facilitation of pollination through very minimal effects to strong competition for pollination, severely limiting seed set in affected individuals. The nature of this interaction varies with the frequency of different plants within a habitat, and can operate either through the direct attraction of pollinators or through the consequences of pollinator fidelity for heterospecific pollen transfer, or through both mechanisms together. An understanding of how these interactions between concurrently flowering plants operate can inform agricultural practice, as well as strategies to conserve and restore species, particularly in response to climate change. However, the purpose of this chapter was simply to ask whether competition for pollinator attention sometimes occurs. If plants compete for pollination services, and those individuals within a species with showier flowers attract more or better pollinators (or make better use of each visit), we can expect showier flowers to become the norm. Indeed, the strength of such natural selection on floral traits may increase in the near future. Recent models of climate change forecast a smaller number of pollinators globally. These models suggest that selection on floral traits will be strongest in species-rich habitats, where competition for pollination services is likely to be strongest if animal species diversity declines. It is therefore predicted that plants in biodiversity hotspots (such as tropical forests or the South African Cape flora) will be at increased risk of extinction if they are not able to specialize on particularly abundant pollinators or substantially enhance their floral attractiveness (Vamosi *et al.* 2006). In the uncertain and constantly changing circumstances experienced by most flowering plants, it is a daunting task to quantify the frequency and importance of the selection imposed by pollination limitation relative to the many other selection pressures that plants face. However, the fact that pollination competition occurs is sufficient to allow flowers to evolve into more elaborate structures and for pollination syndromes to exist, at least in theory. In Chapter 20 we consider another key concept underlying the pollination syndrome theory, before in Chapter 21 we assess whether specialized pollination systems are the rule or the exception.

Do pollinators discriminate between different floral forms?

In Chapter 19 we concluded that some plants, at some times, experience competition for pollinator attention. In such circumstances, natural selection will favour traits that enhance the attractiveness of flowers to their pollinators. In this chapter we investigate whether the different shapes, structures, and colours that flowers produce have the potential to enhance pollinator visitation. To do this, they must fulfil two criteria. First, they must be visible to the appropriate pollinator, or detectable using some other sense. Secondly, the pollinator must discriminate between different floral forms. Simply because a change in floral form is detectable to an animal, it does not necessarily follow that the animal will discriminate between the original and the novel form. Such discrimination will only occur if one form provides an advantage to the animal, either by being easier to detect or handle in some way, or by providing a superior reward of nectar or pollen. In this chapter, then, we shall begin by discussing the current evidence on what different pollinating animals can see and detect in other ways. Then we shall consider the experimental evidence that pollinators do discriminate between different floral forms, focusing on colour, shape, and scent. An understanding of which floral features are discriminated by pollinators will allow us to assess the usefulness of the pollination syndrome concept in Chapter 21.

20.1 What pollinators see

Animal vision depends on two components—the presence of structural features that allow objects to be detected, and the presence of the neural processing ability to translate such signals. Our understanding of animal vision varies greatly between different types of pollinators. In this section we shall consider two comparatively well-studied examples, bees and birds.

20.1.1 Bees

Bees, like all other insects, possess compound eyes. This means that the eye is composed of several thousand independent units, known as ommatidia. Each ommatidium consists of a small lens, around 20 μm in diameter, that focuses light on to a set of photoreceptor cells. Each ommatidium functions as a mini eye in its own right, collecting light from a small cone-shaped section of the world that is slightly offset from the section viewed by the neighbouring ommatidia.

Bumblebee and honeybee colour vision, like human colour vision, is trichromatic, based on three colours. The ommatidia contain photoreceptor cells, containing opsin pigments, which perceive green (540 nm), blue (430 nm), and ultraviolet (UV, 340 nm) light. They do not contain photoreceptors for red, leading to a common misconception that bees cannot see red. However, the spectral sensitivity curve of the green receptor has an extended tail, and reaches to around 650 nm, well beyond the orange/red boundary (Chittka and Waser 1997). This allows bees to perceive red light, although it may be difficult to distinguish from green light of a lower intensity. Bee colour vision is therefore very different from human colour vision, being weaker with regard to one of our primary colours but containing an extra colour, UV, which we cannot see. As a result, it is essential to accurately measure colour using spectrophotometry at a range of wavelengths when considering how a flower looks to a bee, rather than relying on the human eye (see Fig. 20.1). Indeed, bee colour vision has recently been shown to be more complex than was previously thought. All the ommatidia of *Bombus impatiens* contain green receptor cells, but they do not all contain blue and UV receptor cells. There are three types of ommatidia in the eye—those containing green and blue photoreceptors, those containing green and UV photoreceptors, and those containing all three types of photoreceptor, shown by *in situ* hybridization of probes for the UV absorbing pigment to sections of the eye (Spaethe and Briscoe 2005). This means that different ommatidia may perceive a flower as different colours, adding to the complex task of the neural processing systems in decoding the signals they receive. In addition to differences in colour sensitivity, it is thought that different regions of the compound eye are specialized for different functions (Srinivasan 2010). The front and lower part of the eye are predominantly focused on colour vision, while the dorsal region of the eye is specialized for the recognition of polarization of light. As well as these differences within the eye, it is also likely that colour vision varies between individuals within a population, as well as between the many different species of bee.

Compound eyes do not provide the same degree of resolution as the simple eyes of vertebrates, and it has been estimated that the honeybee eye has a resolving ability around 100–170 times worse than that of the human eye (Chittka and Raine 2006; Srinivasan 2010). Combined with this relatively poor resolution, insects also appear to have relatively poor visual processing, and it is necessary for at least 15 ommatidia to perceive a flower for the bee to recognize it by its colour alone (Chittka and Raine 2006). It is clearly the case that larger flowers should be recognizable at greater distances than smaller flowers on the basis of their colour. Spaethe et al. (2001) confirmed that it took longer for bees to find flowers as their size decreased, even though their colour remained constant.

However, bees do not only use colour to identify flowers, but rely at greater distances on the contrast between flowers and green vegetation (Chittka and Raine 2006). This so-called 'green contrast' provides a reliable means of identifying a target (the flower)

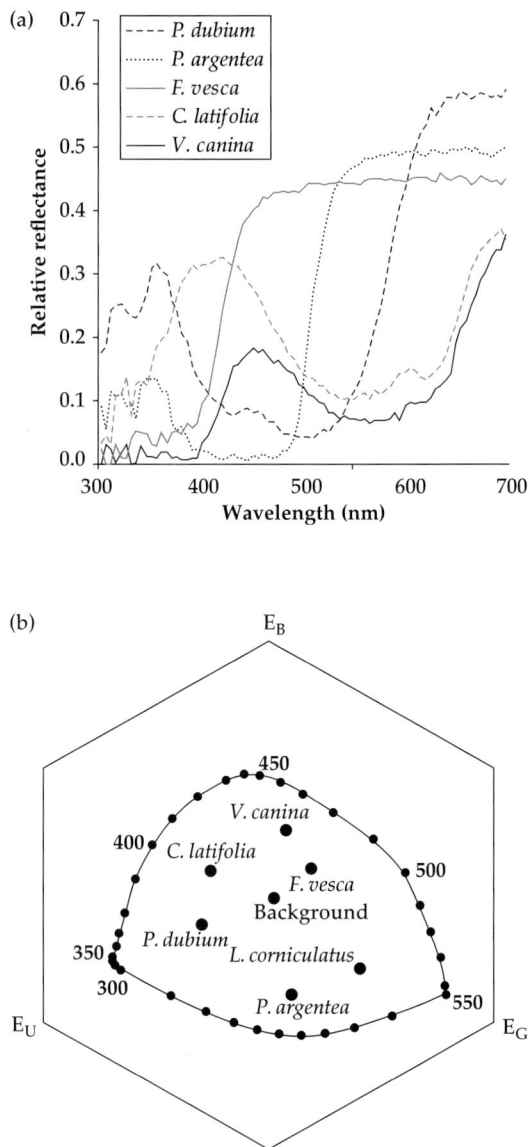

Figure 20.1 Bee colour vision. (a) A set of typical floral spectral reflectance functions. The flowers measured are yellow *Potentilla argentea* (with UV reflectance), red *Papaver dubium* (with UV reflectance), blue *Viola canina* (without UV reflectance), violet *Campanula latifolia* (with UV reflectance), and white *Fragaria vesca* (without UV reflectance). (b) The colour hexagon shows how colours look to a bee. The continuous curve denotes the spectrum locus, in 10-nm steps from 300 to 550 nm. The bottom segment of the spectrum locus connects the loci of 300 and 550 nm in nine mixtures of the two lights in ratios of 0.9:01, 0.8:0.2, and so on. Colour loci for the flowers whose reflectance spectra are given in (a) are marked here. Both parts of this figure kindly provided by Lars Chittka (QMUL).

against the almost universal background colour of plants. It greatly increases the distance at which flowers are visible to bees, although larger flowers will still be visible as coloured objects at greater distances than will smaller flowers. A consequence of the use of green contrast by bees is that the difference between a flower's colour and green can be of more importance in bee discrimination than the intensity or brightness of the colour itself, particularly when the flower is small (Spaethe *et al.* 2001).

In addition to colour, bees can also recognize different patterns and shapes. Many studies over the last century have shown that bees can be trained to recognize a variety of abstract patterns, at a range of angles. Shape and pattern recognition by bees is thought to work through learning abstracts of the patterns, or rules, rather than photographically memorizing particular patterns. Abstracting the key features of a pattern requires less memory capacity and fewer neurons than the sort of photographic memory that humans often use. In particular, a number of studies have shown that bees can detect patterns of symmetry, and the axis of that symmetry, which may be important in accessing the nectar reward in a zygomorphic flower (Giurfa *et al.* 1996; Horridge 1996).

20.1.2 Birds

Birds have camera eyes like those of other vertebrates, including humans. This means that their resolving power at various distances is more similar to our own than is that of insects. Indeed, since bird eyes occupy around 50% of the volume of the skull, compared with 5% in humans, the image projected on to the bird retina is considerably more detailed than that projected on to a human retina (reviewed by Jones *et al.* 2007). However, bird colour vision is quite complex, and may vary significantly between different species of bird. Essentially, bird eyes contain cone cells with pigments, opsins, which absorb light of different colours. Bird vision is tetrachromatic, based on the presence of four different types of cone cell, each of which contains one of the four different photopigments (Bennett and Thery 2007). However, the colours that these absorb are modified by the presence of a drop of oil through which light must pass to reach the pigments. The oil

contains one of a number of different carotenoids, and the combination of different carotenoids in the oil with different opsin pigments results in a wide range of optimum colour sensitivities, varying between species (Varela *et al.* 1993). All birds have red photopigment (peak sensitivity around 565 nm), and many birds are most sensitive to colours at the red end of the spectrum. Most birds also have photopigments with peak sensitivity at around 500 nm and at 470 nm. However, the peak sensitivity of the fourth photopigment varies, peaking at either 405 nm or 365 nm. Therefore, for some birds, colour vision extends into the UV range (Chen *et al.* 1984), and so covers a wider spectrum of colours than that of either humans or insects. This UV sensitivity of the fourth photopigment has evolved repeatedly—for example, in seagulls, many songbirds, and parrots (reviewed by Osorio and Vorobyev 2008). Hummingbirds are not known to have particularly strong sensitivity to red, and do not show a strong preference for red when offered a choice of coloured feeders, although many hummingbird-pollinated flowers produce red pigment (Goldsmith and Goldsmith 1979). One possible explanation for this is that red flowers are harder for bees to handle (because they show little contrast to green in insect vision), and are therefore preferentially visited by birds as their nectar is less likely to be depleted than that of more strongly bee-visible flowers (Raven 1972; Rodriguez-Girones and Santamaria 2004).

20.2 What pollinators sense in other ways

Pollinating animals may also be attracted to flowers over a range of distances by their scent. Scent functions as a long-distance cue for moths, beetles, and some bees, and at more moderate distances for butterflies. It is also known to act as a short-range landing cue or nectar guide for a large array of insects. Once the pollinator is on a flower, tactile cues may be important in allowing rapid handling of the flower, and may therefore make flowers with certain surface properties more attractive than those without them. Taste may also play a part in flower discrimination, if particular nectar qualities can be associated with features that can be discriminated at longer ranges.

The scents produced by flowers can travel very great distances. For example, the amines volatilized by the inflorescence of titan arum (*Amorphophallus titanium*, Alismatales) attract pollinating bees and flies over distances of hundreds of metres, as individual plants are very widely dispersed within the environment. Scent make-up is very complex, with over 1700 volatile organic compounds isolated from hundreds of plant families (Knudsen *et al.* 2006). Different animal species are able to detect different scent components to different degrees, and so will perceive mixes of these compounds differently. They may also draw different information from different odours. Some scents carry very specific information (such as organic decay or sex pheromones), while others are more ubiquitous and can be learned by insects (particularly bees, moths, and butterflies) as generally associated with floral rewards.

Insects are well equipped to perceive odour, usually having large antennae containing neurons that can detect a wide range of chemicals. Insect odour perception has been best studied in the fruit fly, *Drosophila melanogaster*, which produces 62 odour receptor proteins, a family of G protein-coupled receptors with seven transmembrane domains (Ache and Young 2005; Hallem *et al.* 2006). Although the general mechanisms of odour perception, and the types of proteins involved, are comparable in other insects studied, it is likely that *Drosophila* has a simpler olfactory system with regard to floral scents than do pollinating insects, which distinguish between many compounds in their foraging bouts. Although bees have not been studied in as much detail, bee antennae are thought to contain around 130 types of odour receptor (Chittka and Raine 2006), making the understanding of bee response to different floral scents a much more complex problem than that of visual response to colour. Guerrieri *et al.* (2005) tested the ability of bees to distinguish between a range of arbitrarily chosen odours, and observed that alcohols, aldehydes, and ketones were the most important chemicals in bee scent perception, with the majority of odour receptors focused on discriminating between molecules of these types. These compounds are not all commonly released by flowers, and so bee discrimination of floral scents may differ somewhat from this general model. It is also clear that bees can learn to associate

particular scents with food, associating scent cues with food more quickly than they can visual cues, and retaining stronger associations between scents and food than between visual cues and food. Their ability to learn scent-based cues depends both on the concentration of scent compounds present and the ratios of different molecules present in a mixed scent. The perception and learning of olfactory cues has not been well studied in many other animal groups. It is likely to be particularly complex in animals that do not rely heavily on visual signals, such as some beetles, moths, and bats.

The sense of touch is much more similar between different animals than either vision or olfaction. However, differences in body size will significantly influence the range and scale of structures that an animal can perceive tactilely. The great variety of cell types present on petals suggests that insect feet and antennae can distinguish shapes and structures as small as a single cell in size, and behavioural experiments have confirmed this ability (Kevan and Lane 1985; Whitney *et al.* 2009a). Indeed, studies of the Colorado potato beetle in contact with various leaf surfaces suggest that nanoscale sculpturing of the surface of plant epidermal cells can also be detected by insect feet, with beetles finding surfaces covered with wax crystals or folds of cuticle more slippery than smooth surfaces, irrespective of the shape of the underlying cells (Prum *et al.* 2012).

The sense of taste depends on receptors triggered by contact with particular molecules. Most animals can sense sweet and salt, with other taste receptors varying more between species. Electrophysiological techniques have revealed that insects typically have four different types of taste-sensitive neuron, responding to sugar, water, low salt concentrations, and high salt concentrations, and they may also respond to the presence of amino acids (Hallem *et al.* 2006). These are all triggered by the activities of taste receptors, of which *Drosophila melanogaster* has 68 (Hallem *et al.* 2006). Bees themselves detect taste using the forelegs, the antennae, and the mouthparts, with hair receptors localized in hair-like structures called sensilla. Perhaps surprisingly, analysis of the honeybee genome revealed the presence of only 10 genes encoding taste receptors—compared with the 68 receptors (encoded by 60 genes) in *Drosophila* (Robertson and Wanner 2006).

A number of explanations have been put forward for this surprising result, including the suggestion that the more reliable food source of bees, compared with most insects, allows a reduced range of taste perception (reviewed by Sanchez 2011). Bee taste perception is certainly sensitive in some ways—their sweet receptors are sensitive enough to allow discrimination between sucrose at concentrations varying by only 0.5%, although their preferred range is 20–30%. Analysis of hummingbird responses to sucrose feeders at various concentrations has shown that 40% sucrose is preferred over lower concentrations (Stromberg and Johnsen 1990). This discrimination between rewards at different concentrations could be used to select between flowers if it can be linked to a colour or shape difference that is remotely detectable. However, temperature may also influence the perceived taste of a substance if taste receptors are not buffered for temperature. For example, there is evidence that some insect sweet receptors perceive warmer sucrose as sweeter than cooler sucrose of the same concentration (Uehara and Morita 1972). Another effect of temperature on reward relates to the need for warm-blooded animals to heat ingested food to body temperature. For hummingbirds this metabolic cost may have a significant effect on their perception of the quality of reward offered by a flower (Lotz *et al.* 2003). These factors may result in selective pressure to moderate flower temperature as well as nectar quality.

20.3 Discrimination between petals of different colours

Investigation of pollinator discrimination between different floral traits is experimentally difficult. It has traditionally been approached in two different ways. Bee behavioural scientists have tended to use artificial stimuli, often simple pieces of coloured card, to investigate behavioural responses to single factors such as colour, size, or shape. These experiments can be difficult to interpret from a plant evolution standpoint, because they do not necessarily mimic structures and colours that can easily be achieved in nature with the resources available to the plant. Plant evolutionary biologists have tended to compare pollinator responses to, or seed set in,

flowers with different forms. However, unless the differences between the two floral forms are very carefully described and controlled, this can result in equally problematic data, as the flowers may vary in more features than just the one under analysis. Such experiments are also constrained by the availability of individual species with variation in an experimentally interesting trait, such as flower colour.

The literature contains many examples of attempts to assess pollinator discrimination between petals of different colours. In some cases discrimination is clear, in others the animals showed no discrimination, and in many cases some animals discriminated while others did not. For example, experiments with white and violet alfalfa flowers (Fabales) in a growth room indicated that honeybees did not discriminate between flower colours. When a similar trial was conducted in the field, honeybees again showed no discrimination, but pollen-gathering leafcutter bees much preferred the purple flowers (Pedersen 1967). Similarly, experiments with equal numbers of purple- and white-flowered plants of *Vigna sinensis* (cowpea, Fabales) showed that bumblebees visited purple flowers three times as often as white flowers, but that honeybees visited white flowers twice as often as purple flowers (Leleji 1973). These experiments indicate that the attraction of multiple pollinators can result in mixed selective pressures, even where one animal shows very clear discrimination, a key issue that we shall return to in Chapter 21. In another example, *Raphanus raphanistrum* (wild radish, Brassicales) has yellow or white flowers, controlled by a single genetic locus. The frequency of the yellow morph varied from 7% to 60% in the populations studied by Kay (1976), but in all of these populations the butterfly *Pieris rapae* much preferred the yellow form to the white. On the site with 60% yellow flowers, 307 visits to wild radish flowers by the butterflies were observed—and 306 of the 307 visits were to yellow flowers. However, honeybees show no preference for yellow flowers over white ones, maintaining the polymorphism within the population. In 1986, Stanton *et al.* followed up this work by investigating the consequences of pollinator preferences for maternal and paternal reproductive success in *Raphanus raphanistrum*. Since the white allele is dominant over the yellow one, they were able to assess male reproductive success by growing progeny of yellow parents and noting whether they had white or yellow flowers. They discovered that, despite pollinator preference for the yellow morph, female reproductive success of white and yellow flowered plants was equal, suggesting that resource limitation is more important than pollinator limitation in female fitness of this system. However, nearly 75% of the progeny of yellow-flowered plants were themselves yellow-flowered, indicating that their male parent was also yellow. In this example, pollinator discrimination in favour of the yellow morph has clear consequences for the male fitness of yellow-flowered plants. In contrast, Schemske and Bierzychudek (2001) showed that the pollinating beetle, *Trichochorous* sp., showed no preference for blue- or white-flowered forms of the desert annual *Linanthus parryae* (Ericales), suggesting that other factors maintain this polymorphism within the population.

One factor that has been shown to complicate matters when analysing pollinator discrimination between different floral colours is the tendency of many animals to behave in a frequency-dependent manner. In theory, pollinators should preferentially visit more common floral forms, because they are easier to learn to handle and easier to remember. In wild populations, then, particular colour morphs can be selected against simply because they are rare. This has been shown in particular detail with the tall morning glory, *Ipomoea purpurea* (Solanales), and its native American pollinators, mainly bumblebees. The wild type flower is blue-purple, but a white-flowered form commonly arises in natural populations, and is present at a range of frequencies. When white flowers are present at frequencies below around 25%, bumblebees discriminate against white flowers more than would be expected based solely on their frequency. However, when the frequency of white flowers increases to 50%, bumblebees no longer show any colour discrimination (Epperson and Clegg 1987; Clegg and Durbin 2003). This positive-frequency-dependent discrimination may serve to limit the extent to which novel flower colour mutations can spread through populations.

Despite the different preferences shown by different animals, the studies described above, and many similar studies, provide evidence that pollinating animals do sometimes discriminate between flowers of different colours, as we would expect if they can link colour to quality or availability of reward, or if certain colours are easier to see or learn than others. This should be sufficient for flower colour to be under selection in some circumstances, and thus to form part of pollination syndromes. However, these studies all suffer from the classic difficulty of the unknown nature of the polymorphism. The observed difference in flower colour may not be the only difference between two naturally occurring morphs of a plant species, and pollinators may actually be discriminating on the basis of other differences besides colour. This is particularly likely where flower colour is genetically linked to a trait such as nectar quality or flower size. Recent attempts to address this difficulty have focused on using near isogenic lines (NILs) produced by extensive backcrossing, or using molecularly characterized mutant lines. These allow an assessment of the strength of direct selection on isolated phenotypic traits, although such data are only applicable to wild situations where the trait in question can segregate independently of other aspects of pollinator attraction.

Bradshaw and Schemske (2003) provided clear evidence that both bumblebees and hummingbirds distinguish between different coloured forms of *Mimulus* (Lamiales), using NILs. This followed on from their previous work (Schemske and Bradshaw 1999) using a segregating F2 population resulting from a cross between *Mimulus lewisii* and *Mimulus cardinalis*. *M. lewisii* is normally pink, as a result of anthocyanin deposition, and is primarily pollinated by bumblebees (see Fig. 20.2a). *M. cardinalis* is normally orange/red, as a result of both anthocyanin and carotenoid deposition, and is primarily pollinated by hummingbirds (see Fig. 20.2b). These two types of animal rarely visit the 'incorrect' flower type—some observations of pollinator behaviour in mixed plots have shown that hummingbirds only visit *M. cardinalis* and that 98.9% of bee visits are to *M. lewisii* (Ramsey *et al.* 2003), although other studies suggest that the incidence of crossover visits is somewhat higher than this. In the segregating F2 population, bumblebee visits were shown to increase when flower size increased, and to be negatively correlated with carotenoid and anthocyanin

Figure 20.2 Flowers for which pollinators have been shown to discriminate between colour morphs. (a) *Mimulus lewisii*. (b) *Mimulus cardinalis*. Images (a) and (b) kindly provided by Toby Bradshaw (Washington State University). (c) *Antirrhinum majus* wild type and *incolorata* lines. (d) *Antirrhinum majus* wild type and *sulfurea* lines. Images (c) and (d) kindly provided by Enrico Coen (John Innes Centre). See also Plate 24.

levels, whereas hummingbird visits increased with increased anthocyanin levels and increased nectar volume (Schemske and Bradshaw 1999). To analyse these data further, Bradshaw and Schemske (2003) introgressed the *YUP* locus, responsible for carotenoid deposition, from each species into the other background, through four generations, ensuring 97% genetic identity between the new lines and their most similar parent. This resulted in orange-coloured *M. lewisii* flowers and deep pink *M. cardinalis* flowers. Nectar production (a highly significant factor in pollinator attraction) was compared between parent and new lines, and was found not to be significantly different. Mixed plots containing 50 wild type and 50 novel plants of a species were grown, with the two species being investigated separately. Pollinator visits were recorded, and revealed that orange-flowered *M. lewisii* received 68-fold more visits from hummingbirds than the wild type pink, but a significant reduction in bumblebee visits. Similarly, the pink-flowered *M. cardinalis* received 74-fold more visits from bumblebees than the wild type orange (although little reduction in hummingbird visits). These experiments show that both bumblebees and hummingbirds exhibit strong discrimination on the basis of petal colour. The near isogenic nature of the lines used in this study makes

it likely, although not certain, that colour is the only significant factor in the choices made by pollinators. However, since the pollinators were not flower naïve, but accustomed to foraging among wild varieties of both species in which colour is linked with a number of other traits, including nectar availability, it is possible that the animals had learned that colour was a cue that they could usually expect to be associated with a desirable reward.

A similar study by Hopkins and Rausher (2012) demonstrated that shifts in flower colour could influence pollinator behavior not through different choices of different pollinators, but by influencing flower constancy within a single pollinator type. In Texas the usual pale blue flowers of *Phlox drummondii* (Ericales) are replaced by a dark red flowered form when the species grows sympatrically with *Phlox cuspidata*, which also has pale blue flowers. The novel colour morph of *P. drummondii* is the result of loss of the gene encoding flavanone 3'5'hydroxylase (F3'5'H; see Chapter 16), resulting in redder anthocyanin, and gain of an allele conferring stronger expression of a MYB regulator of anthocyanin synthesis. Hopkins and Rausher showed that strong selection favoured in particular the darker pigment allele when the two species were planted together, but that there was no selection acting on either the hue or the intensity alleles when *P. cuspidata* was not present. By exploring pollinator behaviour in mixed field plots they were able to show that flower constancy by butterfly pollinators was the explanation for the selective pressure. When both *Phlox* species produced pale blue flowers, the swallowtail butterfly (*Battus philenor*) moved between the two species without discrimination, generating subfertile interspecific hybrid seed. However, when the dark-flowered forms (dark red or dark blue) of *P. drummondii* were grown with *P. cuspidata*, the butterflies discriminated between flowers of different colours, reducing interspecific pollen flow and thus limiting the rate of hybrid production.

As with the study described above, where the genetic basis of flower colour was well known, the use of mutant or transgenic lines which have been subjected to extensive molecular characterization allows the assessment of pollinator discrimination between different colour morphs with absolute confidence. *Antirrhinum majus* has proved to be an extremely useful model for this kind of work, being both bumblebee-pollinated and a classic molecular genetic model. Jones and Reithel (2001) investigated the responses of wild American bumblebees to experimental plots containing wild type red, white *incolorata* mutant (see Fig. 20.2c) (*INC* encodes flavanone 3-hydroxylase; see Chapter 16), and yellow *sulfurea* mutant lines of Antirrhinum (see Fig. 20.2d). Different bees exhibited different preferences, with individual animals expressing strong preferences for either red or yellow flowers within a foraging bout. Similarly, Glover and Martin (1998) analysed pollinator responses to the *nivea* mutant of Antirrhinum, which contains a deletion of the gene encoding chalcone synthase (the first step of the anthocyanin synthetic pathway; see Chapter 16) and is therefore white. The petals contain no UV-absorbing pigment, as well as no visible pigment, and are therefore white to the bee eye as well as to the human eye. Analysis of seed set of emasculated flowers in mixed plots indicated that the *nivea* mutant flowers set significantly less seed than wild type red flowers, even though they were equally fertile when hand-pollinated. The extent of pollinator discrimination varied with the number of pollinators present in the plot, but in all circumstances *nivea* mutants were discriminated against. Dyer *et al.*

(2007) conducted a detailed analysis of bumblebee behaviour in response to the *nivea* mutant, and concluded that the mutant was significantly less visible than red flowers at a range of distances, as a result of the similarity of the white flower colour to green vegetation when viewed through a bee's visual system. This similarity reduced the 'green contrast' of the *nivea* mutant relative to wild type flowers.

Recently it has even proved possible to use molecularly characterized loci controlling flower colour and scent in genetic model species to compare the relevant importance of these two cues to pollinators. Work by Klahre *et al.* (2011) analysing the interaction between petal colour and scent production in attracting hawkmoths is described in Section 20.7 below.

20.3.1 Discrimination in favour of nectar guides

Many flowers produce markings on their petals that are believed to act as nectar guides, directing pollinators towards the nectar or pollen reward. These are predicted to reduce the time taken to handle a flower, and therefore increase the attractiveness of the plant by improving the energetic efficiency of foraging. A very few studies have investigated pollinator discrimination between flowers with differences in this specialized subset of petal colours.

Scora (1964) showed that bees and wasps use the nectar guides on *Monarda punctata* flowers (Lamiales), consisting of three to five spots or lines of pigment in the back of the throat, to position themselves correctly to access the nectar. Scora used colchicine treatment to produce mutant plants without the nectar guides, but with otherwise normal flowers. These initially appeared to attract bees and wasps to the same degree as wild type flowers, but the pollinators rejected the flowers after landing, without trying to probe for nectar.

A particularly well-studied system in which nectar guides have been shown to be important is that of *Delphinium nelsonii* (Ranunculales), which has two sets of pollinators—hummingbirds and bumblebees (Waser and Price 1981). The flowers are usually deep blue, but occasional albino flowers arise in natural populations. The seed set of white flowers and blue flowers is very similar if they are hand-pollinated, suggesting that they are equally fertile. However, the seed set of open-pollinated albino flowers is as much as 45% lower than that of blue flowers, and white flowers in artificial plots receive 24% fewer visits from pollinators than do blue flowers. Both bees and hummingbirds released into cages containing both blue and white flowers preferred the blue flowers. This appeared to be due to a difference in handling time between the two flower morphs, resulting from the 'disappearance' of the nectar guides in the white flowers. The blue flowers have a nectar guide, consisting of two small white petals that frame the entrance to the concealed nectaries (see Fig. 20.3a). These are not visible against the white background in the albino flowers. It takes up to 1.8 times as long for an animal to extract nectar from the white flowers as from the blue flowers, and up to 2.2 times as long for it to fly between white flowers as between blue flowers. This results in a significant difference in the total time taken to process a white flower and a blue flower, even though they provide identical nectar rewards. In this case, pollinator discrimination between different colours of flower is directly attributable to the visibility of the nectar guides against the different backgrounds, and the consequent effects on handling time (Waser and Price 1981).

Venation patterns (lines of pigmentation overlying the vasculature) are common in many groups of flowering plants, and have been hypothesized to act as nectar guides, directing foraging animals to the reward. A recent study by Leonard and Papaj found that artificial nectar guides consisting of thin radiating lines enhance the speed with which bumblebees find the reward in artificial flowers (Leonard and Papaj 2011). The work of Shang *et al.* (2011) on the developmental control of venation patterning was introduced in Chapter 17. To test the functional significance of venation patterning, the authors also planted out field plots of a set of near isogenic lines of *Antirrhinum majus*, including red, veined, pink, ivory, and white flower colours. They observed three species of bumblebee foraging in their plots, and found that all three species significantly favoured red flowers over pink, ivory, or white ones, but that there was no significant bumblebee preference for red over veined flowers. They concluded that venation patterning, as seen in many wild *Antirrhinum* species, is highly attractive

Figure 20.3 Flowers with significant nectar guides. (a) *Delphinium nelsonii*. Image kindly supplied by Nick Waser (University of California – Riverside). (b) *Clarkia xantiana* subsp. *xantiana*. Image kindly provided by Vince Eckhart (Grinnell College, USA). See also Plate 25.

to bumblebee pollinators. Further work on the same lines, using naïve *Bombus terrestris* foragers, concluded that the bees were not innately attracted to venation-patterned flowers, choosing them at a frequency only slightly higher than that at which they selected ivory flowers. Whitney *et al.* (2013) suggested that bumblebee preference for venation patterning in the field was most likely to be a learned response to the utility of the veins in directing the animal towards the nectar reward.

It is not just bumblebees and hummingbirds that respond to patterns on flowers. The presence of dark spots on the inflorescences of *Daucus carota* subsp. *maxima* (Apiales) attracts *Musca domestica* flies (Eiskowitch 1980). The South African daisy *Gorteria diffusa* (Asterales) has black spots on its ray florets and is pollinated by a small bee-fly, *Megapalpus capensis*. Capitula of *G. diffusa* with black spots received more fly visits than capitula from which the spotted ray florets had been removed, although in some cases this may be the result of sexual deception, with male flies attempting to mate with the spots, rather than of a simple nectar guide effect (Johnson and Midgley 1997; Ellis and Johnson 2010).

Frequency-dependent behaviour may also influence pollinator discrimination between flowers with and without nectar guides. *Clarkia xantiana* subsp. *xantiana* (Myrtales) produces pale pink flowers and is polymorphic for the presence of nectar guides. These consist of dramatic spots of deep red pigment, surrounded by a white area, at the base of each petal (see Fig. 20.3b; the genetic basis of petal spot formation in the related *Clarkia gracilis* is discussed in Chapter 18). Eckhart *et al.* (2006) showed that the pollinators of *C. xantiana* (several species of bee) showed no discrimination between flowers with spots and flowers without them, unless they were presented at different frequencies. When flower frequencies were varied, one bee species preferentially visited the more common form, whether that was spotted or spotless, while two other bee species preferentially visited the rarer form. As with petal colour in general, it appears that frequency-dependent foraging strategies can serve to complicate our understanding of pollinator discrimination between different forms of nectar guide.

20.3.2 Discrimination between colours as a cue to other differences

Pollinating animals may discriminate between different petal colours not only because some colours are more visible, or provide more handling information than others, but also through learned associations. Dyer *et al.* (2006) showed that bumblebees could learn to associate particular flower colours with particular rewards, in this case warmer sucrose. Bees were presented with two colours of artificial flower, pink and purple. The colour difference was very slight, and bees had previously been shown not to discriminate between the two colours when both were equally rewarded. The purple flowers were then heated to several degrees Centigrade above ambient, while the pink flowers were not, and both were provided with sucrose solution. The bees quickly learned only to land on the purple flowers, which contained the warmer reward. This ability to learn an association between flower colour and a particular reward may explain pollinator discrimination between apparently equally visible flower colours, if one colour is associated with other beneficial properties such as enhanced nectar supply, warmth, or reduced handling time.

20.4 Discrimination between corollas of different sizes

There have been fewer experimental demonstrations of pollinator discrimination between flowers of different sizes, despite the prediction, based on our knowledge of bee vision, that insects should strongly favour larger flowers because of their enhanced visibility. Spaethe *et al.* (2001) demonstrated that a decrease in target size from 28 mm to 5 mm diameter was correlated with an increase in time to find the flower from 10 seconds to over 2 minutes. One reason for the shortage of convincing experimental data may be that flower size is not as variable as might be expected, as perturbations in cell division or cell expansion are compensated for by reciprocal changes to the other process (see Chapter 15). However, where experimental systems have been identified in which flower size varies significantly and reproducibly within a population, then it has been possible to show that both bees and hummingbirds discriminate on the basis of various aspects of corolla size. There is some evidence in each case to suggest that reward production increases with flower size, so such discrimination may be due to a learned association between flower size and reward quality or quantity.

Some of the most thorough and convincing work has been conducted by Candace Galen over a number of years on *Polemonium viscosum*, the alpine skypilot (Ericales). This plant produces flowers with different degrees of corolla width or flare at different elevations in Colorado. The flowers of upland plants, growing in tundra habitats, have corollas on average 12% wider than those of lowland plants growing at the timberline. The upland plants are bumblebee-pollinated, whereas the lowland plants are pollinated by a range of different insects, but mostly by flies. Both sets of plants are self-incompatible, so effective pollination is essential for seed set. Bumblebees entering a mixed plot visit the wide-flowered plants first, and wide-flowered plants set more seed than narrow-flowered plants. Pollen removal by bumblebees has also been shown to be correlated with corolla flare (Galen and Stanton 1989), indicating that the male component of fitness is also related to corolla size through bumblebee discrimination. Estimates of the selective advantage of wider corollas in attracting bumblebees predicted that the 12% difference in corolla flare could be achieved in only one to three generations (Galen 1996a,b). Pollinators are not the only animals to distinguish between alpine skypilots with different degrees of corolla flare. Nectar-gathering ants, which often sever the stigma and cause female infertility, preferentially visit short flowers with wide corollas. To test whether reduced corolla flare would protect against this damage, Galen and Cuba (2001) constricted floral growth using glue, so that flowers attained their normal size but had corollas 36–46% narrower than usual. When ants were presented with mixed arrays of untreated control flowers and the manipulated narrower flowers, they chose the wider controls 68% of the time. However, such protection is not selectively advantageous in the wild, because when bumblebees were allowed to pollinate a mixed array of the two flower types, the narrower manipulated flowers set 62% fewer seeds than the controls. These experiments confirm that pollinating bumblebees, as well as predatory ants, can and do discriminate between flowers of the same species with different widths of corolla.

Work with *Ipomopsis aggregata* (Ericales), a hummingbird-pollinated plant that grows in the Rocky Mountains of America, has shown that birds also discriminate between flowers with different widths and lengths. Campbell *et al.* (1991) showed that wider flowers in their study site produced more

Figure 20.4 Flowers for which pollinators have been shown to discriminate on the basis of size. *Ipomopsis aggregata.* Image kindly provided by Nick Waser (University of California – Riverside). See also Plate 26.

nectar than narrower flowers, and that there was a relationship between nectar production, corolla width, and hummingbird visitation. Flowers with wide corollas exported significantly more pollen as a consequence. *I. aggregata* is shown in Fig. 20.4, being visited by its other main pollinator, a large bumblebee.

It is important to note that the selection pressure imposed by pollinators on floral shape is not only by direct floral discrimination and choice. Fulton and Hodges (1999) showed that the hawkmoths which are the major pollinator of *Aquilegia pubescens* (Ranunculales) exert selective pressure on floral position through their choices—when flowers were positioned pendantly, as opposed to their normal upright position, they were visited ten times less frequently than flowers in the normal position. However, artificial shortening of the nectar spurs of *A. pubescens* had no effect on hawkmoth visitation rates—pollinator choice was not related to spur length. Shortening nectar spurs did have a strong effect on pollination efficiency, with hawkmoths removing significantly fewer pollen grains from flowers with short nectar spurs than from flowers with long nectar spurs, explaining the retention of the long spur trait even in the absence of direct pollinator discrimination.

20.5 Discrimination between zygomorphic and actinomorphic flowers

Although zygomorphy (bilateral symmetry) has evolved multiple times in the angiosperms, and is often described as a key specialization to attract a particular subset of pollinators (Cubas 2003), there have been few experimental attempts to investigate pollinator discrimination between floral morphs with different symmetries. This is mainly due to the difficulty of identifying suitable model systems. Very few species exhibit naturally occurring variation in floral symmetry to any great degree. However, the use of actinomorphic mutants, such as those of *Antirrhinum majus*, is also problematic, as their phenotypes may be so extreme as to prohibit pollination at all. For example, the radially symmetrical *cycloidea/dichotoma* double mutant (see Chapter 15, Section 15.2) produces a tightly closed corolla which is so narrow that free-flying bumblebees, the normal pollinator, cannot enter the flower to probe for nectar (B. J. Glover, unpublished data). Many studies have shown that bees can distinguish between artificial shapes with different symmetries (Giurfa *et al.* 1999), but it is only very recently that evidence for pollinator discrimination between differently symmetrical floral forms of the same species has arisen. A recent study using *Erysimum mediohispanicum* (Brassicales) has shown pollinator discrimination in favour of zygomorphy. This plant, growing in the Sierra Nevada, shows continuous variation in flower shape from purely actinomorphic to quite strongly zygomorphic with the two abaxial petals considerably larger than the two adaxial ones. One of its major pollinators in the study site is the beetle *Meligethes maurus*. Gomez *et al.* (2006) studied flower morphology and a number of components of fitness in a wild population of *E. mediohispanicum*, and reported that zygomorphic flowers received more

pollinator visits than actinomorphic flowers and also produced more surviving juvenile offspring.

20.6 Discrimination between flowers with different petal cell shapes

In Section 17.6 we discussed the variety of roles that petal epidermal cell shape can play in modifying flower colour, texture, and temperature. The fact that bumblebees can discriminate between wild type Antirrhinum plants with conical petal epidermal cells and isogenic *mixta* mutants with flat petal cells was shown by Glover and Martin (1998). Fruit set was recorded in mixed field plots of emasculated flowers, and the conical-celled form was shown to set significantly more fruit, reflecting more pollinator visits. These data were confirmed by Comba *et al.* (2000), who observed that bumblebees failed to land on flat-celled *mixta* flowers as often as on conical-celled wild type flowers, and that they also rejected flat-celled flowers without probing them more frequently than they did conical-celled flowers. The current consensus is that bees discriminate between flowers with different shapes of epidermal cells on the basis of the grip that they provide, and can use the effect of conical cells on light capture as a colour cue to guide their choice of easy-to-grip flowers. It has long been known that immobile bees can be trained to recognize the petals of different species (and extend their proboscis) by the shape of the cells (Kevan and Lane 1985). Experiments with epoxy replicas of wild type and *mixta* mutant Antirrhinum petals confirmed that free-foraging bees can discriminate between them using touch alone. Bee preference for conical epidermal surfaces was stronger when artificial flowers were presented vertically rather than horizontally (Whitney *et al.* 2009a). When the same experiment was repeated with Antirrhinum flowers, it was found that all preference for the wild type (conical-celled) line was lost when flowers were presented horizontally so that bees could land on them easily. In additional support of this explanation for bee preference for conical cells, Alcorn *et al.* (2012) showed that bumblebee preference for the wild type conical-celled form of petunia over a flat celled mutant was increased when the flowers were moving.

These recent studies suggest that, although conical cells do visibly enhance flower colour, their primary adaptive function might be to enhance pollinator grip on the flower. It is through this effect on grip and foraging efficiency that some pollinating animals can discriminate between flowers with different shaped petal epidermal cells.

20.7 Discrimination between flowers on the basis of scent

Analysis of pollinator response to floral scent has been considerably more difficult than analysis of visual responses, because variation in scent production is less obvious to the investigator than is variation in appearance. However, a number of studies have established that pollinating animals do distinguish between flowers on the basis of their scent. For example, Dobson *et al.* (1999) showed that the scent released by the pollen of *Rosa rugosa* (Rosales) flowers had a significant effect on their attractiveness to bumblebees. These flowers produce no nectar and attract pollen-gathering bees. Emasculated flowers received fewer bee visits than controls, but attractiveness was significantly enhanced by applying eugenol, an odour normally released by rose pollen, to the emasculated flowers. Similarly, Ashman *et al.* (2005) showed that odour extracts of male flowers, when applied to the female flowers of the wild strawberry *Fragaria virginiana* (Rosales), increased the frequency of approaches by several species of small bees to those flowers. A recent study by Riffell *et al.* (2013) established that moths are also able to distinguish between different floral scents, have an innate preference for the scent mixtures found in flowers from which moths usually forage, and can learn to associate other floral scent mixtures with a nectar reward. Flower-naïve individuals of *Manduca sexta* were introduced to an array of artificial flowers impregnated with volatile organic compounds extracted from a variety of flowers, some of which were regular food sources for hawkmoths, and others of which were not. The moths preferentially selected the artificial flowers impregnated with 'moth flower' volatiles over those impregnated with volatiles from more generalized flowers.

When the moths were trained that the non-innately attractive scent of bat-pollinated *Agave palmeri* (Asparagales) was associated with a reward, they then selected flowers impregnated with *A. palmeri* scent from the artificial flower array, in addition to selecting the 'moth flower' scented targets (Riffell *et al.* 2013).

The ability of hawkmoth pollinators to discriminate between flowers with different scents was also demonstrated by Klahre *et al.* (2011). Red-flowered *Petunia exserta* (Solanales) is bird-pollinated and largely scentless, whereas white-flowered *P. axillaris* is strongly scented and pollinated by *Manduca sexta*. By reciprocally introgressing the locus controlling scent production between these two species, Klahre *et al.* were able to produce red *P. exserta*-like flowers with scent, and white *P. axillaris*-like flowers that were scentless. In a pairwise comparison between scented and non-scented varieties of either *P. exserta* or *P. axillaris*, hawkmoths preferred scented flowers. However, when presented with a scentless white *P. axillaris* flower and a scented red *P. exserta* flower, hawkmoths showed no consistent preference for either type, which suggests that visual and olfactory cues are of equal significance in foraging decisions. Experiments such as this, where near-isogenic lines are used to compare the combined effects of multiple cues, provide great insight into pollinator discrimination.

In this chapter we have considered how pollinators sense various features of flowers, and whether there is convincing evidence that they do sometimes discriminate between such features. Although it is by no means clear from the literature that all diversity in floral form is of significance to pollinating animals, there are certainly data to suggest that polymorphisms in flower size, flower shape, flower colour, and flower scent are all sometimes noted and responded to by pollinators. These data provide the evidence of natural selection operating on floral form that is necessary before any attempt can be made to classify floral diversity with regard to the preferences of particular pollinators. In the next chapter we shall look in more detail at pollination syndromes, considering the other assumptions underlying the concept and the evidence as to whether the concept is useful.

CHAPTER 21

Pollination syndromes: the evidence

In Chapter 14 we discussed the concept of a pollination syndrome—that is, a suite of floral characters such as those described in Chapters 15, 16, and 17, which have evolved together because they enhance pollination by a particular taxonomic or functional group of animals. This concept has underpinned much of floral biology for many years, and our purpose in this chapter is to assess the usefulness of the concept in understanding flowers and flowering. We shall begin by considering why and how the pollination syndrome concept has become so entrenched in the literature on flowering, and then assess whether the key assumptions that underlie it are met. Finally, we shall assess the experimental evidence that pollination syndromes do exist, and the experimental evidence against them—those cases where the major pollinator in the native habitat is not that which the flower's morphology would lead us to predict. This chapter will also provide a brief overview of the relative importance of generalization and specialization in pollination ecology.

21.1 Historical context

The recognition that pollinating animals are essential for reproductive success in many flowering plants came in 1761, in the work of Koelreuter (discussed and partially translated by Waser 2006). By 1793, Sprengel was already assigning likely pollinators to plants on the basis of their floral form, and attempting to classify flowers into groups according to their rewards and the numbers of their floral parts (Waser 2006). A century or so later came the first attempt to classify flowers according to the morphological and structural features that suited them for pollination by different animals. Delpino's scheme was the forerunner of the

pollination syndromes that we know today (Waser 2006). However, much of the scientific enthusiasm for pollination syndromes can be traced to Charles Darwin and the controversy surrounding his theories. Darwin's own work on orchid pollination was one of the earlier essays on the relationships between suites of floral characters and their animal pollinators. In the rush to find evidence to support Darwin's new evolutionary theory, the visually obvious similarities between flower size and pollinator size, flower colour and pollinator preferences, and flower structure and pollinator feeding mechanism were seized upon by many authors as clear examples of the consequences of natural selection. The thorough description of the different pollination syndromes by Faegri and van der Pijl (1966), along with several other authors, with reference to corolla size and shape, colour, scent, and reward production, formally established the pollination syndrome as the paradigm within which pollination ecology was conducted. It was only in the 1980s that experimental approaches were first used to assess whether particular floral traits were under selection by pollinators. Although these experiments (some of which are described in Chapter 20) did reveal the existence of pollinator selection on floral form, they did not address the frequency with which such selection occurs. In 1996, two critical essays were published, both questioning the utility of the pollination syndrome concept, and setting out the difficulties with its entrenchment in the botanical literature (Herrera 1996; Waser *et al.* 1996). Both papers suggested that generalization, with flowers receiving pollinator service from more than one type of animal, was more frequent in nature than the pollination syndrome-dominated literature might lead us to expect. One concern was that positive cases of pollinator selection were over-reported in relation to negative cases, generating a situation where we can say qualitatively that such selection occurs, but nothing quantitative about its relative frequency or importance. Most studies also fail to take account of constraints that might prevent plants from adapting, even when selection by pollinators is strong. This requires an understanding of the genetic control of flower development as well as an overview of the other ecological constraints on a flower, such as the need to avoid predators (for example, the nectar-robbing ants studied in *Polemonium viscosum* by Galen and Cuba 2001). It also requires an appreciation of the relationship between the traits of the flower and the lifetime fitness of the plant. It is only in recent years that a more integrative approach to analysing pollination syndromes has been reported, and the data from such studies now allow us to make initial comments on the utility of the pollination syndrome concept.

21.2 Putting the assumptions together

The pollination syndrome concept rests on a number of assumptions, although they are rarely stated explicitly. These are as follows:

1. Plant reproductive success is limited by a lack of pollinator attention.
2. Pollinators discriminate between different floral morphs within a species, exerting selection on floral traits.
3. Plants can respond to that selective pressure, having the necessary genetic variability with respect to key traits.
4. Plants are not penalized in other ways for responding to that selective pressure.
5. This sequence of events has occurred in a majority, or even a substantial majority, of the plant–pollinator interactions that we see today.
6. This has resulted in a degree of specialization in many or most plant–pollinator interactions.

1. Plant reproductive success is limited by a lack of pollinator attention

We discussed Assumption 1 in Chapter 19, and concluded that plant reproductive success is sometimes limited by lack of pollinator attention (or too general pollinator attention, resulting in interference competition). What is not as easy to assess is how frequently such limitations occur. The best study giving any quantitative data is the review by Knight *et al.* (2005) of published reports of pollen limitation. As many as 63% of published studies did record pollen limitation of control plants, although the variety of methods used in the different studies makes even this conclusion tentative. In addition, it is likely that studies which failed to find pollen limitation are under-represented in the published data set, since negative results are less likely to be published. There are also vastly insufficient data for us to conclude whether pollinator attention limits the fitness of plants in all or a few particular habitats, and whether it occurs occasionally in a plant lifespan, for all of the lifespan, or something in between. In short, Assumption 1 is qualitatively true, but we do not know very much about it at a quantitative level.

2. Pollinators discriminate between different floral morphs within a species, exerting selection on floral traits

We discussed Assumption 2 in Chapter 20, and concluded that some pollinators certainly discriminate between many differences in floral form, including those of colour, size, and scent. Although assessment of pollinator discrimination does require appropriate consideration of pollinator sensory capabilities, there is sufficient evidence that it does occur to allow us to predict its occurrence between many other floral morphs.

It is a great deal harder to demonstrate selective pressure than it is to demonstrate pollinator discrimination, so many authors assume that one follows from the other. It is likely that this is often, although not always, the case. Studies such as those of Candace Galen with the alpine skypilot (Ericales) have demonstrated changes in floral morphology over generations to be consistent with selective pressure imposed by pollinators, making it likely that such selective pressure does occur in other systems. What is not yet known is how significant different floral traits are for different pollinators. For example, it is likely that traits that directly affect reward availability are under stronger selective pressure than those that affect the colour of the flower.

3. Plants can respond to that selective pressure, having the necessary genetic variability with respect to key traits

In several earlier chapters we discussed the genetic control of floral size, shape, symmetry, colour, and pattern. It is clear that many of these traits are under the control of multiple loci, often acting in a coordinated fashion as a result of the activity of transcription factors. Mutations to the coding or regulatory regions of these transcription factors, or to the regulatory regions of the genes they interact with, may result in morphological changes on which selective pressure can act. There is certainly no shortage of genetic material with the potential to be variable, and most of the traits under discussion are quantitative (e.g. spur length, pigment content, petal size) and should therefore be developmentally labile. Indeed, much evidence exists from a number of species that variation within floral traits that influence pollinators is heritable, although it may also have an environmental component (Mitchell and Shaw 1993; Campbell 1996; Galen 1996a,b). In Chapter 18 we considered examples of lability of floral traits within a variety of phylogenetic frameworks, finding evidence that variation in traits was present and differentially fixed in different populations or species. It is therefore clear that plants can respond to selective pressure with shifts in floral morphology.

4. Plants are not penalized in other ways for responding to that selective pressure

In Chapter 20 we discussed the possibility that traits which enhance attractiveness to pollinators might also have detrimental consequences. For example, Galen and Cuba (2001) showed that increased corolla flare in the alpine skypilot made the flowers more attractive to bumblebee pollinators but also more vulnerable to nectar-robbing ants which damaged carpels and reduced female fertility. Similarly, Kessler *et al.* (2010) showed that *Nicotiana attenuata* (Solanales) alters its pollinator-attracting scent production

to reduce the attraction of herbivores. The flowers are usually pollinated by night-flying hawkmoths, attracted by scent released from night-opening flowers. However, when plants are attacked by the hawkmoth caterpillars, scent production is reduced and flowers shift to diurnal opening, enhancing pollination by hummingbirds. In this example, selection has favoured flexibility of floral traits, but the situation provides a good example of why selection would not always favour increased scent production, even if it did enhance moth pollination. Changes to other floral traits might also carry costs. For example, changes to floral pigmentation might be expected to affect floral temperature, heat load, and ultraviolet damage to tissues. The most obvious cost of an increase in floral size, pigmentation, or reward production is that of increased use of resources. Therefore we should not expect that all floral traits will respond to pollinator-imposed selection, as some will be constrained by other parameters. A good deal of further research is needed to establish how frequently and how severely interactions with the abiotic environment, and with predators, constrain floral responses to pollinator-imposed selection.

5. This sequence of events has occurred in a majority, or even a substantial majority, of the plant–pollinator interactions that we see today

Assumption 5 cannot be tested experimentally, as it considers events which have occurred during evolutionary time. At best we can hope to extrapolate from experimental evidence obtained on extant plants. Since we know that this sequence of assumptions can be true for some plants today, we can certainly assume that it was true for some plants in the past. However, we are again faced with the difficulty of making any quantitative statement about the frequency of its occurrence.

6. This has resulted in a degree of specialization in many or most plant–pollinator interactions

This final assumption must be true for the organizing principle of the pollination syndrome to be useful. If it is not true, then pollination syndromes represent tidy ideas that are not reflected by what we see in nature, and should be treated with a great deal of caution. While the statement that a flower exhibits a particular pollination syndrome does not imply that it is only visited by a single species, it does imply that a substantial proportion of its pollination is achieved by a single functional (usually taxonomic) group of animals. Hence, a pollination syndrome implies some degree of specialization and is inconsistent with the idea of flowers as general advertisers of rewards to all passing animals. The assumption that plant–pollinator interactions are specialized has pervaded the literature for many years, but recently more empirical approaches have been taken to test this assumption. These approaches have been based on the assessment of the number of pollinating species per plant species, and the number of plant species visited per pollinator, and they have found that generalization is much more common than expected. These approaches are reviewed extensively elsewhere (see Waser *et al.* 1996; Waser and Ollerton 2006). Waser *et al.* (1996) concluded that moderate generalization was the rule in most published plant–pollinator systems, with the variation in pollinator availability (and, less commonly, efficiency) limiting the extent to which plants could specialize on single pollinators. Similarly, they pointed out that the limited flowering seasons of many plant species, and the variability in numbers of short-lived plants of any one species in each habitat each year, would limit the extent to which pollinators could specialize on single plant species. In all the studies that Waser *et al.* cited, the great majority of plant species were visited by (and, where this was tested, actually pollinated by) more than one animal species. These species were often of multiple genera or even orders.

In contrast, Fenster *et al.* (2004) argued that specialization and pollination syndromes should not be thought of as acting at the level of species, but at the level of 'functional groups.' They argued that the precise species which pollinated a flower was of no relevance, but that what mattered in terms of floral evolution was the selective pressure that this animal would impose, which would depend on its general morphological and behavioural characters. They reassessed some of the literature that Waser *et al.* (1996) had reviewed, but instead of counting the number

of animal species visiting a plant, they counted the number of functional groups (such as long-tongued bees, or pollen-gathering bees) recorded. They concluded that around 75% of flowering plant species for which data were available were specialized on a single functional group, even if that group contained multiple species. For example, *Silene vulgaris* (Caryophyllales) is pollinated by 26 species, but all of these are nocturnal moths, acting as a single functional group (Fenster *et al.* 2004). We do not currently know to what extent behavioural patterns and discriminatory abilities vary between species within a single pollinator functional group, making it difficult to assess how valid such groups are as units of specialization. A further complication in all of these studies is the need to be sure that the recorded flower visitors are actually pollinating. Padysakova *et al.* (2013) identified more than seven bee species, butterflies, three types of fly, and three bird species visiting the flowers of *Hypoestes aristata* (Lamiales) in one location in Cameroon, but found that only three species of large bees were actually pollinating the plant. Specialization in terms of pollinator species is the important consideration for the evolution of pollination syndromes, whereas non-pollinating flower visitors can be highly variable and generalist.

Specialization between plants and pollinators is frequently asymmetric, adding further complexity to the identification of pollination syndromes. Specialist flowers pollinated by only a small number of animals are often using animals which are themselves great generalists, foraging from very many different plant species. Similarly, specialist pollinators which forage on a limited number of plant species are often foraging from plants that experience pollination service from a great many other animal species. This asymmetry in pollination biology can provide increased stability, increasing the survival chances of rare or highly specialist species (Bascompte *et al.* 2003). The complexities presented by asymmetry and by distinguishing feeding from pollinating flower visitors has led to an emerging trend towards visualizing pollination in terms of networks, rather than species-by-species interactions. The number of species in the network and its degree of connectedness can be useful when estimating robustness of a community to abiotic perturbation (reviewed by Vazquez *et al.* 2009).

This final assumption, then, is either true or false depending on how specialization is defined. If it is true that most flowers are specialized at the level of functional groups, then the pollination syndrome concept has the potential to be a useful organizing principle for pollination ecology. We shall next consider some of the published evidence that plants and pollinators are linked in pollination syndromes, and some of the published examples where they are not.

21.3 Evidence for pollination syndromes

Evidence that the pollination syndromes are useful tools comes from studies in which floral morphology is described, the pollinator is predicted from floral morphology, and then the plant is observed in its native habitat to see which animals do provide pollination service. Very few such studies have been reported, and almost all date to the past 20 years. It can be difficult to perform these studies well, because they should ideally measure pollination, not visitation. Many animals may visit a flower without providing much, if any, pollination service. Indeed, some flower visitors rob nectar or damage flowers, and might therefore introduce selective pressures to minimize floral attractiveness to them. Assessing which species are the true pollinators and measuring pollination efficiency is difficult in a wild situation. It is also important to note

the full range of pollinators of a flower, not just the most common or 'expected' ones. A number of such studies have provided support for the pollination syndrome concept.

Wilson *et al.* (2004) described the morphology of and visitors to 49 different species of *Penstemon*, *Keckiella*, and *Nothochelone* (Lamiales), which are collectively thought of as garden penstemons (see Fig. 21.1a and b). Animals were recorded as pollinating if they visited a flower and left with pollen on their body. The authors found that flower colour segregated very neatly with pollinator—bird-pollinated flowers were shades of red or pink whereas bee-pollinated flowers were purple, blue, yellow, or white. They also found that bee-pollinated flowers differed in floral morphology according to the size of their pollinator. Large open corollas were pollinated by large bees, whereas flowers with long thin corollas were pollinated by smaller bees. This distinction was less clear between bird- and bee-pollinated flowers, with some intermediate forms present. Overall, Wilson *et al.* (2004) concluded that it was possible to correctly predict the main pollinator of a penstemon on the basis of its floral morphology, and that the pollination syndromes are therefore useful tools in this system. Similar conclusions were drawn from a smaller-scale study of four species of *Ipomoea* (Solanales) in the southern USA. *I. hederacea* and *I. trichocarpa* have floral morphologies suggestive of bee pollination, with

Figure 21.1 Plants for which the main pollinator type can be accurately predicted from floral morphology. (a) *Penstemon centranthifolius* (hummingbird-pollinated). (b) *Penstemon heterophyllus* (bee-pollinated). Images (a) and (b) kindly supplied by Scott Armbruster (University of Portsmouth). See also Plate 27.

blue and purple flowers, wide corollas, and small volumes of nectar at average concentrations of 41% sucrose. Analysis of animals feeding from these flowers revealed that around 75% of their visitors were bumblebees, as predicted. Similarly, *I. hederifolia* and *I. quamoclit* have red flowers with narrow corollas, and secrete larger volumes of nectar with a concentration of 29% sucrose. Over 80% of the visits to these flowers were made by butterflies.

Machado and Lopes (2004) conducted a similar study of unrelated plants in a particular tropical dry forest habitat in Brazil. They recorded floral morphology and floral visitors for 99 species, and were able to identify trends along the lines of the pollination syndromes. For example, hummingbird-pollinated flowers were usually red, most bat-pollinated flowers were white, and actinomorphy was common in flowers pollinated by small insects. Also in Brazil, Danieli-Silva *et al.* (2011) used a network approach to assess whether floral morphology accurately predicted pollinator type in the flowers of a grassland community. They identified animals as pollinators if their bodies both touched the stigma during flower visitation and had pollen on them when they left the flower. These authors found that 69% of the interactions in their network were between plants and the functional groups of insects that floral morphology predicted would pollinate them.

The true test of pollinator efficiency is reproductive success, and Hargreaves *et al.* (2004) were able to show that seed set was very strongly dependent on visitation by the predicted pollinator, a sunbird, in the African sugarbush (*Protea roupelliae*, Proteales). The flowers have dark pink bracts, no scent, but large quantities of nectar. They are visited by a range of nectarivorous birds, but also by a large number of different insect species. Their floral morphology would suggest that their main pollinator is a bird, but it was possible that any of these visitors was the actual pollinator. Hargreaves *et al.* (2004) excluded birds, but not insects, from inflorescences, using large mesh cages. They found that 75% of caged inflorescences set less than 5% of the maximum possible seed, whereas less than a third of uncaged inflorescences performed as badly, confirming that birds were the major pollinators of these plants.

A different approach to investigating pollination syndromes is to ask which animals pollinate a number of plant species with very similar flowers. The pollination syndrome concept predicts that the pollinator would be the same in such cases. Pauw (2006) observed pollinator visits to six species of South African orchids (Asparagales) with very similar flowers. All of the flowers studied were greenish-yellow, with shallow corollas, had a pungent scent, and secreted oil. The only animal species found to visit these flowers was the female of the oil-collecting bee *Rediviva peringueyi*, and analysis of captured bees confirmed that they carried pollinia from the orchids on their bodies. This is a relatively specialized pollination syndrome, and it is in these specialized cases that the concept is likely to be of most value in predicting pollinator species.

21.4 Evidence against pollination syndromes

Similar sorts of studies have also found against the pollination syndrome concept. By investigating the reproductive behaviour of plant species that supposedly show a particular pollination syndrome, some authors have found that the real pollinator is not the one predicted. Similarly, some studies of groups of plant species have failed to find overall correlations between floral morphology and pollinator. These studies are usually very similar in design to those discussed above, suggesting that their failure to match flowers to pollination syndromes reflects the inconsistency with which the concept is useful, rather than flaws in any particular study.

Hingston and McQuillan (2000) identified the animals visiting 114 native Tasmanian plant species, and found that pollination syndromes did not predict floral visitors. Although they did not record pollination, only visitation, their results were still rather discouraging for the pollination syndrome concept. Some flowers appeared to have many characters associated with a particular syndrome, but were not visited by the appropriate animals. Almost all flowers had visitors from more than one order of animals, and flowers of the same species had different visitors in different habitats. The authors therefore concluded that pollination syndromes were unreliable indicators of pollinating species, at least in the Tasmanian flora.

Figure 21.2 Plants that have cast doubt on the pollination syndrome concept. (a) *Viola cazorlensis.* Image kindly supplied by Carlos Herrera (Seville). (b) *Microloma sagittatum* with its sunbird pollinator. Image kindly supplied by Anton Pauw (Stellenbosch). See also Plate 28.

A detailed analysis of a specialized pollination syndrome was published by Herrera (1993), who studied the floral morphology and female reproductive success of a Spanish endemic violet, *Viola cazorlensis* (Malpighiales), over 5 years (see Fig. 21.2a). This violet is pollinated by a single species, the diurnal hawkmoth, *Macroglossum stellatarum*. The flowers are characterized by a particularly long thin nectar spur (averaging around 25 mm in length), which had been assumed to be part of the plant's pollination syndrome, fitting the long proboscis of the moth. However, Herrera showed that spur length had no effect at all on female reproductive

success. Although most of the other variable flower morphological traits studied did show some correlation with seed set, spur length did not show any correlation at all, indicating that it is not currently under selective pressure by the hawkmoth and is therefore not part of the current pollination syndrome of this plant. Even more strikingly, Herrera showed that fruit production, flower production, plant size, and herbivory all affected seed set to a greater degree than all accumulated differences in floral morphology. He therefore concluded that selection on floral morphology may be virtually irrelevant in comparison with other ecological factors that determine plant fitness.

Analysis of plant reproductive behaviour has undermined confidence in even some of the most specialized of pollination syndromes. Zhang *et al.* (2005) showed that the apparently sapromyophilous inflorescences of *Tacca chantrieri* (Dioscoreales) were predominantly self-pollinated in China, rather than pollinated by flies. The inflorescences contain dark purple bracts and flowers that are nearly black. However, the anthers were found to dehisce before the flowers opened, depositing pollen very effectively on the adjacent stigmas. Selfing rates were estimated using molecular markers (isozymes) and found to vary from 76% to 94%. Therefore, although this species has a highly specialist floral morphology strongly suggestive of one particular pollination syndrome, in its native habitat it is primarily self-pollinated. A similarly surprising result was reported by Pauw (1998), who discovered that the supposedly insect-pollinated milkweed, *Microloma sagittatum* (Gentianales), was in fact pollinated by sunbirds (see Fig. 21.2b). The plant produces pollen in pollinaria, which had previously only been thought to attach to insect bodies. However, observation in the native habitat revealed that sunbirds collect nectar from the flower, and in the process the pollinaria attach to their tongues, from where they are transferred to the next flower. The pollinaria are attached using a microscopic groove which clips to the edge of the tongue. Pauw (1998) also showed that caged plants, visited by a single sunbird but no insects, were successfully pollinated. This is another striking example in which the floral morphology would have predicted an entirely different pollinator from the one observed (Ollerton 1998).

A recent and extremely wide-ranging study has concluded that pollination syndromes do not allow us to correctly predict pollinators for the majority of plants. Ollerton *et al.* (2009) conducted a multivariate analysis of the floral traits associated with traditional pollination syndromes, generating three-dimensional spaces associated with each syndrome. They then recorded the same floral traits from flowers from six different communities on different continents, and tested each species for fit to the positions of known pollination syndromes. Their results showed that the great majority of plant species (479 of the 482 species they analysed) did not produce flowers which fit within one of the spaces associated with the traditional pollination syndromes. Real flowers were generally more variable in morphology than idealized ones, expressing combinations of traits associated with multiple pollination syndromes. The authors then recorded the most frequent effective pollinator of each of the plants under study, using pollen removal and contact between animal body and floral reproductive organs to ensure only effective pollinators were recorded. Averaged across the six communities, in only 29% of species was the main pollinator the one that would have been predicted from floral morphology, basing this prediction on the animal type associated with the classical pollination syndrome nearest in space to the position occupied by the actual flower. The authors concluded that the traditional pollination syndromes largely failed to describe the real variation in floral morphology and also failed to identify the true associations between plants and their pollinators.

21.5 The most effective pollinator?

The literature tells us that the assumptions which underlie the pollination syndrome concept are sometimes true, and that authors in some cases have found evidence for the existence of pollination syndromes—that is, that plants are sometimes pollinated by the animals that their floral morphology would lead us to predict. However, there are also many recorded cases where pollinator identity does not match floral morphology, or where floral morphology cannot be assigned to a specific traditional syndrome, and the recent data of Ollerton

et al. (2009) indicate that this may be the common situation globally. The arguments about the relative importance of generalization and specialization in pollination ecology are likely to continue for some time, but their resolution may depend on a change in terminology so that flowers are described with reference to their full range of pollinators, rather than to one 'ideal' pollinator. Indeed, many authors have pointed out the importance of considering the effects of multiple pollinators on a species. Flowers have often been described as having one 'major' pollinator and one or more 'minor' pollinators. However, even this terminology is insufficiently precise, and other authors have proposed the term 'most effective pollinator.' Stebbins (1970) proposed that pollination ecologists should focus on those pollinators that visit a flower 'most frequently and effectively', as they would be the likeliest to impose selective pressure. Mayfield *et al.* (2001) reported that these two elements were not necessarily linked. They showed that the most frequent visitors to flowers of *Ipomopsis aggregata* (Ericales) at a site in western Colarado were hummingbirds, but that bumblebees were more effective both at pollen removal and at ensuring seed production. A single bumblebee visit resulted in the removal of an average of 112.7 pollen grains, and the production of 4.36 seeds, whereas a single hummingbird visit resulted in the removal of 38.1 pollen grains and the production of only 1.08 seeds. However, bumblebees visited *Ipomopsis* flowers only one-third to one-quarter as frequently as hummingbirds did. The flowers have the morphology supposedly typical of bird-pollinated flowers, being red with long corollas, but bumblebees visit these flowers particularly when hummingbirds are rare and the nectar available is abundant. It may be that in years when few hummingbirds visit a population, bumblebees learn to associate the colour and shape of *I. aggregata* flowers with a rich reward. This would imply that the plant is under selective pressure in two different directions. Its most *frequent* pollinator, the hummingbird, may be expected to impose one set of pressures, while its most *effective* pollinator, the bumblebee, will impose another. The floral form seen in the wild might be thought of as a compromise, and one which must ensure that it does not exclude either pollinator. It would be interesting to observe floral morphology in a population pollinated by only one of these two animals over several generations. More importantly, it would be extremely useful to have similar data for other species. If it is often the case that one pollinator is more efficient but another is more abundant, it is hardly surprising that many flowers fail to match the expectations of traditional pollination syndromes. Indeed, it may be that in some of those studies that recorded the most frequent pollinator as not being that expected by the flower's morphology, a less frequent visitor was more efficient at pollination, and the flower was actually responding to selective pressure imposed by the more effective pollinator.

Aigner (2004, 2006) pointed out that it is not only pollination frequency and pollinator effectiveness that determine what selective pressure is imposed on a flower, but also whether trade-offs occur between fitness for pollination by different animals. A situation where two different pollinators are applying opposing selective pressures, and increased fitness for one reduces the effectiveness of the other may, in fact, be quite rare. Perhaps more common would be a situation in which one pollinator applies pressure in one direction, but this has no consequences for other pollinators. In Aigner's own work with species of *Dudleya* (Saxifragales), hummingbirds were shown to remove dye (as a marker for pollen) more effectively and deposit pollen more successfully in flowers with corollas which had been artificially narrowed. However, although bumblebees, the other major pollinator of these plants, were more effective than hummingbirds at pollinating wider flowers, they were not themselves more effective with flowers of any particular width. In this case, hummingbirds might be expected to

apply selective pressure in favour of reduced corolla flare, but such a reduction in corolla flare has no consequence for the other major pollinators, bumblebees. If the same is true for the *Ipomopsis* example described above, it may be that the flowers have the appearance of hummingbird-pollinated flowers simply because the traits favoured by the hummingbirds are selectively neutral with regard to the bumblebees. Indeed, it is not difficult to imagine situations in which a pollinator which is very clearly neither the most common nor the most effective applies selective pressure that does result in a specialized floral morphology, simply because the trait under selection does not affect pollination by the major pollinators. Such examples would certainly cause deviations from the classical pollination syndromes.

This chapter has considered the evidence that pollination syndromes are real—that is, that flowers are pollinated by the animal that their morphology would lead us to predict. It is difficult to draw firm quantitative conclusions, because the literature is patchy and the quality of the observations varies greatly. However, it is clear that in some cases the pollination syndrome is a useful concept, particularly if applied with caution. It is also clear that much more work remains to be done on the frequency of pollination by different animals, the effectiveness of that pollination, and the selective pressures imposed by each animal. Only then can we hope to fully describe floral morphology with respect to pollinating animals, and only then can we really link the interactions of plants with their pollinators to the genes that control flower development and floral form, the major aim of this book.

Epilogue

When I decided to write this book I had a single aim that seemed very clear in my mind—to link what we know of the molecular and genetic control of *how* flowers come to look as they do with what we know from evolutionary and ecological perspectives about *why* they look as they do. My impetus was a firm belief that the traditional analysis of plant–pollinator interactions would be enhanced by a better understanding of the molecular processes underpinning floral morphology, and that the traditional molecular and developmental analysis of floral morphology would also benefit from an appreciation of the consequences of that morphology for a real plant in its natural habitat. Twenty-one chapters and two editions later I am no less convinced of the need for an integrative approach. But I have also come to realize that such an approach is not straightforward when so much ambiguity remains in the various specialist fields of floral biology. In particular, the current debates about the utility of the pollination syndrome concept, and the relative frequencies and symmetry of specialization and generalization in interactions between plants and pollinators, make it far from easy for a molecular developmentalist to contextualize their own research. It becomes tempting in these circumstances to oversimplify to a degree that will only antagonize the true pollination ecologist. I hope that, if nothing else, this book will serve to show each sort of specialist the complexity of the rest of the field, and perhaps encourage collaboration and discussion as to the way forward.

Despite the difficulties in presenting an integrated picture of floral biology, some major themes have come to the fore. In particular, flexibility seems to be as significant in plant development as it is in plant breeding system. Despite the apparent complexity of the developmental pathways leading to floral induction, floral organ development, and final flower morphogenesis, plant development nevertheless has the flexibility to respond to environmental inputs and is equally labile in evolutionary time. The characteristic flexibility of plant development is usually interpreted as an adaptation to survival as an immobile organism in a world populated by mobile animals (although it could equally be argued that the mobility of animals is an adaptation to survival in the absence of flexible development!). In the same way, most plants exhibit great flexibility of breeding system, tolerating self-fertilization under marginal conditions, varying their reproductive output as males or females in response to environmental pressures, and recruiting different pollen vectors in different habitats. This flexibility can also be interpreted as an adaptation to a sessile life cycle in a changeable world populated by mobile animals.

In the same way that developmental flexibility is mirrored by breeding system flexibility, the presence of redundant mechanisms operating as fail-safes in development is mirrored by the likely general nature of plant–pollinator interactions. Almost uniquely among the eukaryotes, plant development is characterized by an enormous degree of redundancy. The proportion of genes in the Arabidopsis genome that encode transcriptional regulators is several times the proportion of similar genes found in the Drosophila genome. These transcriptional regulators act in overlapping cascades to regulate the expression of redundant structural genes found in large multigene families. The resulting redundancy acts as a safety net, ensuring successful plant development even in the absence of one or several component factors. Redundancy may be a necessary adaptation of an organism that is continually exposed to damaging ultraviolet light, ensuring survival and reproduction even when cells and the genes they contain are damaged by daily exposure to sunlight. In the same way, the generalist nature of plant–pollinator interactions also provides safety nets, ensuring plant reproduction in the absence of the usual or most effective pollinator. This generalism may be at the level of functional groups, with multiple species of animal functioning in the same way with regard to pollen deposition and removal, or it may be at a higher level, with several clearly distinct types of animal all attracted to subsets of the same set of floral characters. The ability to self-fertilize, or reproduce asexually through vegetative organs, provides the ultimate safety net in plant reproduction.

The importance of both flexibility and fail-safe mechanisms in floral development and pollination underlines the vital nature of both processes to plant reproductive success. The angiosperms are the most speciose plant group by some distance, and they also represent the bulk of terrestrial biomass (both plant and animal). They have colonized more habitats and occupied more niches than any other plant group. And this success can be traced to the reproductive innovations made by the first angiosperms—the development of flowers. Flowers, and the flexibility and redundancy inbuilt in their development and function, are responsible for the astonishing radiation of the angiosperms and their continued success today. To fully understand that success we will need to continue our exploration of flower development, evolution, and pollination ecology in an open and integrative spirit, encouraging the disparate fields of floral biology to cross-fertilize each other with ideas just as the angiosperms encourage cross-fertilization through flowers themselves.

References

Aarts, M., R. Hodge, K. Kalantidis, D. Florack, Z. Wilson, B. Mulligan, W. Stiekema, R. Scott, and A. Pereira. (1997). The Arabidopsis *MALE STERILITY 2* protein shares similarity with reductases in elongation/condensation complexes. *Plant Journal* **12**:615–623.

Abbott, R. J., P. A. Ashton, and D. G. Forbes. (1992). Introgressive origin of the radiate groundsel, *Senecio vulgaris* L. var. *hibernicus* Syme: *Aat3* evidence. *Heredity* **60**:295–299.

Abe, M., Y. Kobayashi, S. Yamamoto, Y. Daimon, A. Yamaguchi, Y. Ikeda, H. Ichinoki, A. Notaguchi, K. Goto, and T. Araki. (2005). FD, a bZIP protein mediating signals from the floral pathway integrator FT at the shoot apex. *Science* **309**:1052–1056.

Achard, P., H. Cheng, L. de Grauwe, J. Decat, H. Schoutteten, T. Moritz, D. V. D. Straeten, J. Peng, and N. P. Harberd. (2006). Integration of plant responses to environmentally activated phytohormonal signals. *Science* **311**:91–94.

Ache, B., and J. Young. (2005). Olfaction: diverse species, conserved principles. *Neuron* **48**:417–430.

Ackerman, J. (2000). Abiotic pollen and pollination: ecological, functional and evolutionary perspectives. *Plant Systematics and Evolution* **222**:167–185.

Ackerman, J., and A. Montalvo. (1990). Short- and long-term limitations to fruit production in a tropical orchid. *Ecology* **71**:263–272.

Ahmad, M., and A. R. Cashmore. (1993). *HY4* gene of *A. thaliana* encodes a protein with characteristics of a blue-light photoreceptor. *Nature* **366**:162–166.

Ahn, J. H., D. Miller, V. J. Winter, M. J. Banfield, J. H. Lee, S. Y. Yoo, S. R. Henz, R. L. Brady, and D. Weigel. (2006). A divergent external loop confers antagonistic activity on floral regulators FT and TFL1. *EMBO Journal* **25**:605–614.

Aigner, P. (2004). Floral specialisation without trade-offs: optimal corolla flare in contrasting pollination environments. *Ecology* **85**:2560–2569.

Aigner, P. (2006). The evolution of specialised floral phenotypes in a fine-grained pollination environment. In N. Waser and J. Ollerton (eds), *Plant–Pollinator Interactions*, pp. 23–46. University of Chicago Press, London.

Ainsworth, C. (2000). Boys and girls come out to play: the molecular biology of dioecious plants. *Annals of Botany* **86**:211–221.

Ainsworth, C., S. Crossley, V. Buchanan-Wollaston, M. Thangavelu, and J. Parker. (1995). Male and female flowers of the dioecious plant sorrel show different patterns of MADS box gene expression. *Plant Cell* **7**:1583–1598.

Ainsworth, C., A. Rahman, J. Parker, and G. Edwards. (2005). Intersex inflorescences of *Rumex acetosa* demonstrate that sex determination is unique to each flower. *New Phytologist* **165**:711–720.

Airoldi, C., S. Bergonzi, and B. Davies. (2010). Single amino acid change alters the ability to specify male or female organ identity. *Proceedings of the National Academy of Sciences of the USA* **107**:18898–18902.

Aizen, M., and D. Vázquez. (2006). Flowering phonologies of hummingbird plants from the temperate forest of southern South America: is there evidence of competitive displacement? *Ecogeography* **29**:357–366.

Alabadi, D., T. Oyama, M. Yanovsky, F. Harmon, P. Mas, and S. Kay. (2001). Reciprocal regulation between *TOC1* and *LHY/CCA1* within the *Arabidopsis* circadian clock. *Science* **293**:880–883.

Al-Babili, S., J. Lintig, H. Haubruck, and P. Beyer. (1996). A novel, soluble form of phytoene desaturase from *Narcissus pseudonarcissus* chromoplasts is Hsp70-complexed and competent for flavinylation, membrane association and enzymatic activation. *Plant Journal* **9**:601–612.

Albert, N., D. Lewis, H. Zhang, K. Schwinn, P. Jameson, and K. Davies. (2011). Members of an R2R3-MYB transcription factor family in Petunia are developmentally and environmentally regulated to control complex floral and vegetative pigmentation patterning. *Plant Journal* **65**:771–784.

Alcorn, K., H. Whitney, and B. J. Glover. (2012). Flower movement increases pollinator preference for flowers with better grip. *Functional Ecology* **26**:941–947.

Alexandre, C., and L. Hennig. (2008). FLC or not FLC: the other side of vernalization. *Journal of Experimental Botany* **59**:1127–1135.

Alfenito, M., E. Souer, C. Goodman, R. Buell, J. Mol, R. Koes, and V. Walbot. (1998). Functional complementation of anthocyanin sequestration in the vacuole by widely divergent glutathione S-transferases. *Plant Cell* **10**:1135–1149.

Allen, A., C. Thorogood, M. Hegarty, C. Lexer, and S. Hiscock. (2011). Pollen-pistil interactions and self-incompatibility in the Asteraceae: new insights from studies of *Senecio squalidus*. *Annals of Botany* **108**:687–698.

Almeida, J., R. Carpenter, T. P. Robbins, C. Martin, and E. S. Coen. (1989). Genetic interactions underlying flower colour patterns in *Antirrhinum majus*. *Genes and Development* **3**:1758–1767.

Alvarez-Buylla, E., S. Liljegren, S. Pelaz, S. Gold, C. Burgeff, G. Ditta, F. Vergara-Silva, and M. Yanofsky. (2000). MADS-box gene evolution beyond flowers: expression in pollen, endosperm, guard cells, roots and trichomes. *Plant Journal* **24**:457–466.

Alvarez-Buylla, E., B. Garcia-Ponce, and A. Garay-Arroyo. (2006). Unique and redundant functional domains of APETALA1 and CAULIFLOWER, two recently duplicated Arabidopsis thaliana floral MADS-box genes. *Journal of Experimental Botany* **57**:3099–3107.

Ambrose, B., D. Lerner, P. Ciceri, C. Padilla, M. Yanofsky, and R. Schmidt. (2000). Molecular and genetic analyses of the silky1 gene reveal conservation in floral organ specification between eudicots and monocots. *Molecular Cell* **5**:569–579.

Anastasiou, E., S. Kenz, M. Gerstung, D. Maclean, J. Timmer, C. Fleck, and M. Lenhard. (2007). Control of plant organ size by *KLUH/CYP78A5*-dependent intercellular signalling. *Developmental Cell* **13**:843–856.

Andersen, C. H., C. S. Jensen, and K. Petersen. (2004). Similar genetic switch systems might integrate the floral inductive pathways in dicots and monocots. *Trends in Plant Science* **9**:105–107.

Andersen, O., and M. Jordheim. (2006). The anthocyanins. In: Anderson O.M., Markham K.R. (eds) *Flavonoids: Chemistry, biochemistry and applications*, pp. 471–530. CRC Press, Boca Raton, FL.

Anderson, M., E. Cornish, S. Mau, E. Williams, R. Hoggart, A. Atkinson, I. Boenig, B. Grego, R. Simpson, P. Roche, J. Haley, J. Penschow, H. Niall, G. Treager, J. Coughlan, R. Crawford, and A. Clarke. (1986). Cloning of cDNA for a stylar glycoprotein associated with expression of self-incompatibility in *Nicotiana alata*. *Nature* **321**:38–44.

Angenent, G., J. Franken, M. Busscher, A. Vandijken, J. Vanwent, H. Dons, and A. van Tunen. (1995). A novel class of MADS box genes is involved in ovule development in petunia. *Plant Cell* **7**:1569–1582.

Angiosperm Phylogeny Group. (2009). An update of the Angiosperm Phylogeny Group classification for the orders and families of flowering plants: APG III. *Botanical Journal of the Linnean Society* **161**:105–121.

Anon. (1921). Letters to Editor. *Nature* **107**:301.

Arber, A. (1937). The interpretation of the flower: a study of some aspects of morphological thought. *Biological Reviews* **12**:157–184.

Arber, A. (1946). Goethe's botany. *Chronica Botanica* **10**:63–126.

Armbruster, W. S., M. Edwards, and E. Debevec. (1994). Floral character displacement generates an assemblage structure of Western Australian triggerplants (*Stylidium*). *Ecology* **75**:315–329.

Ashman, T., T. Knight, J. Steets, P. Amarasekare, M. Burd, D. Campbell, M. Dudash, M. Johnston, S. Mazer, R. Mitchell, M. Morgan, and W. Wilson. (2004). Pollen limitation of plant reproduction: ecological and evolutionary causes and consequences. *Ecology* **85**:2408–2421.

Ashman, T., M. Bradburn, D. Cole, B. Blaney, and R. Raguso. (2005). The scent of a male: the role of floral volatiles in pollination of a gender dimorphic plant. *Ecology* **86**:2099–2105.

Aubert, D., L. Chen, Y. Moon, D. Martin, C. Yang, and Z. R. Sung. (2001). EMF1, a novel protein involved in the control of shoot architecture and flowering in Arabidopsis. *Plant Cell* **13**:1865–1875.

Auckerman, M. J., I. Lee, D. Weigel, and R. M. Amasino. (1999). The Arabidopsis flowering-time gene *LUMINIDEPENDENS* is expressed primarily in regions of cell proliferation and encodes a nuclear protein that regulates *LEAFY* expression. *Plant Journal* **18**:195–203.

Ausin, I., C. Alonso-Blanco, J. Jarillo, L. Ruiz-Garcia, and J. Martinez-Zapater. (2004). Regulation of flowering time by FVE, a retinoblastoma-associated protein. *Nature Genetics* **36**:162–166.

Ayasse, M., F. P. Schiestl, H. F. Paulus, F. Ibarra, and W. Francke. (2003). Pollinator attraction in a sexually deceptive orchid by means of unconventional chemicals. *Proceedings of the Royal Society B: Biological Sciences* **270**:517–522.

Ayasse, M., J. Stoekl, and W. Franke. (2011). Chemical ecology and pollinator-driven speciation in sexually deceptive orchids. *Phytochemistry* **72**:1667–1677.

Baker, C., P. Sieber, F. Wellmer, and E. Meyerowitz. (2005). The *early extra petals 1* mutant uncovers a role for microRNA miR164c in regulating petal number in Arabidopsis. *Current Biology* **15**:303–315.

Balasubramanian, S., S. Sureshkumar, J. Lempe, and D. Weigel (2006). Potent induction of *Arabidopsis thaliana* flowering by elevated growth temperature. *PLoS Genetics* **2**:e106.

Barak, S., E. Tobin, C. Andronis, S. Sugano, and R. Green. (2000). All in good time: the *Arabidopsis* circadian clock. *Trends in Plant Science* **5**:517–522.

Barrett, P. M., and K. J. Willis. (2001). Did dinosaurs invent flowers? Dinosaur-angiosperm coevolution revisited. *Biological Reviews of the Cambridge Philosophical Society* **76**:411–447.

Barrett, S. C. H. (2002). The evolution of plant sexual diversity. *Nature Reviews Genetics* **3**:274–284.

Barrett, S. C. H., L. K. Jesson, and A. M. Baker. (2000). The evolution and function of stylar polymorphisms in flowering plants. *Annals of Botany* **85**:A253–A265.

Bartlett, M., and C. Specht. (2011). Changes in expression pattern of the *TEOSINTE BRANCHED1*-line genes in the Zingiberales provide a mechanism for evolutionary shifts in symmetry across the order. *American Journal of Botany* **98**:1–17.

Bascompte, J., P. Jordano, C. Melian, and J. Olesen. (2003). The nested assembly of plant–animal mutualistic networks. *Proceedings of the National Academy of Sciences of the USA* **100**:9383–9387.

Bastow, R., J. S. Mylne, C. Lister, Z. Lippman, R. A. Martienssen, and C. Dean. (2004). Vernalization requires epigenetic silencing of *FLC* by histone methylation. *Nature* **427**:164–167.

Battey, N. H., and F. Tooke. (2002). Molecular control and variation in the floral transition. *Current Opinion in Plant Biology* **5**:62–68.

Baum, D. A. (1998). The evolution of plant development. *Current Opinion in Plant Biology* **1**:79–86.

Baurle, I., L. Smith, D. Baulcombe, and C. Dean. (2007). Widespread role for the flowering time regulators FCA and FPA in RNA-mediated chromatin silencing. *Science* **318**:109–112.

Becker, A., and G. Theissen. (2003). The major clades of MADS-box genes and their role in the development and evolution of flowering plants. *Molecular Phylogenetics and Evolution* **29**:464–489.

Beld, M., C. Martin, H. Huits, A. Stuitje, and A. G. M. Gerats. (1989). Flavonoid synthesis in *Petunia*; partial characterisation of dihydroflavonol 4-reductase genes. *Plant Molecular Biology* **13**:491–502.

Bell, C., D. Soltis, D, and P. Soltis. (2005a). The age of the angiosperms: A molecular timescale without a clock. *Evolution* **59**:1245–1258.

Bell, C., D. Soltis, and P. Soltis. (2010). The age and diversification of the angiosperms re-revisited. *American Journal of Botany* **97**:1296–1303.

Bell, J., J. Karron, and R. Mitchell. (2005b). Interspecific competition for pollination lowers seed production and outcrossing in *Mimulus ringens*. *Ecology* **86**:762–771.

Bennett, A., and M. Thery. (2007). Avian colour vision and colouration: multidisciplinary evolutionary biology. *American Naturalist* **169**:S1–S6.

Bereterbide, A., M. Hernould, S. Castera, and A. Mouras. (2001). Inhibition of cell proliferation, cell expansion and differentiation by the *Arabidopsis SUPERMAN* gene in transgenic tobacco plants. *Planta* **214**:22–29.

Berger, F., and D. Twell. (2011). Germline specification and function in plants. *Annual Reviews in Plant Biology* **62**:461–484.

Bernier, G., and C. Perilleux. (2005). A physiological overview of the genetics of flowering time control. *Plant Biotechnology Journal* **3**:3–16.

Bertin, R. (1993). Incidence of monoecy and dichogamy in relation to self-fertilization in angiosperms. *American Journal of Botany* **80**:557–560.

Bertin, R., and C. Newman. (1993). Dichogamy in angiosperms. *Botanical Review* **59**:112–152.

Beveridge, C., and I. Murfet. (1996). The *gigas* mutant in pea is deficient in the floral stimulus. *Physiologia Plantarum* **96**:637–645.

Bey, M., K. Stuber, K. Fellenberg, Z. Schwarz-Sommer, H. Sommer, H. Saedler, and S. Zachgo. (2004). Characterisation of Antirrhinum petal development and identification of target genes of the class B MADS box gene *DEFICIENS*. *Plant Cell* **16**:3197–3215.

Blazquez, M., and D. Weigel. (2000). Integration of floral inductive signals in *Arabidopsis*. *Nature* **404**:889–892.

Blazquez, M., R. Green, O. Nilsson, M. Sussman, and D. Weigel. (1998). Gibberellins promote flowering of Arabidopsis by activating the *LEAFY* promoter. *Plant Cell* **10**:791–800.

Blazquez, M., J. Ahn, and D. Weigel. (2003). A thermosensory pathway controlling flowering time in *Arabidopsis thaliana*. *Nature Genetics* **33**:168–171.

Blazquez, M., C. Ferrandiz, F. Madueno, and F. Parcy. (2006). How floral meristems are built. *Plant Molecular Biology* **60**:855–870.

Bodson, M. (1985). Changes in adenine nucleotide content in the apical bud of *Sinapis alba* L. during floral transition. *Planta* **163**:34–37.

Bodson, M., and W. Outlaw (1985). Elevation in the sucrose content of the shoot apical meristem of *Sinapis alba* at floral evocation. *Plant Physiology* **79**:420–424.

Boggs, N., K. Dwyer, M. Nasrallah, and J. Nasrallah. (2009). *In vivo* detection of residues required for ligand-selective activation of the S-locus receptor in Arabidopsis. *Current Biology* **19**:786–791.

Böhlenius, H., T. Huang, L. Charbonnel-Campaa, A. Brunner, S. Jansson, S. Strauss, and O. Nilsson. (2006). CO/FT regulatory module controls timing of flowering and seasonal growth cessation in trees. *Science* **312**:1040–1043.

Bond, D., E. Dennis, and J. Finnegan. (2011). The low temperature response pathways for cold acclimation and vernalization are independent. *Plant Cell and Environment* **34**:1737–1748.

Borner, R., G. Kampnann, J. Chandler, R. Gleissner, E. Wisman, K. Apel, and S. Melzer. (2000). A MADS domain gene involved in the transition to flowering in *Arabidopsis*. *Plant Journal* **24**:591–599.

Borthwick, H. A., Hendricks, S. B., Parker, M. W., Toole, E. H. and Toole, V. K. (1952). A reversible photoreaction controlling seed germination. *Proceedings of the National Academy of Sciences of the USA* **38**:662–666.

Boualem, A., C. Troadec, I. Kovalski, M. Sari, R. Perl-Treves, and A. Bendahmane. (2009). A conserved ethylene biosynthesis enzyme leads to andromonoecy in two *Cucumis* species. *PLoS One* **4**:e6144.

Bowman, J., D. Smyth, and E. Meyerowitz. (1991). Genetic interactions among floral homeotic genes of Arabidopsis. *Development* **112**:1–20.

Bowman, J., H. Sakai, T. Jack, D. Weigel, U. Mayer, and E. Myerowitz. (1992). Superman, a regulator of floral homeotic genes in Arabidopsis. *Development* **114**:599–615.

Bowman, J., J. Alvarez, D. Weigel, E. Meyerowitz, and D. Smyth. (1993). Control of flower development in *Arabidopis thaliana* by *APETALA1* and interacting genes. *Development* **119**:721–743.

Box, M., R. Bateman, B. Glover, and P. Rudall. (2008). Floral ontogenetic evidence of repeated speciation via paedomorphosis in subtribe Orchidinae (Orchidaceae). *Botanical Journal of the Linnean Society* **157**:429–454.

Box, M., and B. Glover. (2010). A plant developmentalist's guide to paedomorphosis. *Trends in Plant Science* **15**:241–246.

Box, M. S., S. Dodsworth, P. Rudall, R. Bateman, and B. Glover. (2011). Characterisation of *Linaria KNOX* genes suggests a role in petal spur development. *Plant Journal* **608**:703–714.

Box, M., S. Dodsworth, P. Rudall, R. Bateman, and B. Glover. (2012). Flower-specific KNOX phenotype in the orchid *Dactylorhiza fuchsia. Journal of Experimental Botany* **63**:4811–4819.

Bradley, D., R. Carpenter, H. Sommer, N. Hartley, and E. Coen. (1993). Complementary floral homeotic phenotypes result from opposite orientations of a transposon at the *PLENA* locus of Antirrhinum. *Cell* **72**:85–95.

Bradley, D., R. Carpenter, L. Copsey, C. Vincent, S. Rothstein, and E. Coen. (1996). Control of inflorescence architecture in Antirrhinum. *Nature* **379**:791–797.

Bradley, D., O. Ratcliffe, C. Vincent, R. Carpenter, and E. Coen. (1997). Inflorescence commitment and architecture in *Arabidopsis. Science* **275**:80–83.

Bradshaw, E., P. Rudall, D. Devey, M. Thomas, B. Glover, and R. Bateman. (2010). Comparative labellum micromorphology of the sexually deceptive temperate orchid genus *Ophrys*: diverse epidermal cell types and multiple origins of structural colour. *Botanical Journal of the Linnean Society* **162**:504–540.

Bradshaw, H., and D. Schemske. (2003). Allele substitution at a flower colour locus produces a pollinator shift in monkeyflowers. *Nature* **426**:176–178.

Brenner, G. J. (1996). Evidence for the earliest stages of angiosperm pollen evolution: a paleoequatorial section from Israel. In D. W. Taylor and L. J. Hickey (eds) *Flowering Plant Origin, Evolution and Phylogeny*, pp. 91–115. Chapman and Hall, New York.

Breuil-Broyer, S., P. Morel, J. Almeida-Engler, V. Coustham, I. Negrutiu, and C. Trehin. (2004). High-resolution boundary analysis during *Arabidopsis thaliana* flower development. *Plant Journal* **38**:182–192.

Brioudes, F., C. Joly, J. Szecsi, E. Varaud, J. Leroux, F. Bellvert, C. Bertrand, and M. Bendahmane. (2009). Jasmonate controls late development stages of petal growth in *Arabidopsis thaliana. Plant Journal* **60**:1070–1080.

Brockington, S., R. Walker, B. Glover, P. Soltis, and D. Soltis. (2011). Complex evolution of pigmentation in the Caryophyllales. *New Phytologist* **190**:854–864.

Brockington, S., P. Rudall, M. Frohlich, D. Oppenheimer, P. Soltis, and D. Soltis. (2012). "Living stones" reveal alternative petal identity programmes within the core eudicots. *Plant Journal* **69**:193–203.

Brockington, S., R. Alvarez-Fernandez, J. Landis, K. Alcorn, R. Walker, M. Thomas, L. Hileman, and B. Glover. (2013). Evolutionary analysis of the MIXTA gene family highlights potential targets for the study of cellular differentiation. *Molecular Biology and Evolution* **30**:526–540.

Broholm, S., S. Tahtiharju, R. Laitinen, V. Albert, T. Teeri, and P. Elomaa. (2008). A TCP domain transcription factor controls flower type specification along the radial axis of the *Gerbera* (Asteraceae) inflorescence. *Proceedings of the National Academy of Sciences of the USA* **105**:9117–9122.

Brown, B. J., R. J. Mitchell, and S. A. Graham. (2002). Competition for pollination between an invasive species (purple loosestrife) and a native congener. *Ecology* **83**:2328–2336.

Brugliera, F., G. Barr-Rewell, T. Holton, and J. Mason. (1999). Isolation and characterisation of a flavonoid 3′-hydroxylase cDNA clone corresponding to the *Ht1* locus of *Petunia hybrida. Plant Journal* **19**:441–451.

Bumrungsri, S., E. Sripaoraya, T. Chongsiri, K. Sridith, and P. A. Racey. (2009). The pollination ecology of durian (*Durio zibethinus*, Bombacaceae) in southern Thailand. *Journal of Tropical Ecology* **25**:85–92.

Busch, A., and S. Zachgo. (2007). Control of corolla monosymmetry in the Brassicaceae *Iberis amara. Proceedings of the National Academy of Sciences of the USA* **104**: 16714–16719.

Busch, M., K. Bomblies, and D. Weigel. (1999). Activation of a floral homeotic gene in Arabidopsis. *Science* **285**:585–587.

Buzgo, M., P. Soltis, and D. Soltis. (2004). Floral development and morphology of *Amborella trichopoda* (Amborellaceae). *International Journal of Plant Science* **165**:925–947.

Byzova, M., J. Franken, M. Aarts, J. de Almeida-Engler, G. Engler, C. Mariani, M. Campagne, and G. Angenent. (1999). Arabidopsis STERILE APETALA, a multifunctional gene regulating inflorescence, flower and ovule development. *Genes and Development* **13**:1002–1014.

Cabrillac, D., J. Cock, C. Dumas, and T. Gaude. (2001). The S-locus receptor kinase is inhibited by thioredoxins and activated by pollen coat proteins. *Nature* **410**:220–223.

Campbell, D. (1996). Evolution of floral traits in a hermaphroditic plant: field measurements of heritabilities and genetic correlations. *Evolution* **50**:1442–1453.

Campbell, D., and K. Halama. (1993). Resource and pollen limitations to lifetime seed production in a natural plant population. *Ecology* **74**:1043–1051.

Campbell, D., N. Waser, M. Price, E. Lynch, and R. Mitchell. (1991). Components of phenotypic selection: pollen export and flower corolla width in *Ipomopsis aggregata*. *Evolution* **45**:1458–1467.

Carlsbecker, A., K. Tandre, U. Johanson, M. Englund, and P. Engstrom. (2004). The MADS box gene DAL1 is a potential mediator of the juvenile-to-adult transition in Norway spruce (*Picea abies*). *Plant Journal* **40**: 546–557.

Carpenter, R., and E. Coen. (1990). Floral homeotic mutations produced by transposon-mutagenesis in *Antirrhinum majus*. *Genes and Development* **4**:1483–1493.

Cartolano, M., R. Castillo, N. Efremova, M. Kuckenberg, J. Zethof, T. Gerats, Z. Schwarz-Sommer, and M. Vandenbussche. (2007). A conserved microRNA module exerts homeotic control over *Petunia hybrida* and *Antirrhinum majus* floral organ identity. *Nature Genetics* **39**:901–905.

Cashmore, A., J. Jarillo, Y. Wu, and D. Liu. (1999). Cryptochromes: blue light receptors for plants and animals. *Science* **284**:760–765.

Castillejo, C., and S. Pelaz. (2008). The balance between CONSTANS and TEMPRANILLO activities determines FT expression to trigger flowering. *Current Biology* **18**:1338–1343.

Causier, B., R. Catillo, J. Zhou, R. Ingram, Y. Xue, Z. Schwarz-Sommer, and B. Davies. (2005). Evolution in action: following function in duplicated floral homeotic genes. *Current Biology* **15**:1508–1512.

Causier, B., D. Bradley, H. Cook, and B. Davies. (2009). Conserved intragenic elements were critical for the evolution of the floral C-function. *Plant Journal* **58**:41–52.

Chae, E., Q. Tan, T. Hill, and V. Irish. (2008). An Arabidopsis F-box protein acts as a transcriptional co-factor to regulate floral development. *Development* **135**:1235–1245.

Chanderbali, A., V. Albert, J. Leebens-Mack, N. Altman, D. Soltis, and Soltis, P. (2009). Transcriptional signatures of ancient floral developmental genetics in avocado (*Persea americana*, Laureaceae). *Proceedings of the National Academy of Sciences of the USA* **106**:8929–8934.

Chandler, J., J. M. Martinez-Zapater, and C. Dean. (2000). Mutations causing defects in the biosynthesis and response to gibberellins, abscisic acid and phytochrome B do not inhibit vernalization in *Arabidopsis fca-1*. *Planta* **210**:677–682.

Chang, F., Y. Wang, S. Wang, and H. Ma. (2011). Molecular control of microsporogenesis in Arabidopsis. *Current Opinion in Plant Biology* **14**:66–73.

Chanvivattana, Y., A. Bishopp, D. Schubert, C. Stock, Y. Moon, Z. R. Sung, and J. Goodrich. (2004). Interaction of Polycomb-group proteins controlling flowering in *Arabidopsis*. *Development* **131**:5263–5276.

Chapman, M., S. Tang, D. Draeger, S. Nambeesan, H. Shaffer, J. Barb, S. Knapp, and J. Burke. (2012). Genetic analysis of floral symmetry in Van Gogh's sunflowers reveals independent recruitment of *CYCLOIDEA* genes in the Asteraceae. *PLoS Genetics* **8**:e1002628.

Chappell, J. (1995). Biochemistry and molecular biology of the isoprenoid biosynthetic pathway in plants. *Anuual Review of Plant Physiology and Plant Molecular Biology* **46**:521–547.

Charlesworth, D. (2002). Self-incompatibility: how to stay incompatible. *Current Biology* **12**:R424–R426.

Charlesworth, D. (2006). Evolution of plant breeding systems. *Current Biology* **16**:R726–R735.

Chaw, S., C. Parkinson, Y. Cheng, T. Vincent, and J. Palmer. (2000). Seed plant phylogeny inferred from all three plant genomes: Monophyly of extant gymnosperms and origin of Gnetales from conifers. *Proceedings of the National Academy of Sciences of the USA* **97**:4086–4091.

Chen, D., J. Collins, and T. Goldsmith. (1984). The ultraviolet receptor of bird retinas. *Science* **285**:337–340.

Chen, D., B. Guo, S. Hexige, T. Zhang, D. Shen, and F. Ming. (2007). *SQUA*-like genes in the orchid *Phalaenopsis* are expressed in both vegetative and reproductive tissues. *Planta* **226**:369–380.

Chen, H., Y. Shen, X. Tang, L. Yu, J. Wang, L. Guo, Y. Zhang, H. Zhang, S. Feng, E. Strickland, N. Zeng, and X. Deng. (2006). Arabidopsis CULLIN4 forms an E3 ubiquitin ligase with RBX1 and the CDD complex in mediating light control of development. *Plant Cell* **18**: 1991–2004.

Chen, H., X. Huang, G. Gusmaroli, W. Terzhagi, O. Lau, Y. Yanagawa, Y. Zhang, J. Li, J. Lee, D. Zhu, and X. Deng. (2010). *Arabidopsis* CULLIN4-damaged DNA binding protein 1 interacts with CONSTITUTIVELY PHOTOMORPHOGENIC1-SUPPRESSOR OF PHYA complexes to regulate photomorphogenesis and flowering time. *Plant Cell* **22**:108–123.

Chen, M., and J. Chory. (2011). Phytochrome signalling mechanisms and the control of plant development. *Trends in Cell Biology* **21**:664–671.

Chen, X. (2004). A microRNA as a translational repressor of APETALA2 in Arabidopsis flower development. *Science* **303**:2022–2025.

Chen, Z. J., and L. Tian. (2007). Roles of dynamic and reversible histone acetylation in plant development and polyploidy. *Biochimica et Biophysica Acta* **1769**:295–307.

Chittka, L. (1996). Does bee colour vision predate the evolution of flower colour? *Naturwissenschaften* **83**:136–138.

Chittka, L., and N. M. Waser. (1997). Why red flowers are not invisible to bees. *Israel Journal of Plant Sciences* **45**:169–183.

Chittka, L., and S. Schürkens. (2001). Successful invasion of a floral market. An exotic Asian plant has moved in on Europe's river-banks by bribing pollinators. *Nature* **146**:653.

Chittka, L., and N. Raine. (2006). Recognition of flowers by pollinators. *Current Opinion in Plant Biology* **9**:428–435.

Choi, K., J. Kim, H. Hwang, S. Kim, C. Park, S. Kim, and I. Lee. (2011). The FRIGIDA complex activates transcription of FLC, a strong flowering repressor in Arabidopsis, by recruiting chromatin modification factors. *The Plant Cell* **23**:289–303.

Christinet, L., F. Burdet, M. Zaiko, U. Hinz, and J. Zryd. (2004). Characterization and functional identification of a novel plant 4,5-extradiol dioxygenase involved in betalain pigment biosynthesis in *Portulaca grandiflora*. *Plant Physiology* **134**:265–274.

Chuck, G., R. Meeley, H. Irish, H. Sakai, and S. Hake. (2007). The maize tasselseed4 microRNA controls sex determination and meristem cell fate by targeting Tasselseed6/indeterminate spikelet1. *Nature Genetics* **39**:1517–1521.

Ciaffi, M., A. Paolacci, O. Tanzarella, and E. Porceddu. (2011). Molecular aspects of flower development in grasses. *Sexual Plant Reproduction* **24**:247–282.

Citerne, H., R. T. Pennington, and Q. C. B. Cronk. (2006). An apparent reversal in floral symmetry in the legume *Cadia* is a homeotic transformation. *Proceedings of the National Academy of Sciences of the USA* **103**:12017–12020.

Clack, T., A. Shokry, M. Moffet, P. Liu.M. Faul, and R. Sharrock. (2009). Obligate heterodimerization of Arabidopsis phytochromes C and E and interaction with the PIF3 basic helix-loop-helix transcription factor. *Plant Cell* **21**:786–799.

Clegg, M., and M. Durbin. (2003). Tracing floral adaptations from ecology to molecules. *Nature Reviews Genetics* **4**:206–215.

Clough, S. J., and A. F. Bent. (1998). Floral dip: a simplified method for Agrobacterium-mediated transformation of *Arabidopsis thaliana*. *Plant Journal* **16**:735–743.

Coe, E. H., S. McCormick, and S. A. Modena. (1981). White pollen in maize. *Journal of Heredity* **72**:318–320.

Coen, E., and E. Meyerowitz. (1991). The war of the whorls: genetic interactions controlling flower development. *Nature* **353**:31–37.

Coen, E. S., R. Carpenter, and C. Martin. (1986). Transposable elements generate novel patterns of gene expression in *Antirrhinum majus*. *Cell* **47**:285–296.

Coen, E., J. M. Romero, S. Doyle, R. Elliott, G. Murphy, and R. Carpenter. (1990). *floricaula*: a homeotic gene required for flower development in *Antirrhinum majus*. *Cell* **63**:1311–1322.

Colasanti, J., and V. Sundaresan. (2000). "Florigen" enters the molecular age: long-distance signals that cause plants to flower. *Trends in Biochemical Sciences* **25**:236–240.

Colcombet, J., A. Boisson-Cernier, R. Ros-Palau, C. Vera, and J. Schroeder. (2005). Arabidopsis SOMATIC EMBRYOGENESIS RECEPTOR KINASES 1 and 2 are essential for tapetum development and microspore maturation. *Plant Cell* **17**:3350–3361.

Colombo, L., J. Franken, E. Koetje, J. van Went, H. Dons, G. Angenent, and A. van Tunen. (1995). The petunia MADS box gene FBP11 determines ovule identity. *Plant Cell* **7**:1859–1868.

Comba, L., S. A. Corbet, H. Hunt, S. Outram, J. S. Parker, and B. J. Glover. (2000). The role of genes influencing the corolla in pollination of *Antirrhinum majus*. *Plant, Cell and Environment* **23**:639–647.

Conner, J., and Z. Liu. (2000). LEUNIG, a putative transcriptional corepressor that regulates AGAMOUS expression during flower development. *Proceedings of the National Academy of Sciences of the USA* **97**:12902–12907.

Corbesier, L., and G. Coupland. (2005). Photoperiodic flowering of *Arabidopsis*: integrating genetic and physiological approaches to characterisation of the floral stimulus. *Plant, Cell and Environment* **28**:54–66.

Corbesier, L., C. Vincent, S. Jang, F. Fornara, Q. Fan, I. Searle, A. Giakountis, S. Farrona, L. Gissot, C. Turnbull, and Coupland, G. (2007). FT protein movement contributes to long-distance signaling in floral induction of *Arabidopsis*. *Science* **316**:1030–1033.

Corley, S. B., R. Carpenter, L. Copsey, and E. Coen. (2005). Floral asymmetry involves an interplay between TCP and MYB transcription factors in *Antirrhinum*. *Proceedings of the National Academy of Sciences of the USA* **102**:5068–5073.

Coupland, G. (1995). Genetic and environmental control of flowering time in *Arabidopsis*. *Trends in Genetics* **11**:393–397.

Cox, P. A. (1988). Hydrophilous pollination. *Annual Reviews in Ecology and Systematics* **19**:261–280.

Cox, P. (1991). Abiotic pollination—an evolutionary escape for animal-pollinated angiosperms. *Philosophical Transactions of the Royal Society of London Series B* **333**:217–224.

Crane, P. (1985). Phylogenetic analysis of seed plants and the origin of angiosperms. *Annals of the Missouri Botanical Garden* **72**:716–793.

Crane, P. R., E. M. Friis, and K. R. Pedersen. (1995). The origin and early diversification of angiosperms. *Nature* **374**:27–33.

Crawford, B. C. W., U. Nath, R. Carpenter, and E. S. Coen. (2004). *CINCINNATA* controls both cell differentiation and growth in petal lobes and leaves of Antirrhinum. *Plant Physiology* **135**:244–253.

Cronk, Q. C. B. (2001). Plant evolution and development in a post-genomic context. *Nature Reviews Genetics* **2**:607–619.

Cubas, P. (2003). Floral zygomorphy, the recurring evolution of a successful trait. *BioEssays* **26**:1175–1184.

Cubas, P., C. Vincent, and E. S. Coen. (1999). An epigenetic mutation responsible for natural variation in floral symmetry. *Nature* **401**:157–161.

Culley, T., S. Weller, and A. Sakai. (2002). The evolution of wind pollination in angiosperms. *Trends in Ecology and Evolution* **17**:361–369.

Danieli-Silva, A., J. de Souza, A. Donatti, R. Campos, J. Vicente-Silva, L. Freitas, and I. Varassin. (2011). Do pollination syndromes cause modularity and predict interactions in a pollination network in tropical high-altitude grasslands? *Oikos* **121**:35–43.

Danyluk, J., N. Kane, G. Breton, A. Limin, D. B. Fowler, and F. Sarhan. (2003). TaVRT-1, a putative transcription factor associated with vegetative to reproductive transition in cereals. *Plant Physiology* **132**:1849–1860.

Darwin, C. (1859). *On the Origin of Species by Means of Natural Selection*. Murray, London.

Das, P., T. Ito, F. Wellmer, T. Vernoux, A. Dedieu, J. Traas, and E. Meyerowitz. (2009). Floral stem cell termination involves the direct regulation of AGAMOUS by PERIANTHIA. *Development* **136**:1605–1611.

Davis, C., P. Endress, and D. Baum. (2008). The evolution of floral gigantism. *Current Opinion in Plant Biology* **11**:49–57.

Davis, S., J. Kurepa, and R. Vierstra. (1999). The *Arabidopsis thaliana* HY1 locus, required for phytochrome-chromophore biosynthesis, encodes a protein related to heme oxygenases. *Proceedings of the National Academy of Sciences of the USA* **96**:6541–6546.

de Bodt, S., J. Raes, K. Florquin, S. Rombauts, P. Rouze, G. Theissen, and Y. Van de Peer. (2003). Genomewide structural annotation and evolutionary analysis of the type I MADS-box genes in plants. *Journal of Molecular Evolution* **56**:573–586.

Delgado-Vargas, F., and O. Paredes-Lopez. (2002). Anthocyanins and betalains. In: *Natural Colorants for Food and Nutraceutical Uses*, pp. 167–219. CRC Press, Boca Raton, FL.

Dellaporta, S., and A. Calderon-Urrea. (1994). The sex determination process in maize. *Science* **266**:1501–1505.

Deng, X., M. Matsui, N. Wei, D. Wagner, A. Chu, K. Feldmann, and P. Quail. (1992). COP1, an Arabidopsis regulatory gene, encodes a protein with both a zinc-binding motif and a G beta homologous domain. *Cell* **71**:791–801.

de Vetten, N., F. Quattrocchio, J. Mol, and R. Koes. (1997). The *an11* locus controlling flower pigmentation in petunia encodes a novel WD-repeat protein conserved in yeast, plants and animals. *Genes and Development* **11**:1422–1434.

de Vlaming, P., A. W. Scram, and H. Wiering. (1983). Genes affecting flower colour and pH of flower limb homogenates in *Petunia hybrida*. *Theoretical and Applied Genetics* **66**:271–278.

Dewitte, W., C. Riou-Khamlichi, S. Scofield, J. M. S. Healy, A. Jacqmard, N. J. Kilby, and J. A. H. Murray. (2003). Altered cell cycle distribution, hyperplasia, and inhibited differentiation in Arabidopsis caused by the D-type cyclin CYCD3. *Plant Cell* **15**:79–92.

Diggle, P., V. Di Stilio, A. Gschwend, E. Golenberg, R. Moore, J. Russell, and J. Sinclair. (2011). Multiple developmental processes underlie sex differentiation in angiosperms. *Trends in Genetics* **27**:368–376.

Dilcher, D. (2000). Towards a new synthesis: major evolutionary trends in the angiosperm fossil record. *Proceedings of the National Academy of Sciences of the USA* **97**:7030–7036.

Dinh, T., T. Girke, X. Liu, L. Yant, M. Schmid, and X. Chen. (2012). The floral homeotic protein APETALA2 recognizes and acts through an AT-rich sequence element. *Development* **139**:1978–1986.

Disch, S., E. Anastasiou, V. Sharma, T. Laux, J. Fletcher, and M. Lenhard. (2006). The E3 ubiquitin ligase BIG BROTHER controls Arabidopsis organ size in a dosage-dependent manner. *Current Biology* **16**:272–279.

Distelfeld A., C. Li, and J. Dubcovsky (2009). Regulation of flowering in temperate cereals. *Current Opinion in Plant Biology* **12**:178–184.

Di Stilio, V., E. Kramer, and D. Baum. (2005). Floral MADS box genes and homeotic gender dimorphism in *Thalictrum dioicum* (Ranunculaceae)—a new model for the study of dioecy. *The Plant Journal* **41**:755–766.

Di Stilio, V., C. Martin, A. Schulfer, and C. Connelly. (2009). An ortholog of *MIXTA-like 2* controls epidermal cell shape in flowers of *Thalictrum*. *New Phytologist* **183**:718–728.

Ditta, G., A. Pinyopich, P. Robles, S. Pelaz, and M. Yanofsky. (2004). The *SEP4* gene of *Arabidopsis thaliana* functions in floral organ and meristem identity. *Current Biology* **14**:1935–1940.

Dobson, H., E. Danielson, and I. van Wesep. (1999). Pollen odor chemicals as modulators of bumble bee foraging on *Rosa rugosa* Thunb. (Rosaceae). *Plant Species Biology* **14**:153–166.

Doi, K., T. Izawa, T. Fuse, U. Yamanouchi, T. Kubo, Z. Shimatani, M. Yano, and A. Yoshimura. (2004). Ehd1, a B-type response regulator in rice, confers short-day promotion of flowering and controls FT-like gene expression independently of Hd1. *Genes and Development* **18**:926–936.

Dong, C., M. Agarwal, Y. Zhang, Q. Xie, and J. Zhu. (2006). The negative regulator of plant cold responses, HOS1, is a ring E3 ligase that mediates the ubiquitination and degradation of ICE1. *Proceedings of the National Academy of Sciences of the USA* **103**:8281–8286.

Dooner, H. K. (1983). Coordinate genetic regulation of flavonoid biosynthetic enzymes in maize. *Molecular and General Genetics* **189**:136–141.

Dornelas, M., and A. Rodriguez. (2005). A FLORICAULA/ LEAFY gene homolog is preferentially expressed in developing female cones of the tropical pine *Pinus caribaea* var. *caribaea*. *Genetics and Molecular Biology* **28**:299–307.

Doyle, J., and M. Donoghue. (1986). Seed plant phylogeny and the origin of angiosperms: an experimental cladistic approach. *Botanical Review* **52**:321–431.

Dresselhaus, T. (2006). Cell–cell communication during double fertilization. *Current Opinion in Plant Biology* **9**:41–47.

Drews, G., J. Bowman, and E. Meyerowitz. (1991). Negative regulation of the Arabidopsis homeotic gene AGAMOUS by the APETALA2 product. *Cell* **65**:991–1002.

Drews, G., D. Wang, J. Steffen, K. Schumaker, and R. Yadegari. (2011). Identification of genes expressed in the angiosperm female gametophye. *Journal of Experimental Botany* **62**:1593–1599.

Dudareva, N., and E. Pichersky. (2006). *Biology of Floral Scent*. Taylor and Francis, Boca Raton, FL.

Duek, P., M. Elmer, V. Van Oosten, and C. Fankhauser. (2004). The degradation of HFR1, a putative bHLH class transcription factor involved in light signaling, is regulated by phosphorylation and requires COP1. *Current Biology* **14**:2296–2301.

Dyer, A. G., H. M. Whitney, S. E. J. Arnold, B. J. Glover, and L. Chittka. (2006). Bees associate warmth with floral colour. *Nature* **442**:525.

Dyer, A. G., H. M. Whitney, S. E. J. Arnold, B. J. Glover, and L. Chittka. (2007). Mutations perturbing petal cell shape and anthocyanin synthesis influence bumblebee perception of *Antirrhinum majus* flower colour. *Arthropod Plant Interactions* **1**:45–55.

Ebel, C., L. Mariconti, and W. Gruissem. (2004). Plant retinoblastoma homologues control nuclear proliferation in the female gametophyte. *Nature* **429**:776–780.

Eckardt, N. (2003). A component of the cryptochrome blue light signalling pathway. *Plant Cell* **15**:1051–1052.

Eckhart, V., N. Rushing, G. Hart, and J. Hansen. (2006). Frequency-dependent pollinator foraging in polymorphic *Clarkia xantiana* ssp. *xantiana* populations: implications for flower colour evolution and pollinator interactions. *Oikos* **112**:412–421.

Egea-Cortines, M., H. Saedler, and H. Sommer. (1999). Ternary complex formation between the MADS-box proteins SQUAMOSA, DEFICIENS and GLOBOSA is involved in the control of floral architecture in *Antirrhinum majus*. *EMBO Journal* **18**:5370–5379.

Eiskowitch, D. (1980). The role of dark flowers in the pollination of certain Umbelliferae. *Journal of Natural History* **14**:737–742.

Ellis, A., and S. Johnson. (2010). Floral mimicry enhances pollen export: the evolution of pollination by sexual deceit outside of the Orchidaceae. *American Naturalist* **176**:E143–E151.

Emborg, T., J. Walker, B. Noh, and R. Viestra. (2006). Multiple heme oxygenase family members contribute to the biosynthesis of the phytochrome chromophore in Arabidopsis. *Plant Physiology* **140**:856–868.

Endress, P. K. (1987). The early evolution of the angiosperm flower. *Trends in Ecology and Evolution* **2**:300–304.

Endress, P. K. (1996). Structure and function of female and bisexual organ complexes in Gnetales. *International Journal of Plant Science* **157**:S113–S125.

Endress, P. K. (2001). Evolution of floral symmetry. *Current Opinion in Plant Biology* **4**:86–91.

Endress, P., and A. Igersheim. (2000). The reproductive structures of the basal angiosperm *Amborella trichopoda* (Amborellaceae). *International Journal of Plant Science* **161**:S237–S248.

Epperson, B., and M. Clegg. (1987). Frequency-dependent variation for outcrossing rate among flower colour morphs of *Ipomoea purpurea*. *Evolution* **41**:1302–1311.

Eriksson, S., H. Bohlenius, T. Moritz, and O. Nilsson. (2006). GA$_4$ is the active gibberellin in the regulation of *LEAFY* transcription and *Arabidopsis* floral initiation. *Plant Cell* **18**:2172–2181.

Eriksson, S., L. Stransfield, N. Adamski, H. Breuninger, and M. Lenhard. (2010). *KLUH/CYP78A5* dependent growth signalling coordinates floral organ growth in *Arabidopsis*. *Current Biology* **20**:527–532.

Faegri, K., and L. van der Pijl. (1966). *The Principles of Pollination Ecology*. Pergamon Press, Oxford.

Fan, H., Y. Hu, M. Tudor, and H. Ma. (1997). Specific interactions between the K domains of AG and AGLs, members of the MADS domain family of DNA binding proteins. *Plant Journal* **12**:999–1010.

Fankhauser, C., K. Yeh, J. C. Lagarias, H. Zhang, T. Elich, and J. Chory. (1999). PKS1, a substrate phosphorylated by phytochrome that modulates light signalling in *Arabidopsis*. *Science* **284**:1539–1541.

Farre, G., C. Bai, R. Twyman, T. Capell, P. Christou, and C. Zhu. (2011). Nutritious crops producing multiple carotenoids—a metabolic balancing act. *Trends in Plant Science* **16**:532–540.

Farzad, M., R. Griesbach, and M. R. Weiss. (2002). Floral colour change in *Viola cornuta* L. (Violaceae): a model system to study regulation of anthocyanin production. *Plant Science* **162**:225–231.

Favaro, R., A. Pinyopich, R. Battaglia, M. Kooiker, L. Borghi, G. Ditta, M. Yanofsky, M. Kater, and L. Colombo. (2003). MADS-box protein complexes control carpel and ovule development in Arabidopsis. *Plant Cell* **15**:2603–2611.

Feinsinger, P. (1987). Effects of plant species on each other's pollination: is community structure influenced? *Trends in Ecology and Evolution* **2**:123–126.

Feng, C., X. Liu, Y. Yu, D. Xie, R. Franks, and J. Wang. (2012). Evolution of bract development and B-class MADS box gene expression in petaloid bracts of *Cornus* s.l. (Cornaceae). *New Phytologist* **196**:631–643.

Feng, W., Y. Jacob, K. Veley, L. Ding, X. Yu, G. Choe, and S. Michaels. (2011). Hypomorphic alleles reveal FCA-independent roles for FY in the regulation of FLOWERING LOCUS C. *Plant Physiology* **155**:1425–1434.

Feng, X., Z. Zhao, Z. Tian, *et al.* (2006). Control of petal shape and floral zygomorphy in *Lotus japonicus*. *Proceedings of the National Academy of Sciences of the USA* **103**:4970–4975.

Fenster, C., W. S. Armbruster, P. Wilson, M. Dudash, and J. Thomson. (2004). Pollination syndromes and floral specialization. *Annual Reviews in Ecology, Evolution and Systematics* **35**:375–403.

Feyissa, D., T. Lovdal, K. Olsen, R. Slimestad, and C. Lillo. (2009). The endogenous GL3, but not EGL3, gene is necessary for anthocyanin accumulation as induced by nitrogen depletion in Arabidopsis rosette stage leaves. *Planta* **230**:747–754.

Figueiredo, P., M. Elhabiri, K. Toki, N. Saito, O. Dangles, and R. Brouillard. (1996). New aspects of anthocyanin complexation. Intramolecular copigmentation as a means for colour loss? *Phytochemistry* **41**:301–308.

Finnegan, E. J., R. K. Genger, K. Kovac, W. J. Peacock, and E. S. Dennis. (1998). DNA methylation and the promotion of flowering by vernalization. *Proceedings of the National Academy of Sciences of the USA* **95**:5824–5829.

Finnegan, E. J., K. A. Kovac, E. Jaligot, C. C. Sheldon, W. J. Peacock, and E. S. Dennis. (2005). The downregulation of *FLOWERING LOCUS C (FLC)* expression in plants with low levels of DNA methylation and by vernalization occurs by distinct mechanisms. *Plant Journal* **44**:420–432.

Finnegan, J., and E. Dennis. (2007). Vernalization-induced trimethylation of histone H3 lysine 27 at FLC is not maintained in mitotically quiescent cells. *Current Biology* **17**:1978–1983.

Flanagan, C., and H. Ma (1994). Spatially and temporally regulated expression of the MADS box gene AGL2 in wild type and mutant Arabidopsis flowers. *Plant Molecular Biology* **26**:581–595.

Fleming, T. (2006). Reproductive consequences of early flowering in organ pipe cactus, *Stenocereus thurberi*. *International Journal of Plant Sciences* **167**:473–481.

Flores-Renteria, L., A. Vazquez-Lobo, A. Whipple, D. Pinero, J. Marquez-Guzman, and C. Dominguez. (2011). Functional bisporangiate cones in *Pinus johannis* (Pinaceae): implications for the evolution of bisexuality in seed plants. *American Journal of Botany* **98**:130–139.

Foote, H., J. Ride, V. Franklin-Tong, E. Walker, M. Lawrence, and C. Franklin. (1994). Cloning and expression of a distinctive class of self-incompatibility (S) gene from *Papaver rhoeas* L. *Proceedings of the National Academy of Sciences of the USA* **91**:2265–2269.

Fowler, S., K. Lee, H. Onouchi, A. Samach, K. Richardson, B. Morris, G. Coupland, and J. Putterill. (1999). *GIGANTEA*: a circadian clock-controlled gene that regulates photoperiodic flowering in *Arabidopsis* and encodes a protein with several possible membrane-spanning domains. *EMBO Journal* **18**:4679–4688.

Franklin-Tong, V., and C. Franklin. (2003). Gametophytic self-incompatibility inhibits pollen tube growth using different mechanisms. *Trends in Plant Science* **8**: 598–605.

Franks, R., C. Wang, J. Levin, and Z. Liu. (2002). SEUSS, a member of a novel family of plant regulatory proteins, represses floral homeotic gene expression with LEUNIG. *Development* **129**:253–263.

Franks, R. G., Z. Liu, and R. L. Fischer. (2006). *SEUSS* and *LEUNIG* regulate cell proliferation, vascular development and organ polarity in *Arabidopsis* petals. *Planta* **224**:801–811.

Free, J. (1968). Dandelion as a competitor to fruit trees for bee visits. *Journal of Applied Ecology* **5**:169–178.

Freeman, D. C., E. D. McArthur, K. Harper, and A. Blauer. (1981). Influence of environment on the floral sex ratio of monoecious plants. *Evolution* **35**:194–197.

Freeman, D., J. Doust, A. El-Keblawy, K. Miglia, and E. McArthur. (1997). Sexual specialization and inbreeding avoidance in the evolution of dioecy. *Botanical Review* **63**:65–92.

Friedman, J., and S. Barrett. (2008). A phylogenetic analysis of the evolution of wind pollination in the angiosperms. *International Journal of Plant Sciences* **169**:49–58.

Friedman, W. (1990). Double fertilisation in *Ephedra*, a nonflowering seed plant: its bearing on the origin of angiosperms. *Science* **247**:951–954.

Friedman, W. E. (2006). Embryological evidence for developmental lability during early angiosperm evolution. *Nature* **441**:337–340.

Friedman, W., and K. Ryerson. (2009). Reconstructing the ancestral female gametophyte of angiosperms: insights from *Amborella* and other ancient lineages of flowering plants. *American Journal of Botany* **96**:129–143.

Friis, E. M., K. R. Pedersen, and P. R. Crane. (1999). Early angiosperm diversification: the diversity of pollen associated with angiosperm reproductive structures in Early

Cretaceous floras from Portugal. *Annals of the Missouri Botanical Garden* **86**:259–296.

Friis, E., K. Pedersen, and P. Crane. (2001). Fossil evidence of water lilies (Nymphaeales) in the Early Cretaceous. *Nature* **410**:357–360.

Friis, E., K. Pedersen, and P. Crane. (2005). When Earth started blooming: insights from the fossil record. *Current Opinion in Plant Biology* **8**:5–12.

Frohlich, M. W. (2002). The Mostly Male theory of flower origins: summary and update regarding the Jurassic pteridosperm *Pteroma*. In: Q. C. B. Cronk, R. M. Bateman, and J. A. Hawkins (eds) *Developmental Genetics and Plant Evolution*, pp. 85–108. Taylor and Francis, London.

Frohlich, M. W., and D. S. Parker. (2000). The mostly male theory of flower evolutionary origins: from genes to fossils. *Systematic Botany* **25**:155–170.

Fukada-Tanaka, S., Y. Inagaki, T. Yamaguchi, N. Saito, and S. Iida. (2000). Colour-enhancing protein in blue petals. *Nature* **407**:581.

Fulton, M., and S. Hodges. (1999). Floral isolation between *Aquilegia formosa* and *Aquilegia pubescens*. *Proceedings of the Royal Society of London Series B* **266**:2247–2252.

Furner, I. J., and M. Matzke. (2010). Methylation and de-methylation of the Arabidopsis genome. *Current Opinion in Plant Biology* **14**:137–141.

Furner, I. J., J. Ainscough, J. Pumfrey, and L. Petty. (1996). Clonal analysis of the late flowering *fca* mutant of *Arabidopsis thaliana*: cell fate and cell autonomy. *Development* **122**:1041–1050.

Furness, C., P. Rudall, and F. B. Sampson. (2002). Evolution of microsporogenesis in angiosperms. *International Journal of Plant Science* **163**:235–260.

Galego, L., and J. Almeida. (2002). Role of DIVARICATA in the control of dorsoventral asymmetry in Antirrhinum flowers. *Genes and Development* **16**:880–891.

Galen, C. (1996a). Rates of floral evolution: adaptation to bumblebee pollination in an alpine wildflower, *Polemonium viscosum*. *Evolution* **50**:120–125.

Galen, C. (1996b). The evolution of floral form: insights from an alpine wildflower, *Polemonium viscosum* (Polemoniaceae). In: D. Lloyd and S. Barrett (eds), *Floral Biology*, pp. 273–291. Chapman and Hall, New York.

Galen, C., and M. Stanton. (1989). Bumble bee pollination and floral morphology: factors influencing pollen dispersal in the alpine sky pilot, *Polemonium viscosum* (Polemoniaceae). *American Journal of Botany* **76**: 419–426.

Galen, C., and J. Cuba (2001). Down the tube: pollinators, predators and the evolution of flower shape in the alpine skypilot, *Polemonium viscosum*. *Evolution* **55**:1963–1971.

Galizia, C. G., J. Kunze, A. Gumbert, A. K. Borg-Karlson, S. Sachse, C. Markl, and R. Menzel. (2005). Relationship of visual and olfactory signal parameters in a food-deceptive flower mimicry system. *Behavioural Ecology* **16**:159–168.

Garner, W., and H. Allard. (1920). Effect of the relative length of day and night and other factors of the environment on growth and reproduction in plants. *Journal of Agricultural Research* **18**:553–606.

Garray-Arroyo, A., A. Pineyro-Nelson, B. Garcia-Ponce, M. Sanchez, and E. Alvarez-Buylla. (2012). When ABC becomes ACB. *Journal of Experimental Botany* **63**:2377–2395.

Gendall, A. R., Y. Y. Levy, A. Wilson, and C. Dean. (2001). The *VERNALIZATION 2* gene mediates the epigenetic regulation of vernalization in *Arabidopsis*. *Cell* **107**:525–535.

Ghazoul, J. (2006). Floral diversity and the facilitation of pollination. *Journal of Ecology* **94**:295–304.

Gigord, L., M. Macnair, and A. Smithson. (2001). Negative frequency-dependent selection maintains a dramatic flower colour polymorphism in the rewardless orchid *Dactylorhiza sambucina* (L.) Soo. *Proceedings of the National Academy of Sciences of the USA* **98**:6253–6255.

Giurfa, M., B. Eichmann, and R. Menzel. (1996). Symmetry perception in an insect. *Nature* **382**:458–461.

Giurfa, M., A. Dafni, and P. Neal. (1999). Floral symmetry and its role in plant-pollinator systems. *International Journal of Plant Sciences* **160**:S41–S50.

Glover, B. J., and C. Martin. (1998). The role of petal cell shape and pigmentation in pollination success in *Antirrhinum majus*. *Heredity* **80**:778–784.

Glover, B. J., M. Perez-Rodriguez, and C. Martin. (1998). Development of several epidermal cell types can be specified by the same MYB-related plant transcription factor. *Development* **125**:3497–3508.

Gocal, G., C. Sheldon, F. Gubler, T. Moritz, D. Bagnall, C. MacMillan, S. Li, R. Parish, E. Dennis, D. Weigel, and R. King. (2001). *GAMYB-like* genes, flowering, and gibberellin signalling in Arabidopsis. *Plant Physiology* **127**:1682–1693.

Goldraij, A., K. Kondo, C. Lee, C. Hancock, M. Sivaguru, S. Vazquez-Santana, S. Kim, T. Phillips, F. Cruz-Garcia, and B. McClure. (2006). Compartmentalization of S-RNase and HT-B degradation in self-incompatible *Nicotiana*. *Nature* **439**:805–810.

Goldsmith, T., and K. Goldsmith. (1979). Discrimination of colours by the black-chinned hummingbird, *Archilochus alexandri*. *Journal of Comparative Physiology A* **130**:209–220.

Golz, J. F., E. J. Keck, and A. Hudson. (2002). Spontaneous mutations in *KNOX* genes give rise to a novel floral structure in Antirrhinum. *Current Biology* **12**:515–522.

Gomez, J., F. Perfectti, and J. Camacho. (2006). Natural selection on *Erysimum mediohispanicum* flower shape: insights into the evolution of zygomorphy. *American Naturalist* **168**:531–545.

Gomi, K., and M. Matsuoka. (2003). Gibberellin signalling pathway. *Current Opinion in Plant Biology* **6**:489–493.

Gonthier, R., A. Jacqmard, and G. Bernier. (1987). Changes in cell cycle duration and growth fraction in the shoot meristem of *Sinapis* during floral transition. *Planta* **170**: 55–59.

Goodman, C., P. Casati, and V. Walbot. (2004) A multidrug resistance-associated protein involved in anthocyanin transport in *Zea mays*. *Plant Cell* **16**:1812–1826.

Goodrich, J., R. Carpenter, and E. S. Coen. (1992). A common gene regulates pigmentation pattern in diverse plant species. *Cell* **68**:955–964.

Goodwin, T. W. (1980). *The Biochemistry of the Carotenoids. Volume 1*. Chapman and Hall, New York.

Goremykin, V. V., K. I. Hirsch-Ernst, S. Wolfl, and F. H. Hellwig. (2003). Analysis of the *Amborella trichopoda* chloroplast genome sequence suggests that *Amborella* is not a basal angiosperm. *Molecular Biology and Evolution* **20**:1499–1505.

Gorton, H. L., and T. C. Vogelmann. (1996). Effects of epidermal cell shape and pigmentation on optical properties of *Antirrhinum* petals at visible and ultraviolet wavelengths. *Plant Physiology* **112**:879–888.

Goto, K., and E. Meyerowitz. (1994). Function and regulation of the Arabidopsis floral homeotic gene *PISTILLATA*. *Genes and Development* **8**:1548–1560.

Goto, K., J. Kyozuka, and J. Bowman. (2001). Turning floral organs into leaves, leaves into floral organs. *Current Opinion in Genetics and Development* **11**:449–456.

Goto, N., T. Kumagai, and M. Koornneef. (1991). Flowering responses to light breaks in photomorphogenic mutants of *Arabidopsis thaliana*, a long-day plant. *Physiologia Plantarum* **83**:209–215.

Gramzow, L., M. Ritz, and G. Theissen. (2010). On the origin of MADS-domain transcription factors. *Trends in Genetics* **26**:149–153.

Grant, S., B. Hunkirchen, and H. Saedler. (1994). Developmental differences between male and female flowers in the dioecious plant *Silene latifolia*. *Plant Journal* **6**:471–480.

Grant-Downton, R., S. Hafidh, D. Twell, and H. Dickinson. (2009). Small RNA pathways are present and functional in the Arabidopsis male gametophyte. *Molecular Plant* **2**:500–512.

Gray, J., B. McClure, I. Boenig, M. Anderson, and A. Clarke. (1991). Action of the style product of the self-incompatibility gene of *Nicotiana alata* (S-RNase) on in vitro-grown pollen tubes. *Plant Cell* **3**:271–283.

Green, R., and E. Tobin. (1999). Loss of the circadian clock-associated protein 1 in *Arabidopsis* results in altered clock-regulated gene expression. *Proceedings of the National Academy of Sciences of the USA* **96**:4176–4179.

Grigorova, B., C. Mara, C. Hollender, P. Sijacic, X. Chen, and Z. Liu. (2011). LEUNIG and SEUSS co-repressors regulate *miR172* expression in *Arabidopsis* flowers. *Development* **138**:2451–2456.

Gross, C. (2005). A comparison of the sexual systems in the trees from the Australian tropics with other tropical biomes—more monoecy but why? *American Journal of Botany* **92**:907–919.

Grotewold, E. (2006). The genetics and biochemistry of floral pigments. *Annual Review of Plant Biology* **57**:761–780.

Gu, T., M. Mazzurco, W. Sulaman, D. Matias, and D. Goring. (1998). Binding of an arm-repeat protein to the kinase domain of the S-locus receptor kinase. *Proceedings of the National Academy of Sciences of the USA* **95**: 382–387.

Guerrieri, F., M. Schubert, J. Sandoz, and M. Giurfa. (2005). Perceptual and neural olfactory similarity in honeybees. *PLoS Biology* **3**:e60.

Gustafson-Brown, C., B. Savidge, and M. Yanofsky. (1994). Regulation of the Arabidopsis floral homeotic gene *Apetala1*. *Cell* **76**:131–143.

Gyula, P., E. Schafer, and F. Nagy. (2003). Light perception and signalling in higher plants. *Current Opinion in Plant Biology* **6**:446–452.

Hallem, E., A. Dahanukar, and J. Carlson. (2006). Insect odor and taste receptors. *Annual Review of Entomology* **51**:113–135.

Hamès, C., D. Ptchelkine, C. Grimm, E. Thevenon, E. Moyroud, F. Gerard, J. Martiel, R. Benlloch, F. Parcy, and C. Müller. (2008). Structural basis for LEAFY floral switch function and similarity with helix-turn-helix proteins. *EMBO Journal* **27**:2628–2637.

Hamilton, A.J., and D.C. Baulcombe. (1999). A species of small antisense RNA in posttranscriptional gene silencing in plants. *Science* **286**:950–952.

Hanano, S., and K. Goto. (2011). Arabidopsis TERMINAL FLOWER1 is involved in the regulation of flowering time and inflorescence development through transcriptional repression. *Plant Cell* **23**:3172–3184.

Hanzawa, Y., T. Money, and D. Bradley. (2005). A single amino acid converts a repressor to an activator of flowering. *Proceedings of the National Academy of Sciences of the USA* **102**:7748–7753.

Hargreaves, A., S. Johnson, and E. Nol. (2004). Do floral syndromes predict specialisation in plant pollination systems? An experimental test in an 'ornithophilous' African *Protea*. *Oecologia* **140**:295–301.

Harmer, S. (2009). The circadian system in higher plants. *Annual Review of Plant Biology* **60**:357–377.

Haseloff, J., and B. Amos. (1995). GFP in plants. *Trends in Genetics* **11**:328–329.

Hayama, R., T. Izawa, and K. Shimamoto. (2002). Isolation of rice genes possibly involved in the photoperiodic control of flowering by a fluorescent differential display method. *Plant Cell Physiology* **43**:494–504.

Hayama, R., S. Yokoi, S. Tamaki, M. Yano, and K. Shimamoto. (2003). Adaptation of photoperiodic control pathways produces short-day flowering in rice. *Nature* **422**:719–722.

Hazebroek, J. P., J. D. Metzger, and E. R. Mansager. (1993). Thermoinductive regulation of gibberellin metabolism in *Thlaspi arvensei*. *Plant Physiology* **102**:547–552.

He, Y., and R. M. Amasino. (2005). Role of chromatin modification in flowering-time control. *Trends in Plant Science* **10**:30–35.

He, Y., S. D. Michaels, and R. M. Amasino. (2003). Regulation of flowering time by histone acetylation in *Arabidopsis*. *Science* **302**:1751–1754.

He, Y., M. Doyle, and R. M. Amasino. (2004). PAF1-complex-mediated histone methylation of *FLOWERING LOCUS C* chromatin is required for the vernalization-responsive, winter-annual habit in Arabidopsis. *Genes and Development* **18**:2774–2784.

Hecht, V., R. Laurie, J. Vander Schoor, S. Ridge, C. Knowles, L. Liew, F. Sussmilch, I. Murfet, R. Macknight, and J. Weller. (2011). The pea GIGAS gene is a FLOWERING LOCUS T homolog necessary for graft-transmissible specification of flowering but not for responsiveness to photoperiod. *Plant Cell* **23**:147–161.

Heidmann, I., N. Efremova, H. Saedler, and Z. Schwarz-Sommer. (1998). A protocol for transformation and regeneration of *Antirrhinum majus*. *Plant Journal* **13**:723–728.

Heijmans, K., P. Morel, and M. Vandenbussche. (2012a). MADS-box genes and floral development: the dark side. *Journal of Experimental Botany* **63**:5397–5404.

Heijmans, K., K. Ament, A. Rijpkema, J. Zethof, M. Wolters-Arts, T. Gerats, and M. Vandenbussche. (2012b). Redefining C and D in the Petunia ABC. *Plant Cell* **24**:2305–2317.

Helliwell, C. A., C. C. Wood, M. Robertson, W. J. Peacock, and E. S. Dennis. (2006). The Arabidopsis FLC protein interacts directly *in vivo* with SOC1 and FT chromatin and is part of a high-molecular-weight protein complex. *Plant Journal* **46**:183–192.

Henderson, I. R., F. Liu, S. Drea, G. G. Simpson, and C. Dean. (2005). An allelic series reveals essential roles for FY in plant development in addition to flowering-time control. *Development* **132**:3597–3607.

Henschel, K., R. Kofuji, M. Hasebe, H. Saedler, T. Munster, and G. Theissen. (2002). Two ancient classes of MIKC-type MADS-box genes are present in the moss *Physcomitrella patens*. *Molecular Biology and Evolution* **19**:801–814.

Hepworth, S. R., F. Valverde, D. Ravenscroft, A. Mouradov, and G. Coupland. (2002). Antagonistic regulation of flowering-time gene *SOC1* by CONSTANS and FLC via separate promoter motifs. *EMBO Journal* **21**:4327–4337.

Hepworth, S. R., J. E. Klenz, and G. W. Haughn. (2006). UFO in the *Arabidopsis* inflorescence apex is required for floral-meristem identity and bract suppression. *Planta* **223**:769–778.

Herrera, C. (1993). Selection on floral morphology and environmental determinants of fecundity in a hawk moth-pollinated violet. *Ecological Monographs* **63**:251–275.

Herrera, C. (1996). Floral traits and plant adaptation to insect pollinators: a devil's advocate approach. In: D. Lloyd and S. Barrett (eds), *Floral Biology*, pp. 65–87. Chapman and Hall, New York.

Heyer, A., M. Raap, B. Schroeer, B. Marty, and L. Willmitzer. (2004). Cell wall invertase expression at the apical meristem alters floral, architectural and reproductive traits in Arabidopsis thaliana. *Plant Journal* **39**:161–169.

Hicks, K., A. Millar, I. Carre, D. Somers, M. Straume, D. Meeks-Wagner, and S. Kay. (1996). Conditional circadian dysfunction of the *Arabidopsis early-flowering 3* mutant. *Science* **274**:790–792.

Hicks, K., T. Albertson, and D. Wagner (2001). *EARLY FLOWERING 3* encodes a novel protein that regulates circadian clock function and flowering in Arabidopsis. *Plant Cell* **13**:1281–1292.

Hileman, L., E. Kramer, and D. Baum (2003). Differential regulation of symmetry genes and the evolution of flower morphologies. *Proceedings of the National Academy of Sciences of the USA* **100**:12814–12819.

Hingston, A., and P. McQuillan. (2000). Are pollination syndromes useful predictors of floral visitors in Tasmania? *Austral Ecology* **25**:600–609.

Hirschberg, J. (1999). Production of high-value compounds: carotenoids and vitamin E. *Current Opinion in Biotechnology* **10**:186–191.

Hisamatsu, T., and R. King. (2008). The nature of floral signals in Arabidopsis. II. Roles for FLOWERING LOCUS T (FT) and gibberellins. *Journal of Experimental Botany* **59**:3821–3829.

Hiscock, S., and S. McInnis. (2003). Pollen recognition and rejection during the sporophytic self-incompatibility response: *Brassica* and beyond. *Trends in Plant Science* **8**:606–613.

Hiscock, S., and D. Tabah. (2003). The different mechanisms of sporophytic self-incompatibility. *Philosophical Transactions of the Royal Society of London, Series B* **358**:1037–1045.

Hiscock, S., S. McInnis, D. Tabah, C. Henderson, and A. Brennan. (2003). Sporophytic self-incompatibility in *Senecio squalidus* L. (Asteraceae)—the search for S. *Journal of Experimental Botany* **54**:169–174.

Hoballah, M., T. Gubitz, J. Stuurman, L. Broger, M. Barone, T. Mandel, A. Dell'Olivo, M. Arnold, and C. Kuhlemeier. (2007). Single gene-mediated shift in pollinator attraction in Petunia. *Plant Cell* **19**:779–790.

Hodges, S., and N. Derieg. (2009). Adaptive radiations: from field to genomic studies. *Proceedings of the National Academy of Sciences of the USA* **106**:9947–9954.

Hoffmann-Benning, S., D. Gage, L. McIntosh, H. Kende, and J. Zeevaart. (2002). Comparison of peptides in the phloem sap of flowering and non-flowering *Perilla* and lupine plants using microbore HPLC followed by matrix-assisted laser desorption/ionization time-of-flight mass spectrometry. *Planta* **216**:140–147.

Holton, T. A., F. Brugliera, and Y. Tanaka. (1993). Cloning and expression of flavonol synthase from *Petunia hybrida*. *Plant Journal* **4**:1003–1010.

Honma, T., and K. Goto. (2001). Complexes of MADS-box proteins are sufficient to convert leaves into floral organs. *Nature* **409**:525–529.

Honys, D., and D. Twell. (2003). Comparative analysis of the Arabidopsis pollen transcriptome. *Plant Physiology* **132**:640–652.

Hopkins, R., and M. Rausher. (2011). Identification of two genes causing reinforcement in the Texas wildflower *Phlox drummondii*. *Nature* **469**:411–414.

Hopkins, R., and M. Rausher. (2012). Pollinator-mediated selection on flower colour alleles drives reinforcement. *Science* **335**:1090–1092.

Hord, C., C. Chen, B. Deyoung, S. Clark, and H. Ma. (2006). The BAM1/BAM2 receptor-like kinases are important regulators of Arabidopsis early anther development. *Plant Cell* **18**:1667–1680.

Hornyik, C., L. Terzi, and G. Simpson. (2010). The Spen family protein FPA controls alternative cleavage and polyadenylation of RNA. *Developmental Cell* **18**:203–213.

Horridge, G. (1996). The honeybee (*Apis mellifera*) detects bilateral symmetry and discriminates its axis. *Journal of Insect Physiology* **42**:755–764.

Horvitz, C., and D. Schemske. (1988). A test of the pollinator limitation hypothesis for a neotropical herb. *Ecology* **69**:200–206.

Howarth, D., and M. Donoghue. (2006). Phylogenetic analysis of the "ECE" (CYC/TB1) clade reveals duplications predating the core eudicots. *Proceedings of the National Academy of Sciences of the USA* **103**:9101–9106.

Howarth, D., T. Martins, E. Chimney, and M. Donoghue. (2011). Diversification of *CYCLOIDEA* expression in the evolution of bilateral flower symmetry in Caprifoliaceae and *Lonicera* (Dipsacales). *Annals of Botany* **107**:1521–1532.

Hsu, C-Y., J. Adams, H. Kim, K. No, C. Ma, S. Strauss, J. Drnevich, L. Vandervelde, J. Ellis, B. Rice, N. Wickett, L. Gunter, G. Tuskan, A. Brunner, G. Page, A. Barakat, J. Carlson, C. dePamphilis, D. Luthe, and C. Yuceer. (2011). FLOWERING LOCUS T duplication coordinates reproductive and vegetative growth in perennial poplar. *Proceedings of the National Academy of Sciences of the USA* **108**:10756–10761.

Hu, Y., Q. Xie, and N. Chua. (2003). The Arabidopsis auxin-inducible gene *ARGOS* controls lateral organ size. *Plant Cell* **15**:1951–1961.

Hu, Y., H. Poh, and N. Chua. (2006). The Arabidopsis *ARGOS-LIKE* gene regulates cell expansion during organ growth. *Plant Journal* **47**:1–9.

Huanca-Mamani, W., M. Garcia-Aguilar, G. Leon-Martinez, U. Grossniklaus, and J. Vielle-Calzada. (2005). CHR11, a chromatin-remodeling factor essential for nuclear proliferation during female gametogenesis in *Arabidopsis thaliana*. *Proceedings of the National Academy of Sciences of the USA* **102**:17231–17236.

Huang, T., H. Boehlenius, S. Eriksson, F. Parcy, and O. Nilsson. (2005). The mRNA of the *Arabidopsis* gene *FT* moves from leaf to shoot apex and induces flowering. *Science* **309**:1694–1696.

Huang, T., F. Lopez-Giraldez, J. Townsend, and V. Irish. (2012). RBE controls microRNA164 expression to effect floral organogenesis. *Development* **139**:2161–2169.

Hughes, N. F. (1994). *The Enigma of Angiosperm Origins*. Cambridge University Press, Cambridge.

Huijser, P., J. Klein, W. Lonnig, H. Meijer, H. Saedler, and H. Sommer. (1992). Bracteomania, an inflorescence anomaly, is caused by the loss of function of the MADS-box gene *squamosa* in *Antirrhinum majus*. *EMBO Journal* **11**:1239–1249.

Hun, S., I. Shogo, and I. Takato. (2010). Similarities in the circadian clock and photoperiodism in plants. *Current Opinion in Plant Biology* **13**:594–603.

Hunaca-Mamani, W., M. Garcia-Aguilar, G. Leon-Martinez, U. Grossniklaus, and J. Vielle-Calzada. (2005). CHR11, a chromatin-remodeling factor essential for nuclear proliferation during female gametogenesis in *Arabidopsis thaliana*. *Proceedings of the National Academy of Sciences of the USA* **102**:17231–17236.

Huq, E., J. Tepperman, and P. Quail. (2000). GIGANTEA is a nuclear protein involved in phytochrome signalling in Arabidopsis. *Proceedings of the National Academy of Sciences of the USA* **97**:9789–9794.

Immink, R., I. Tonaco, S. de Folter, A. Schchennikova, A. van Dijk, J. Busscher-Lange, J. Borst, and G. Angenent. (2009). SEPALLATA3: the "glue" for MADS box transcription factor complex formation. *Genome Biology* **10**:R24.

Ingram, G., J. Goodrich, M. Wilkinson, R. Simon, G. Haughn, and E. Coen. (1995). Parallels between *UNUSUAL FLORAL ORGANS* and *FIMBRIATA*, genes controlling flower development in Arabidopsis and Antirrhinum. *Plant Cell* **7**:1501–1510.

Ingram, G., S. Doyle, R. Carpenter, E. Schultz, R. Simon, and E. Coen. (1997). Dual role for *fimbriata* in regulating floral homeotic genes and cell division in *Antirrhinum*. *EMBO Journal* **16**:6521–6534.

Internicola, A., and L. Harder. (2012). Bumble-bee learning selects for both early and long flowering in food-deceptive plants. *Proceedings of the Royal Society B: Biological Sciences* **279**:1538–1543.

Irish, V., and Y. Yamamoto. (1995). Conservation of floral homeotic gene function between Arabidopsis and Antirrhinum. *Plant Cell* **7**:1635–1644.

Ishiguro, K., M. Taniguchi, and Y. Tanaka. (2012). Functional analysis of *Antirrhinum kelloggii* flavonoid 3′-hydroxylase and flavonoid 3′,5′-hydroxylase genes; critical role in flower colour and evolution in the genus *Antirrhinum*. *Journal of Plant Research* **125**:451–456.

Ishikawa, R., S. Tamaki, S. Yokoi, N. Inagaki, T. Shinomura, M. Takano, and K. Shimamoto. (2005). Suppression of the floral activator *Hd3a* is the principle cause of the night break effect in rice. *Plant Cell* **17**:3326–3336.

Ishikawa, R., T. Shinomura, M. Takano, and K. Shimamoto. (2009). Phytochrome dependent quantitative control of Hd3a transcription is the basis of the night break effect in rice flowering. *Genes and Genetic Systems* **84**: 179–184.

Ito, T., H. Sakai, and E. Meyerowitz. (2003). Whorl-specific expression of the *SUPERMAN* gene of *Arabidopsis* is mediated by *cis* elements in the transcribed region. *Current Biology* **13**:1524–1530.

Itoh, H., Y. Nonoue, M. Yano, and T. Izawa (2010). A pair of floral regulators sets critical day length for Hd3a florigen expression in rice. *Nature Genetics* **42**:635–638.

Iwano, M., and S. Takayama. (2012). Self/non-self discrimination in angiosperm self-incompatibility. *Current Opinion in Plant Biology* **15**:78–83.

Iwata, H., A. Gaston, A. Remay, T. Thouroude, J. Jeauffre, K. Kawamura, L. Oyant, T. Araki, B. Denoyes, and F. Foucher. (2012). The TFL1 homologue KSN is a regulator of continuous flowering in rose and strawberry. *Plant Journal* **69**:116–125.

Izawa, T., T. Oikawa, S. Tokutomi, K. Okuno, and K. Shimamoto. (2000). Phytochromes confer the photoperiodic control of flowering in rice (a short-day plant). *Plant Journal* **22**:391–399.

Izawa, T., T. Oikawa, N. Sugiyama, T. Tanisaka, M. Yano, and K. Shimamoto. (2002). Phytochrome mediates the external light signal to repress *FT* orthologues in photoperiodic flowering of rice. *Genes and Development* **16**: 2006–2020.

Izawa, T., Y. Takahashi, and Yano, M. (2003). Comparative biology comes into bloom: genomic and genetic comparison of flowering pathways in rice and *Arabidopsis*. *Current Opinion in Plant Biology* **6**:113–120.

Jack, T. (2001). Relearning our ABCs: new twists on an old model. *Trends in Plant Science* **6**:310–316.

Jack, T. (2004). Molecular and genetic mechanisms of floral control. *Plant Cell* **16**:S1–S17.

Jack, T., L. Brockman, and E. Meyerowitz. (1992). The homeotic gene APETALA3 of *Arabidopsis thaliana* encodes a MADS box and is expressed in petals and stamens. *Cell* **68**:683–697.

Jackson, D. (1992). In situ hybridization in plants. In: D.J. Bowles, S.J. Gurr and M. McPhereson (eds) *Molecular Plant Pathology: a practical approach*, pp. 163–174. Oxford University Press, Oxford.

Jacobsen, S. E., and N. E. Olszewski. (1993). Mutations at the *SPINDLY* locus of Arabidopsis alter gibberellin signal transduction. *Plant Cell* **5**:887–896.

Jacobsen, S., K. Binkowski, and N. Olszewski. (1996). SPINDLY, a tetratricopeptide repeat protein involved in gibberellin signal transduction in *Arabidopsis*. *Proceedings of the National Academy of Sciences of the USA* **93**:9292–9296.

Jacqmard, A., J. Miksche, and G. Bernier. (1972). Quantitative study of nucleic acids and proteins in the shoot apex of *Sinapis alba* during transition from the vegetative to the reproductive condition. *American Journal of Botany* **59**:714–721.

Jaeger, K., and Wigge, P. (2007). FT protein acts as a long-range signal in *Arabidopsis*. *Current Biology* **17**:1050–1054.

Jefferson, R., T. Kavanagh, and M. Bevan. (1987). GUS fusions: beta-glucuronidase as a sensitive and versatile gene fusion marker in higher plants. *EMBO Journal* **6**:3901–3907.

Jeon, J., S. Jang, S. Lee, J. Nam, C. Kim, S. Lee, Y. Chung, S. Kim, Y. Lee, Y. Cho, and G. An. (2000). *leafy hull sterile 1* is a homeotic mutation in a rice MADS box gene affecting flower development. *Plant Cell* **12**:871–884.

Jesson, L. K., and S. C. H. Barrett. (2002). Solving the puzzle of mirror-image flowers. *Nature* **417**:707.

Jesson, L., and S. Barrett. (2005). Experimental tests of the function of mirror-image flowers. *Biological Journal of the Linnean Society* **85**:167–179.

Jia, G., X. Liu, H. Owen, and D. Zhao. (2008). Signaling of cell fate determination by the TPD1 small protein and EMS1 receptor kinase. *Proceedings of the National Academy of Sciences of the USA* **105**:2220–2225.

Jofuku, K., B. Denboer, M. Van Montagu, and J. Okamuro. (1994). Control of Arabidopsis flower and seed development by the homeotic gene APETALA2. *Plant Cell* **6**:1211–1225.

Johanson, U., J. West, C. Lister, S. Michaels, R. Amasino, and C. Dean,. (2000). Molecular analysis of *FRIGIDA*, a major determinant of natural variation in *Arabidopsis* flowering time. *Science* **290**:344–347.

Johnson, S. D., and J. J. Midgley. (1997). Fly pollination in *Gorteria diffusa* (Asteraceae) and a possible mimetic function for dark spots on the capitulum. *American Journal of Botany* **84**:429–436.

Johnson, S., and K. Steiner. (1997). Long-tongued fly pollination and evolution of floral spur length in the *Disa draconis* complex (Orchidaceae). *Evolution* **51**:45–53.

Johnson S., and A. Jürgens. (2010). Convergent evolution of carrion and faecal scent mimicry in fly-pollinated angiosperm flowers and a stinkhorn fungus. *South African Journal of Botany* **76**:796–807.

Jones, A. M., K. H. Im, M. A. Savka, M. J. Wu, N. G. DeWitt, R. Shillito, and A. N. Binns. (1998). Auxin-dependent cell expansion mediated by overexpressed auxin binding protein 1. *Science* **282**:1114–1117.

Jones, K. N., and J. S. Reithel. (2001). Pollinator-mediated selection on a flower colour polymorphism in experimental populations of Antirrhinum (Scrophulariaceae). *American Journal of Botany* **88**:447–454.

Jones, M., K. Pierce, and D. Ward (2007). Avian vision: a review of form and function with special consideration to birds of prey. *Journal of Exotic Pet Medicine* **16**:69–87.

Joy, R., M. Sugiyama, H. Fukuda, and A. Komamine. (1995). Cloning and characterization of polyphenol oxidase cDNAs of *Phytolacca americana*. *Plant Physiology* **107**:1083–1089.

Judd, W., C. Campbell, E. Kellogg, P. Stevens, and M. Donoghue. (2007). *Plant Systematics: A phylogenetic approach*, 3rd edition. Sinauer Associates, Sunderland, MA.

Juergens, A., S. Doetterl, and U. Meve. (2006). The chemical nature of fetid floral odors in stapeliads (Apocynaceae-Asclepiadoideae-Ceropegieae). *New Phytologist* **172**:452–468.

Jung, J., Y. Ju, P. Seo, J. Lee, and C. Park. (2012). The SOC1-SPL module integrates photoperiod and gibberellic acid signals to control flowering time in Arabidopsis. *Plant Journal* **69**:577–588.

Kajiwara, S., P. D. Fraser, K. Kondo, and N. Misawa. (1997). Expression of an exogenous isopentenyl diphosphate isomerase gene enhances isoprenoid biosynthesis in *Escherichia coli*. *Biochemical Journal* **324**:421–426.

Kalisz, S., A. Randle, D. Chaiffetz, M. Faigeles, A. Butera, and C. Beight. (2012). Dichogamy correlates with outcrossing rate and defines the selfing syndrome in the mixed-mating genus *Collinsia*. *Annals of Botany* **109**:571–582.

Kandasamy, M., D. Paolillo, C. Faraday, J. Nasrallah, and M. Nasrallah. (1989). The S-locus specific glycoproteins of *Brassica* accumulate in the cell wall of developing stigma papillae. *Developmental Biology* **134**:462–472.

Kandori, I., T. Hirao, S. Matsunaga, and T. Kurosaki. (2009). An invasive dandelion unilaterally reduces the reproduction of a native congener through competition for pollination. *Oecologia* **159**:559–569.

Kang, H., J. Jeon, S. Lee, and G. An. (1998). Identification of class B and class C floral organ identity genes from rice plants. *Plant Molecular Biology* **38**:1021–1029.

Kanno, A., H. Saeki, T. Kameya, H. Saedler, and G. Theissen. (2003). Heterotropic expression of class B floral homeotic genes supports a modified ABC model for tulip (*Tulipa gesneriana*). *Plant Molecular Biology* **52**:831–841.

Kardailsky, I., V. Shukla, J. Ahn, N. Dagenais, S. Christensen, J. Nguyen, J. Chory, M. Harrison, and D. Weigel. (1999). Activation tagging of the floral inducer *FT*. *Science* **286**:1962–1965.

Kaufmann, K., R. Melzer, and G. Theissen. (2005). MIKC-type MADS-domain proteins: structural modularity, protein interactions and network evolution in land plants. *Gene* **347**:183–198.

Kaufmann, K., F. Wellmer, J. Muino, T. Ferrier, S. Wuest, V. Kumar, A. Serrano-Mislata, F. Madueno, P. Krajewski, E. Meyerowitz, G. Angenent, and J. Riechmann. (2010). Orchestration of floral initiation by APETALA1. *Science* **328**:85–89.

Kay, Q. O. N. (1976). Preferential pollination of yellow-flowered morphs of *Raphanus raphanistrum* by *Pieris* and *Ersistralis* spp. *Nature* **261**:230–232.

Kay, Q. O. N., H. S. Daoud, and C. H. Stirton. (1981). Pigment distribution, light reflection and cell structure in petals. *Botanical Journal of the Linnean Society* **83**:57–84.

Kazama, Y., M. Fujiwara, A. Koizumi, K. Nishihara, R. Nishiyama, E. Kifune, T. Abe, and S. Kawano. (2009). A SUPERMAN-like gene is exclusively expressed in female flowers of the dioecious plant *Silene latifolia*. *Plant and Cell Physiology* **50**:1127–1141.

Keck, E., P. McSteen, R. Carpenter, and E. Coen. (2003). Separation of genetic functions controlling organ identity in flowers. *EMBO Journal* **22**:1058–1066.

Kempin, S., B. Savidge, and M. Yanovsky. (1995). Molecular basis of the *cauliflower* phenotype in *Arabidopsis*. *Science* **267**:522–525.

Kenrick, P. (1999). The family tree flowers. *Nature* **402**:358–359.

Kessler, D., C. Diezel, and I. Baldwin. (2010). Changing pollinators as a means of escaping herbivores. *Current Biology* **20**:237–242.

Kevan, P. G., and M. A. Lane. (1985). Flower petal microtexture is a tactile cue for bees. *Proceedings of the National Academy of Sciences of the USA* **82**:4750–4752.

Kevan, P., M. Giurfa, and L. Chittka. (1996). Why are there so many and so few white flowers? *Trends in Plant Science* **1**:280–284.

Khanna, R., E. Huq, E. Kikis, B. Al-Sady, C. Lanzatella, and P. Quail. (2004). A novel molecular recognition motif necessary for targeting photoactivated phytochrome signaling to specific basic helix-loop-helix transcription factors. *Plant Cell* **16**:3033–3044.

Kim, M., M. Cui, P. Cubas, A. Gillies, K. Lee, M. Chapman, R. Abbott, and E. Coen. (2008). Regulatory genes control a key morphological and ecological trait transferred between species. *Science* **322**:1116–1119.

Kim, S., J. Koh, M. Yoo, H. Kong, Y. Hu, H. Ma, P. Soltis, and D. Soltis. (2005). Expression of floral MADS-box genes in basal angiosperms: implications for the evolution of floral regulators. *Plant Journal* **43**:724–744.

Kim, S., K. Choi, C. Park, H. Hwang, and I. Lee. (2006). *SUPPRESSOR OF FRIGIDA 4*, encoding a C2H2-type zinc finger protein, represses flowering by transcriptional activation of Arabidopsis FLOWERING LOCUS C. *Plant Cell* **18**:2985–2998.

King, J. (1997). *Reaching for the Sun.* Cambridge University Press, Cambridge.

Klahre, U., A. Gurba, K. Hermann, M. Saxenhofer, E. Bossolini, P. Guerin, and C. Kuhlemeier. (2011). Pollinator choice in Petunia depends on two major genetic loci for floral scent production. *Current Biology* **21**:730–739.

Knight, T., J. Steets, J. Vamosi, S. Mazer, M. Burd, D. Campbell, M. Dudash, M. Johnston, R. Mitchell, and T. Ashman. (2005). Pollen limitation of plant reproduction: pattern and process. *Annual Reviews of Ecology, Evolution and Systematics* **36**:467–497.

Knudsen, J., R. Eriksson, J. Gershenzon, and B. Stahl. (2006). Diversity and distribution of floral scent. *Botanical Review* **72**:1–120.

Kobayashi, Y., H. Kaya, K. Goto, M. Iwabuchi, and T. Araki. (1999). A pair of related genes with antagonistic roles in mediating flowering signals. *Science* **286**:1960–1962.

Koelewijn, H., and J. van Damme. (1996). Gender variation, partial male sterility and labile sex expression in gynodioecious *Plantago coronopus. New Phytologist* **132**: 67–76.

Koes, R. E., C. E. Spelt, H. J. Reif, P. van den Elzen, E. Veltkamp, and J. N. M. Mol. (1986). Floral tissue of *Petunia hybrida* (V30) expresses only one member of the chalcone synthase multigene family. *Nucleic Acids Research* **14**:5229–5239.

Kohchi, T., K. Mukougawa, N. Frankenberg, M. Masuda, A. Yokota, and J. Lagarias. (2001). The Arabidopsis HY2 gene encodes phytochromobilin synthase, a ferredoxin-dependent biliverdin reductase. *Plant Cell* **13**:425–436.

Kojima, S., Y. Takahashi, Y. Kobayashi, L. Monna, T. Sasaki, T. Araki, and M. Yano. (2002). *Hd3a*, a rice ortholog of the *Arabidopsis FT* gene, promotes transition to flowering downstream of *Hd1* under short-day conditions. *Plant Cell Physiology* **43**:1096–1105.

Komiya, R., S. Yokoi, and K. Shimamoto (2009). A gene network for long-day flowering activates RFT1 encoding a mobile flowering signal in rice. *Development* **136**: 3443–3450.

Kondo, T., Y. Toyama-Kato, and K. Yoshida. (2005). Essential structure of co-pigment for blue sepal-color development of hydrangea. *Tetrahedron Letters* **46**:6645–6649.

Kotilainen, M., P. Elomaa, A. Uimari, V. Albert, D. Yu, and T. Teeri. (2000). GRCD1, an AGL2-like MADS box gene, participates in the C function during stamen development in *Gerbera hybrida. Plant Cell* **12**:1893–1902.

Koornneef, M., C. J. Hanhart, and J. H. van der Veen. (1991). A genetic and physiological analysis of late flowering mutants in *Arabidopsis thaliana. Molecular and General Genetics* **229**:57–66.

Koornneef, M., C. Alonso-Blanco, H. Blankestijn-DeVries, C. J. Hanhart, and A. J. Peeters. (1998). Genetic interactions among late-flowering mutants of *Arabidopsis. Genetics* **148**:885–892.

Kowyama, Y., K. Kakeda, K. Kondo, T. Imada, and T. Hattori. (1996). A putative receptor protein kinase gene in *Ipomoea trifida. Plant Cell Physiology* **37**:681–685.

Kramer, E. M., and V. F. Irish. (1999). Evolution of genetic mechanisms controlling petal development. *Nature* **399**:144–148.

Kramer, E. M., M. A. Jaramillo, and V. S. Di Stilio. (2004). Patterns of gene duplication and functional evolution during the diversification of the *AGAMOUS* subfamily of MADS box genes in angiosperms. *Genetics* **166**: 1011–1023.

Kramer, E. M., H. Su, C. Wu, and J. Hu. (2006). A simplified explanation for the frameshift mutation that created a novel C-terminal motif in the *APETALA3* gene lineage. *BMC Evolutionary Biology* **6**:30.

Kramer, E., L. Holappa, B. Gould, M. Jaramillo, D. Setnikov, and P. Santiago. (2007). Elaboration of B gene function to include the identity of novel floral organs in the lower eudicot *Aquilegia* (Ranunculaceae). *Plant Cell* **19**:750–766.

Krassilov, V. A., and I. A. A. Dobruskina. (1995). Angiosperm fruit from the Lower Cretaceous of Israel and origins in rift valleys. *Paleontological Journal* **29**: 110–115.

Krizek, B., and E. Meyerowitz. (1996). The Arabidopsis homeotic genes APETALA3 and PISTILLATA are sufficient to provide the B class organ identity function. *Development* **122**:11–22.

Krizek, B. A., M. W. Lewis, and J. C. Fletcher. (2006). *RABBIT EARS* is a second-whorl repressor of *AGAMOUS* that maintains spatial boundaries in Arabidopsis flowers. *Plant Journal* **45**:369–383.

Kubo, K., T. Entani, A. Takara, N. Wang, A. Fields, Z. Hua, M. Toyoda, S. Kawashima, T. Ando, A. Isogai, T. Kao, and S. Takayama. (2010). Collaborative non-self recognition system in S-RNase-based self-incompatibility. *Science* **330**:796–799.

Kunin, W. (1993). Sex and the single mustard: population density and pollinator behaviour effects on seed set. *Ecology* **74**:2145–2160.

Kusaba, M., C. Tung, M. Nasrallah, and J. Nasrallah. (2002). Monoallelic expression and dominance interactions in anthers of self-incompatible *Arabidopsis lyrata. Plant Physiology* **128**:17–20.

Kwantes, M., D. Liebsch, and W. Verelst. (2012). How MIKC* MADS-box genes originated and evidence for their conserved function throughout the evolution of vascular plant gametophytes. *Molecular Biology and Evolution* **29**:293–302.

Labandeira, C. C. (1997). Insect mouthparts: ascertaining the paleobiology of insect feeding strategies. *Annual Reviews in Ecology and Systematics* **28**:153–193.

Lai, Z., W. Ma, B. Han, L. Liang, Y. Zhang, G. Hng, and Y. Xue. (2002). An F-box gene linked to the self-incompatibility (S) locus of *Antirrhinum* is expressed specifically in pollen and tapetum. *Plant Molecular Biology* **50**:29–42.

Lariguet, P., I. Schepens, D. Hodgson, U. V. Pedmale, M. Trevisan, C. Kami, M. Carbonnel, J. M. Alonso, J. R. Ecker, E. Liscum, and C. Fankhauser. (2006). PHYTOCHROME KINASE SUBSTRATE 1 is a phototropin 1 binding protein required for phototropism. *Proceedings of the National Academy of Sciences of the USA* **103**:10134–10139.

Larson, R. L., and E. H. Coe. (1977). Gene-dependent flavonoid glucosyltransferase in maize. *Biochemical Genetics* **15**:153–156.

Laurie, R., P. Diwadkar, M. Jaudal, L. Zhang, V. Hecht, J. Wen, M. Tadege, K. Mysore, J. Putterill, J. Weller, and R. Macknight. (2011). The Medicago *FLOWERING LOCUS T* homolog, *MtFTa1*, is a key regulator of flowering time. *Plant Physiology* **156**: 2207–2224.

Laux, T., K. Mayer, J. Berger, and G. Jurgens. (1996). The *WUSCHEL* gene is required for shoot and floral meristem integrity in *Arabidopsis*. *Development* **122**:87–96.

Lázaro, A., A. Gómez-Zambrano, L. López-González, M. Piñeiro, and J. Jarillo. (2008). Mutations in the *Arabidopsis SWC6* gene, encoding a component of the SWR1 chromatin remodelling complex, accelerate flowering time and alter leaf and flower development. *Journal of Experimental Botany* **59**:653–666.

Le Corre, V., F. Roux, and X. Reboud. (2002). DNA polymorphism at the *FRIGIDA* gene in *Arabidopsis thaliana*: extensive nonsynonymous variation is consistent with local selection for flowering time. *Molecular Biology and Evolution* **19**:1261–1271.

Lee, H., S. Huang, and T. Kao. (1994a). S proteins control rejection of incompatible pollen in *Petunia inflata*. *Nature* **367**:560–563.

Lee, H., S. Suh, E. Park, E. Cho, J. Ahn, S. Kim, J. Lee, Y. Kwon, and I. Lee. (2000). The AGAMOUS-LIKE 20 MADS domain protein integrates floral inductive pathways in *Arabidopsis*. *Genes and Development* **14**:2366–2376.

Lee, H., L. Xiong, Z. Gong, M. Ishitani, B. Stevenson, and J. Zhu. (2001). The *Arabidopsis HOS1* gene negatively regulates cold signal transduction and encodes a RING finger protein that displays cold-regulated nucleo-cytoplasmic partitioning. *Genes and Development* **15**:912–924.

Lee, I., and R. Amasino. (1995). Effect of vernalization, photoperiod, and light quality on the flowering phenotype

of *Arabidopsis* plants containing the *FRIGIDA* gene. *Plant Physiology* **108**:157–162.

Lee, I., M. Auckerman, S. Gore, K. Lohman, S. Michaels, L. Weaver, M. John, K. Feldmann, and R. Amasino. (1994b). Isolation of *LUMINIDEPENDENS*: a gene involved in the control of flowering time in Arabidopsis. *Plant Cell* **6**:75–83.

Lee, I., D. Wolfe, O. Nilsson, and D. Weigel. (1997). A LEAFY co-regulator encoded by UNUSUAL FLORAL ORGANS. *Current Biology* **7**:95–104.

Lee, J., S. Yoo, S. Park, I. Hwang, J. Lee, and J. Ahn. (2007). Role of *SVP* in the control of flowering time by ambient temperature in *Arabidopsis*. *Genes and Development* **21**:397–402.

Lee, S., J. Kim, J. Han, M. Han, and G. An. (2004). Functional analyses of the flowering time gene *OsMADS50*, the putative *SUPRESSOR OF OVEREXPRESSION OF CO 1/AGAMOUS LIKE 20 (SOC1/AGL20)* ortholog in rice. *Plant Journal* **38**:754–764.

Leivar, P., and P. Quail. (2011). PIFs: pivotal components in a cellular signalling hub. *Trends in Plant Science* **16**: 19–28.

Leleji, O. (1973). Apparent preference by bees for different flower colours in cowpeas (*Vigna sinensis* (L) Savi ex Hassk. *Euphytica* **22**:150–153.

Lenhard, M., A. Bohnert, G. Jürgens, and T. Laux. (2001). Termination of stem cell maintenance in Arabidopsis floral meristems by interactions between WUSCHEL and AGAMOUS. *Cell* **105**:805–814.

Leonard, A., and D. Papaj. (2011). 'X' marks the spot: the possible benefits of nectar guides to bees and plants. *Functional Ecology* **25**:1293–1301.

Leseberg, C., A. Li, H. Kang, M. Duvall, and L. Mao. (2006). Genome-wide analysis of the MADS-box gene family in *Populus trichocarpa*. *Gene* **378**:84–94.

Levin, J., and E. Meyerowitz. (1995). UFO: an Arabidopsis gene involved in both floral meristem and floral organ development. *Plant Cell* **7**:529–548.

Levy, Y. Y., S. Mesnage, J. S. Mylne, A. R. Gendall, and C. Dean. (2002). Multiple roles of *Arabidopsis VRN1* in vernalization and flowering time control. *Science* **297**: 243–246.

Li, C., and J. Dubcovsky. (2008). Wheat FT protein regulates *VRN1* transcription through interactions with FDL2. *Plant Journal* **55**:543–554.

Li, D., C. Liu, L. Shen, Y. Wu, H. Chen, M. Robertson, C. Helliwell, T. Ito, E. Meyerowitz, and H. Yu. (2008). A repressor complex governs the integration of flowering signals in *Arabidopsis*. *Developmental Cell* **15**:110–120.

Li, J., M. Webster, M. Smith, and P. Gilmartin. (2011). Floral heteromorphy in *Primula vulgaris*: progress towards isolation and characterisation of the *S* locus. *Annals of Botany* **108**:715–726.

Li, Z., B. Li, W. Shen, H. Huang, and A. Dong. (2012). TCP transcription factors interact with AS2 in the repression of class-I KNOX genes in *Arabidopsis thaliana*. *Plant Journal* **71**:99–107.

Liang, H., and L. Mahadevan. (2011). Growth, geometry and mechanics of a blooming lily. *Proceedings of the National Academy of Sciences of the USA* **108**:5516–5521.

Liljegren, S., C. Gustafson-Brown, A. Pinyopich, G. Ditta, and M. Yanofsky. (1999). Interactions among APETALA1, LEAFY and TERMINAL FLOWER 1 specify meristem fate. *Plant Cell* **11**:1007–1018.

Lim, M., J. Kim, Y. Kim, K. Chung, Y. Seo, I. Lee, J. Kim, C. Hong, H. Kim, and C. Park. (2004). A new *Arabidopsis* gene, *FLK*, encodes an RNA binding protein with K homology motifs and regulates flowering time via *FLOWERING LOCUS C*. *Plant Cell* **16**:731–740.

Lin, C. (2000). Plant blue-light receptors. *Trends in Plant Science* **5**:337–342.

Lin, C., H. Yang, H. Guo, T. Mockler, J. Chen, and A. Cashmore. (1998). Enhancement of blue-light sensitivity of Arabidopsis seedlings by a blue light receptor cryptochrome 2. *Proceedings of the National Academy of Sciences of the USA* **95**:2686–2690.

Litt, A. (2007). An evaluation of A function: evidence from the APETALA1 and APETALA2 gene lineages. *International Journal of Plant Sciences* **168**:73–91.

Liu, C., F. Lu, X. Cui, and X. Cao. (2010a). Histone methylation in higher plants. *Annual Review of Plant Biology* **61**:395–420.

Liu, F., V. Quesada, P. Crevillen, I. Baurle, S. Swiezewski, and C. Dean. (2007). The Arabidopsis RNA binding protein FCA requires a lysine-specific demethylase 1 homolog to downregulate FLC. *Molecular Cell* **28**:398–407.

Liu, F., S. Marquardt, C. Lister, S. Swiezewski, and C. Dean. (2010b). Targeted 3′ processing of antisense transcripts triggers Arabidopsis FLC chromatin silencing. *Science* **327**:94–97.

Liu, H., X. Yu, K. Li, J. Klejnot, H. Yang, D. Lisiero, and C. Lin. (2008). Photoexcited CRY2 interacts with CIB1 to regulate transcription and floral initiation in Arabidopsis. *Science* **322**:1535–1539.

Liu, H., B. Liu, C. Zhao, M. Pepper, and C. Lin. (2011a). The action mechanisms of plant cryptochromes. *Trends in Plant Science* **16**:684–691.

Liu, J., S. Gilmour, M. Thomashow, and S. van Nocker. (2002). Cold signalling associated with vernalization in *Arabidopsis thaliana* does not involve CBF1 or abscisic acid. *Physiologia Plantarum* **114**:125–134.

Liu, X., Y. Kim, R. Mueller, R. Yumul, C. Liu, Y. Pan, X. Cao, J. Goodrich, and X. Chen. (2011b). AGAMOUS terminates floral stem cell maintenance in Arabidopsis by directly repressing WUSCHEL through recruitment of Polycomb Group proteins. *Plant Cell* **23**:3654–3670.

Liu, Z., and E. Meyerowitz. (1995). LEUNIG regulates AGAMOUS expression in Arabidopsis flowers. *Development* **121**:975–991.

Liu, Z., and C. Mara. (2010). Regulatory mechanisms for floral homeotic gene expression. *Seminars in Cell and Developmental Biology* **21**:80–86.

Lobo, J., M. Quesada, K. Stoner, E. Fuchs, Y. Herrerias-Diego, J. Rojas, and G. Saborio. (2003). Factors affecting phonological patterns of bombacaceous trees in seasonal forests in Costa Rica and Mexico. *American Journal of Botany* **90**:1054–1063.

Locke, J. C. W., L. Kozma-Bognar, P. D. Gould, B. Fehér, É. Kevei, F. Nagy, M. S. Turner, A. Hall, and A. J. Millar. (2006). Experimental validation of a predicted feedback loop in the multi-oscillator clock of *Arabidopsis thaliana*. *Molecular Systems Biology* **2**:59.

Lockhart, P., and D. Penny.(2005). The place of Amborella within the radiation of angiosperms. *Trends in Plant Science* **10**:201–202.

Lohmann, J., R. Hong, M. Hobe, M. Busch, F. Parcy, R. Simon, and D. Weigel. (2001). A molecular link between stem cell regulation and floral patterning in Arabidopsis. *Cell* **105**:793–803.

Lotz, C., C. Martinez del Rio, and S. Nicolson. (2003). Hummingbirds pay a high cost for a warm drink. *Journal of Comparative Physiology B* **173**:455–462.

Loukoianov, A., L. Yan, A. Blechl, A. Sanchez, and J. Dubcovsky. (2005). Regulation of VRN-1 vernalization genes in normal and transgenic polyploid wheat. *Plant Physiology* **138**:2364–2373.

Lowman, A., and M. Purugganan. (1999). Duplication of the *Brassica oleracea* APETALA1 floral homeotic gene and the evolution of domesticated cauliflower. *Journal of Heredity* **90**:514–520.

Luo, D., R. Carpenter, C. Vincent, L. Copsey, and E. Coen. (1996). Origin of floral asymmetry in Antirrhinum. *Nature* **383**:794–799.

Luo, Y., and Widmer, A. (2013). Herkogamy and its effects on mating patterns in *Arabidopsis thaliana*. *PLoS ONE* **8**:e57902.

Luu, D., X. Qin, D. Morse, and M. Cappadocia. (2000). S-RNase uptake by compatible pollen tubes in gametophytic self-incompatibility. *Nature* **407**:649–651.

Ma, H. (2005). Molecular genetic analyses of microsporogenesis and microgametogenesis in flowering plants. *Annual Reviews of Plant Biology* **56**:393–434.

McClure, B., and V. Franklin-Tong. (2006). Gametophytic self-incompatibility: understanding the cellular mechanisms involved in 'self' pollen tube inhibition. *Planta* **224**:233–245.

McClure, B., F. Cruz-Garcia, and C. Romero. (2011). Compatibility and incompatibility in S-RNase-based systems. *Annals of Botany* **108**:647–658.

McCormick, S. (2004). Control of male gametophyte development. *Plant Cell* **16**:S142–S153.

Macknight, R., I. Bancroft, T. Page, C. Lister, R. Schmidt, K. Love, L. Westphal, G. Murphy, S. Sherson, C. Cobbett, and C. Dean. (1997). *FCA*, a gene controlling flowering time in Arabidopsis, encodes a protein containing RNA-binding domains. *Cell* **89**:737–745.

Macknight, R., M. Duroux, R. Laurie, P. Dijkwel, G. Simpson, and C. Dean. (2002). Functional significance of the alternative transcript processing of the Arabidopsis floral promoter *FCA*. *Plant Cell* **14**:877–888.

McSteen, P., C. Vincent, S. Doyle, R. Carpenter, and E. Coen. (1998). Control of floral homeotic gene expression and organ morphogenesis in *Antirrhinum*. *Development* **125**:2359–2369.

Machado, I., and A. Lopes. (2004). Floral traits and pollination systems in the Caatinga, a Brazilian tropical dry forest. *Annals of Botany* **94**:365–376.

Magnani, E., K. Sjolander, and S. Hake. (2004). From endonucleases to transcription factors: evolution of the AP2 DNA binding domain in plants. *Plant Cell* **16**:2265–2277.

Maier, A., S. Stehling-Sun, H. Wollmann, M. Demar, R. Hong, S. Haubeiss, D. Weigel, and J. Lohmann. (2009). Dual roles of the bZIP transcription factor PERIANTHIA in the control of floral architecture and homeotic gene expression. *Development* **136**:1613–1620.

Maizel, A., M. A. Busch, T. Tanahashi, J. Perkovic, M. Kato, M. Hasebe, and D. Weigel. (2005). The floral regulator LEAFY evolves by substitutions in the DNA binding domain. *Science* **308**:260–263.

Makino, S., T. Kiba, A. Imamura, N. Hanaki, A. Nakamura, T. Suzuki, M. Taniguchi, C. Ueguchi, T. Sugiyama, and T. Mizuno. (2000). Genes encoding pseudo-response regulators: insight into his-to-asp phosphorelay and circadian rhythm in *Arabidopsis thaliana*. *Plant Cell Physiology* **41**:791–803.

Makkena, S., E. Lee, F. Sack, and R. Lamb. (2012). The R2R3 MYB transcription factors FOUR LIPS and MYB88 regulate female reproductive development. *Journal of Experimental Botany* **63**:5545–5558.

Malcomber, S., and E. Kellog. (2005). *SEPALLATA* gene diversification: brave new whorls. *Trends in Plant Science* **10**:427–435.

Mallory, A., D. Dugas, D. Bartel, and B. Bartel. (2004). MicroRNA regulation of NAC-domain targets is required for proper formation and separation of adjacent embryonic, vegetative and floral organs. *Current Biology* **14**:1035–1046.

Mandel, M. A., and M. Yanofsky. (1995). A gene triggering flower formation in *Arabidopsis*. *Nature* **377**:522–524.

Mandel, M., and M. Yanofsky. (1998). The Arabidopsis AGL9 MADS box gene is expressed in young flower primordia. *Sexual Plant Reproduction* **11**:22–28.

Mandel, M. A., C. Gustafson-Brown, B. Savidge, and M. Yanofsky. (1992). Molecular characterisation of the *Arabidopsis* floral homeotic gene *APETALA1*. *Nature* **360**:273–277.

Manzano, D., S. Marquardt, A. Jones, I. Baurle, F. Liu, and C. Dean. (2009). Altered interactions within FY/AtCPSF complexes required for Arabidopsis FCA-mediated chromatin silencing. *Proceedings of the National Academy of Sciences of the USA* **106**:8772–8777.

Mara, C., T. Huang, and V. Irish. (2010). The Arabidopsis floral homeotic proteins APETALA3 and PISTILLATA negatively regulate the BANQUO genes implicated in light signaling. *Plant Cell* **22**:690–702.

Marinova, K., L. Pourcel, B. Weder, M. Schwarz, D. Barron, J. Routaboul, I. Debeaujon, and M. Klein. (2007). The *Arabidopsis* MATE transport *TT12* acts as a vacuolar flavonoid/H^+-antiporter active in proanthocyanidin-accumulating cells of the seed coat. *Plant Cell* **19**:2023–2038.

Markham, K. R., K. S. Gould, C. S. Winefield, K. A. Mitchell, S. J. Bloor, and M. R. Bloase. (2000). Anthocyanic vacuolar inclusions—their nature and significance in flower colouration. *Phytochemistry* **55**:327–336.

Markovic, J., N. Petranovic, and J. Baranac. (2005). The copigmentation effect of sinapic acid on malvin: a spectroscopic investigation on colour enhancement. *Journal of Photochemistry and Photobiology B* **78**:223–228.

Marrs, K. A., M. R. Alfenito, A. M. Lloyd, and V. Walbot. (1995). A glutathione S-transferase involved in vacuolar transfer encoded by the maize gene *Bronze-2*. *Nature* **375**:397–400.

Martin, C., and T. Gerats. (1993). The control of flower coloration. In: B. R. Jordan (ed.), *The Molecular Biology of Flowering*, pp. 219–255. CAB International, Wallingford.

Martin, C., R. Carpenter, H. Sommer, H. Saedler, and E. Coen. (1985). Molecular analysis of instability in flower pigmentation of *Antirrhinum majus* following isolation of the *PALLIDA* locus by transposon tagging. *EMBO Journal* **4**:1625–1630.

Martin, C., A. Prescott, S. Mackay, J. Bartlett, and E. Vrijlandt. (1991). The control of anthocyanin biosynthesis in flowers of *Antirrhinum majus*. *Plant Journal* **1**:37–49.

Martin, W., A. Gierl, and H. Saedler. (1989). Molecular evidence for pre-Cretaceous angiosperm origins. *Nature* **339**:46–48.

Martin, W., O. Deusch, N. Stawski, N. Grunheit, and V. Goremykin. (2005). Chloroplast genome phylogenetics: why we need independent approaches to plant molecular evolution. *Trends in Plant Science* **10**:203–209.

Martin-Trillo, M., A. Lazaro, R. S. Poethig, C. Gomez-Mena, M. A. Pineiro, J. M. Martinez-Zapater, and J. A. Jarillo. (2006). *EARLY IN SHORT DAYS 1 (ESD1)* encodes ACTIN-RELATED PROTEIN 6 (AtARP6), a putative component of chromatin remodelling complexes that positively regulates *FLC* accumulation in *Arabidopsis*. *Development* **133**:1241–1252.

Martinez, M., J. Jørgensen, M. Lawton, C. Lamb, and P. Doerner. (1992). Spatial pattern of cdc2 expression in relation to meristem activity and cell proliferation during plant development. *Proceedings of the National Academy of Sciences of the USA* **89**:7360–7364.

Martinez-Garcia, J., E. Huq, and P. Quail. (2000). Direct targeting of light signals to a promoter element-bound transcription factor. *Science* **288**:859–863.

Martinez-Garcia, J., J. Garcia-Martinez, J. Bou, and S. Prat. (2001). The interaction of gibberellins and photoperiod in the control of potato tuberization. *Journal of Plant Growth Regulation* **20**:377–386.

Martins, T., J. Berg, S. Blinka, M. Rausher, and D. Baum. (2013). Precise spatio-temporal regulation of the anthocyanin biosynthetic pathway leads to petal spot formation in *Clarkia gracilis*. *New Phytologist* **197**:958–969.

Mas, P., F. Devlin, S. Panda, and S. A. Kay. (2000). Functional interaction of phytochrome B and cryptochrome 2. *Nature* **408**:207–211.

Mas, P., W. Kim, D. Somers, and S. Kay. (2003). Targeted degradation of TOC1 by ZTL modulates circadian function in *Arabidopsis thaliana*. *Nature* **426**:567–570.

Masiero, S., L. Colombo, P. Grini, A. Schnittger, and M. Kater. (2011). The emerging importance of type I MADS box transcription factors for plant reproduction. *Plant Cell* **23**:865–872.

Massinga, P., S. Johnson, and L. Harder. (2005). Heteromorphic incompatibility and efficiency of pollination in two distylous *Pentanisia* species (Rubiaceae). *Annals of Botany* **95**:389–399.

Mast, A., and E. Conti. (2006). The primrose path to heterostyly. *New Phytologist* **171**:439–442.

Mathews, S. (2010). Evolutionary studies illuminate the structural-functional model of plant phytochromes. *Plant Cell* **22**:4–16.

Mathieu, J., N. Warthmann, F. Kuttner, and Schmid, M. (2007). Export of FT protein from phloem companion cells is sufficient for floral induction in *Arabidopsis*. *Current Biology* **17**:1055–1060.

Matias-Hernandez, L., R. Battaglia, F. Galbiati, M. Rubes, C. Eichenberger, U. Grossniklaus, M. Kater, and L. Colombo. (2010). *VERDANDI* is a direct target of the MADS domain ovule identity complex and affects embryo sac differentiation in Arabidopsis. *Plant Cell* **22**:1702–1715.

Mayfield, M., N. Waser, and M. Price. (2001). Exploring the 'most effective pollinator principle' with complex flowers: bumblebees and *Ipomopsis aggregata*. *Annals of Botany* **88**:591–596.

Mazzurco, M., W. Sulaman, H. Elina, J. M. Cock, and D. Goring. (2001). Further analysis of the interactions between the *Brassica* S receptor kinase and three interacting

proteins (ARC1, THL1 and THL2) in the yeast two-hybrid system. *Plant Molecular Biology* **45**:365–376.

Medrano, M., C. Herrera, and S. Barrett. (2005). Herkogamy and mating patterns in the self-compatible daffodil *Narcissus longispathus*. *Annals of Botany* **95**:1105–1111.

Mello, C.C., and D. Conte. (2004). Revealing the world of RNA interference. *Nature* **431**:338–342.

Melzer, R., Y. Wang, and G. Theissen. (2010). The naked and the dead: the ABCs of gymnosperm reproduction and the origin of the angiosperm flower. *Seminars in Cell and Developmental Biology* **21**:118–128.

Meyerowitz, E. M. (1996). Plant development: local control, global patterning. *Current Opinion in Genetics and Development* **6**:475–479.

Michaels, S. D., and R. M. Amasino. (1999). *FLOWERING LOCUS C* encodes a novel MADS domain protein that acts as a repressor of flowering. *Plant Cell* **11**:949–956.

Michaels, S. D., and R. M. Amasino. (2001). Loss of *FLOWERING LOCUS C* activity eliminates the late-flowering phenotype of *FRIGIDA* and autonomous pathway mutations but not responsiveness to vernalization. *Plant Cell* **13**:935–941.

Michaels, S. D., I. C. Bezerra, and R. M. Amasino. (2004). *FRIGIDA*-related genes are required for the winterannual habit in *Arabidopsis*. *Proceedings of the National Academy of Sciences of the USA* **101**:3281–3285.

Millar, A. J., I. A. Carre, C. A. Strayer, N. Chua, and S. A. Kay. (1995). Circadian clock mutants in Arabidopsis identified by luciferase imaging. *Science* **267**:1161–1163.

Ming, R., A. Bendahmane, and S. Renner. (2011). Sex chromosomes in land plants. *Annual Reviews in Plant Biology* **62**:485–514.

Mitchell, R., and R. Shaw. (1993). Heritability of floral traits for the perennial wild flower *Penstemon centranthifolius* (Scrophulariaceae): clones and crosses. *Heredity* **71**:185–192.

Mizoguchi, T., K. Wheatley, Y. Hanzawa, L. Wright, M. Mizoguchi, H. Song, I. Carre, and G. Coupland. (2002). *LHY* and *CCA1* are partially redundant genes required to maintain circadian rhythms in *Arabidopsis*. *Developmental Cell* **2**:629–641.

Mizoguchi, T., L. Wright, S. Fujiwara, F. Cremer, K. Lee, H. Onouchi, A. Mouradov, S. Fowler, H. Kamada, J. Putterill, and G. Coupland. (2005). Distinct roles of *GIGANTEA* in promoting flowering and regulating circadian rhythms in Arabidopsis. *Plant Cell* **17**:2255–2270.

Mizukami, Y. (2001). A matter of size: developmental control of organ size in plants. *Current Opinion in Plant Biology* **4**:533–539.

Mizukami, Y., and R. L. Fischer. (2000). Plant organ size control: *AINTEGUMENTA* regulates growth and cell numbers during organogenesis. *Proceedings of the National Academy of Sciences of the USA* **97**:942–947.

Mockler, T., H. Yang, X. Yu, D. Parikh, Y. Cheng, S. Dolan, and C. Lin. (2003). Regulation of photoperiodic flowering by *Arabidopsis* photoreceptors. *Proceedings of the National Academy of Sciences of the USA* **100**: 2140–2145.

Moehs, C. P., L. Tian, K. W. Osteryoung, and D. DellaPenna. (2001). Analysis of carotenoid biosynthetic gene expression during marigold petal development. *Plant Molecular Biology* **45**:281–293.

Moise, A. R., J. von Lintig, and K. Palczewski. (2005). Related enzymes solve evolutionarily recurrent problems in the metabolism of carotenoids. *Trends in Plant Science* **10**:178–186.

Mol, J. N. M., M. P. Robbins, R. A. Dixon, and E. Veltkamp. (1985). Spontaneous and enzymic rearrangement of naringenin chalcone to flavanone. *Phytochemistry* **24**:2267–2269.

Mol, J., E. Grotewold, and R. Koes. (1998). How genes paint flowers and seeds. *Trends in Plant Science* **3**:212–217.

Mondragon-Palomino, M., and G. Theissen. (2008). MADS about the evolution of orchid flowers. *Trends in Plant Science* **13**:51–59.

Mondragon-Palomino, M., and G. Theissen. (2011). Conserved differential expression of paralogous *DEFICIENS*- and *GLOBOSA*-like MADS box genes in the flowers of Orchidaceae: refining the "orchid code". *Plant Journal* **66**: 1008–1019.

Momonoi, K., K. Yoshida, S. Mano, H. Takahashi, C. Nakamori, K. Shoji, A. Nitta, and M. Nishimura. (2009). A vacuolar iron transporter in tulip, TgVit1, is responsible for blue coloration in petal cells through iron accumulation. *Plant Journal* **59**:437–447.

Montalvo, A., and J. Ackermann. (1987). Limitations to fruit production in *Ionopsis utricularioides* (Orchidaceae). *Biotropica* **19**:24–31.

Moon, J., S. Suh, H. Lee, K. Choi, C. Hong, N. Paek, S. Kim, and I. Lee. (2003). The *SOC1* MADS-box gene integrates vernalization and gibberellin signals for flowering in Arabidopsis. *Plant Journal* **35**:613–623.

Moon, J., H. Lee, M. Kim, and I. Lee. (2005). Analysis of flowering pathway integrators in *Arabidopsis*. *Plant Cell Physiology* **46**:292–299.

Mooney, M., T. Desnos, K. Harrison, J. Jones, R. Carpenter, and E. Coen. (1995). Altered regulation of tomato and tobacco pigmentation genes caused by the *delila* gene of *Antirrhinum*. *Plant Journal* **7**:333–339.

Moran, N., and T. Jarvik. (2010). Lateral transfer of genes from fungi underlies carotenoid production in aphids. *Science* **328**:624–627.

Motte, P., H. Saedler, and Z. Schwarz-Sommer. (1998). *STYLOSA* and *FISTULATA*: regulatory components of the homeotic control of *Antirrhinum* floral organogenesis. *Development* **125**:71–84.

Mouradov, A., T. Glassick, B. Hamdorf, L. Murphy, B. Fowler, S. Marla, and R. D. Teasdale. (1998). *NEEDLY*, a *Pinus radiata* ortholog of *FLORICAULA/LEAFY* genes, expressed in both reproductive and vegetative meristems. *Proceedings of the National Academy of Sciences of the USA* **95**:6537–6542.

Mouradov, A., F. Cremer, and G. Coupland. (2002). Control of flowering time: interacting pathways as a basis for diversity. *Plant Cell* **14**:S111–S130.

Moyano, E., J. F. Martinez-Garcia, and C. Martin. (1996). Apparent redundancy in *myb* gene function provides gearing for the control of flavonoid biosynthesis in Antirrhinum flowers. *Plant Cell* **8**:1519–1532.

Moyroud, E., E. Kusters, M. Monniaux, R. Koes, and F. Parcy. (2010). LEAFY blossoms. *Trends in Plant Science* **15**:346–352.

Moyroud, E., E. Minguet, F. Ott, L. Yant, D. Pose, M. Monniaux, S. Blanchet, O. Bastien, E. Thevenon, D. Weigel, M. Schmid, and F. Parcy. (2011). Prediction of regulatory interactions from genome sequences using a biophysical model for the *Arabidopsis* LEAFY transcription factor. *Plant Cell* **23**:1293–1306.

Mudalige, R., A. Kuehnle, and T. Amore. (2003). Pigment distribution and epidermal cell shape in *Dendrobium* species and hybrids. *HortScience* **38**:573–577.

Müller, R., A. Fernández, A. Hiltbrunner, E. Schäfer, and T. Kretsch. (2009). The histidine kinase-related domain of Arabidopsis phytochrome A controls the spectral sensitivity and the subcellular distribution of the photoreceptor. *Plant Physiology* **150**:1297–1309.

Murfett, J., T. Atherton, B. Mou, C. Gasser, and B. McClure. (1994). S-RNase expressed in transgenic *Nicotiana* causes S-allele-specific pollen rejection. *Nature* **367**:563–566.

Munster, T., J. Pahnke, A. DiRosa, J. Kim, W. Martin, H. Saedler, and G. Theissen. (1997). Floral homeotic genes were recruited from homologous MADS-box genes preexisting in the common ancestor of ferns and seed plants. *Proceedings of the National Academy of Sciences of the USA* **94**:2415–2420.

Mylne, J. S., L. Barrett, F. Tessadori, S. Mesnage, L. Johnson, Y. V. Bernatavichute, S. E. Jacobsen, P. Fransz, and C. Dean. (2006). LHP1, the *Arabidopsis* homologue of HETEROCHROMATIN PROTEIN 1, is required for epigenetic silencing of *FLC*. *Proceedings of the National Academy of Sciences of the USA* **103**:5012–5017.

Nagasawa, N., M. Miyoshi, Y. Sano, H. Satoh, H. Hirano, H. Sakai, and Y. Nagato. (2003). *SUPERWOMAN1* and *DROOPING LEAF* genes control floral organ identity in rice. *Development* **130**:705–718.

Nagatani, A. (2010). Phytochrome: structural basis for its functions. *Current Opinion in Plant Biology* **13**:565–570.

Nageli, C. (1884). *Mechanisch-physiologische Theorie der Abstammungslehre*.

Nagy, F., and E. Schafer. (2002). Phytochromes control photomorphogenesis by differentially regulated, interacting signalling pathways in higher plants. *Annual Review of Plant Biology* **53**:329–355.

Nair, S., N. Wang, Y. Turuspekov, M. Pourkheirandish, S. Sinsuwongwat, G. Chen, M. Sameri, A. Tagiri, I. Honda, Y. Watanabe, H. Kanamori, T. Wicker, N. Stein, Y. Nagamura, T. Matsumoto, and T. Komatsuda. (2010). Cleistogamous flowering in barley arises from the suppression of microRNA-guided *HvAP2* mRNA cleavage. *Proceedings of the National Academy of Sciences of the USA* **107**:490–495.

Nakajima, M., A. Shimada, Y. Takashi, Y. Kim, S. Park, M. Ueguchi-Tanaka, H. Suzuki, E. Katoh, M. Iuchi, M. Kobayashi, T. Maeda, M. Matsuoka, and I. Yamaguchi. (2006). Identification and characterisation of Arabidopsis gibberellin receptors. *Plant Journal* **46**:880–889.

Nakayama, T. (2002). Enzymology of aurone biosynthesis. *Journal of Bioscience and Bioengineering* **94**:487–491.

Nakayama, T., K. Yonekura-Sakakibara, T. Sato, S. Kikuchi, Y. Fukui, M. Fukuchi-Mizutani, T. Ueda, M. Nakao, Y. Tanaka, T. Kusumi, and T. Nishino. (2000). Aureusidin synthase: a polyphenol oxidase homolog responsible for flower colouration. *Science* **290**:1163–1166.

Nasrallah, J., T. Kao, M. Goldberg, and M. Nasrallah. (1985). A cDNA clone encoding an S-locus-specific glycoprotein from *Brassica oleracea*. *Nature* **318**:263–267.

Nath, U., B. C. W. Crawford, R. Carpenter, and E. Coen. (2003). Genetic control of surface curvature. *Science* **299**:1404–1407.

Navarro, C., N. Efremova, J. Golz, R. Rubiera, M. Kuckenberg, R. Castillo, O. Tietz, H. Saedler, and Z. Schwarz-Sommer. (2004). Molecular and genetic interactions between *STYLOSA* and *GRAMINIFOLIA* in the control of *Antirrhinum* vegetative and reproductive development. *Development* **131**:3649–3659.

Newman, D. A., and J. D. Thomson. (2005). Effects of nectar robbing on dynamics and bumblebee foraging strategies in *Linaria vulgaris* (Scrophulariaceae). *Oikos* **110**:309–320.

Ni, M., J. Tepperman, and P. Quail. (1998). PIF3, a phytochrome-interacting factor necessary for normal photoinduced signal transduction, is a novel basic helix-loop-helix protein. *Cell* **95**:657–667.

Niklas, K. (1985). The aerodynamics of wind pollination. *Botanical Review* **51**:328–386.

Nobutoshi, Y., Y. Ayako, A. Mitsutomo, D. Wagner, and Y. Komeda. (2012). LEAFY controls Arabidopsis pedicel length and orientation by affecting adaxial-abaxial cell fate. *Plant Journal* **69**:844–856.

Noda, K., B. J. Glover, P. Linstead, and C. Martin. (1994). Flower colour intensity depends on specialised cell shape controlled by a MYB-related transcription factor. *Nature* **369**:661–664.

Noh, Y., and R. Amasino. (2003). *PIE1*, an ISWI family gene, is required for *FLC* activation and floral repression in Arabidopsis. *Plant Cell* **15**:1671–1682.

Oh, S., H. Zhang, P. Ludwig, and S. van Nocker. (2004). A mechanism related to the yeast transcriptional regulator Paf1c is required for expression of the Arabidopsis FLC/MAF MADS box gene family. *Plant Cell* **16**:2940–2953.

Oh, S., A. Johnson A. Smertenko, D. Rahman, S. Park, P. Hussey, and D. Twell. (2005). A divergent cellular role for the FUSED kinase family in the plant-specific cytokinetic phragmoplast. *Current Biology* **15**:2107–2111.

Oh, S., V. Bourdon, M. Pal, H. Dickinson, and D. Twell. (2008). *Arabidopsis* kinesins HINKEL and TETRASPORE act redundantly to control cell plate expansion during cytokinesis in the male gametophyte. *Molecular Plant* **1**:794–799.

Ohme-Takagi, M., and H. Shinshi. (1995). Ethylene-inducible DNA binding proteins that interact with an ethylene-responsive element. *Plant Cell* **7**:173–182.

Ohmori, S., M. Kimizu, M. Sugita, A. Miyao, H. Hirochika, E. Uchida, Y. Nagato, and H. Yoshida. (2009). *MOSAIC FLORAL ORGANS I*, an AGL6-like MADS box gene, regulates floral organ identity and meristem fate in rice. *Plant Cell* **21**:3008–3025.

Ohnishi, M., S. Fukada-Tanaka, A. Hoshino, J. Takada, Y. Inagaki, and S. Iida. (2005). Characterisation of a novel Na$^+$/H$^+$ antiporter gene *InNHX2* and comparison of *InNHX2* with *InNHX1*, which is responsible for blue flower colouration by increasing the vacuolar pH in the Japanese morning glory. *Plant Cell Physiology* **46**:259–267.

Ohshima, S., M. Murata, W. Sakamoto, Y. Ogura, and F. Motoyoshi. (1997). Cloning and molecular analysis of the *Arabidopsis* gene *Terminal Flower 1*. *Molecular and General Genetics* **254**:186–194.

Ojeda, I., J. Francisco-Ortega, and Q. Cronk. (2009). Evolution of petal epidermal micromorphology in Leguminosae and its use as a marker of petal identity. *Annals of Botany* **104**:1099–1110.

Ojeda, I., A. Santos-Guerra, J. Caujape-Castells, R. Jaen-Molina, A. Marrero, and Q. Cronk. (2012). Comparative micromorphology of petals in Macaronesian *Lotus* (Leguminosae) reveals a loss of papillose conical cells during the evolution of bird pollination. *International Journal of Plant Sciences* **173**:365–374.

Okada, K., T. Saito, T. Kakagawa, M. Kawamukai, and Y. Kamiya. (2000). Five geranylgeranyl diphosphate synthases expressed in different organs are localised into three subcellular compartments in Arabidopsis. *Plant Physiology* **122**:1045–1056.

Okamuro, J., B. Caster, R. Villarroel, M. van Montagu, and K. D. Jofuku. (1997). The AP2 domain of *APETALA2* defines a large new family of DNA binding proteins in *Arabidopsis*. *Proceedings of the National Academy of Sciences of the USA* **94**:7076–7081.

Ollerton, J. (1998). Sunbird surprise for syndromes. *Nature* **394**:726–727.

Ollerton, J., A. Stott, E. Allnutt, S. Shove, C. Taylor, and E. Lamborn. (2007). Pollination niche overlap between a parasitic plant and its host. *Oecologia,* **151**:473–485.

Ollerton, J., R. Alarcon, N. Waser, M. Price, S. Watts, L. Cranmer, A. Hingston, C. Peter, and J. Rotenberry. (2009). A global test of the pollination syndrome hypothesis. *Annals of Botany* **103**:1471–1480.

Olmedo-Monfil, V., N. Duran-Figueroa, M. Arteaga-Vazquez, E. Demesa-Arevalo, D. Autran, D. Grimanelli, R. Slotkin, R. Martienssen, and J. Vielle-Calzada. (2010). Control of female gamete formation by a small RNA pathway in Arabidopsis. *Nature* **464**:628–632.

Osorio, D., and M. Vorobyev. (2008). A review of the evolution of animal colour vision and visual communication signals. *Vision Research* **48**:2042–2051.

Osterlund, M., C. Hardtke, N. Wei, and X. Deng. (2000). Targeted destabilization of HY5 during light-regulated development of *Arabidopsis. Nature* **405**:462–466.

Oyama, T., Y. Shimura, and K. Okada. (1997). The *Arabidopsis HY5* gene encodes a bZIP protein that regulates stimulus-induced development of root and hypocotyl. *Genes and Development* **11**:2983–2995.

Pabon-Mora, N., B. Ambrose, and A. Litt. (2012). Poppy APETALA1/FRUITFULL orthologs control flowering time, branching, perianth identity, and fruit development. *Plant Physiology* **158**:1685–1704.

Padysakova, E., M. Bartos, R. Tropek, and S. Janecek. (2013). Generalization versus specialization in pollination systems: visitors, thieves and pollinators of *Hypoestes aristata* (Acanthaceae). *PLoS One* **4**:e59299.

Pagnussat, G., H. Yu, Q. Ngo, S. Rajani, S. Mayalagu, C. Johnson, A. Capron, L. Xie, D. Ye, and V. Sundaresan. (2005). Genetic and molecular identification of genes required for female gametophyte development and function in Arabidopsis. *Development* **132**: 603–614.

Pagnussat, G., M. Alandete-Saez, J. Bowman, and V. Sundaresan. (2009). Auxin-dependent patterning and gamete specification in the Arabidopsis female gametophyte. *Science* **324**:1684–1689.

Pannell, J. (2002). The evolution and maintenance of androdioecy. *Annual Reviews of Ecology and Systematics* **33**:397–425.

Parcy, F., O. Nilsson, O., M. Busch, I. Lee, and D. Weigel. (1998). A genetic framework for floral patterning. *Nature* **395**:561–566.

Parenicova, L., S. de Folter, M. Kieffer, D. Horner, C. Favalli, J. Busscher, H. Cook, R. Ingram, M. Kater, B. Davies, G. Angenent, and L. Colombo. (2003). Molecular and phylogenetic analyses of the complete MADS-box transcription factor family in Arabidopsis: new openings to the MADS world. *Plant Cell* **15**:1538–1551.

Park, D., D. Somers, Y. Kim, Y. Choy, H. Lim, M. Soh, H. Kim, S. Kay, and H. Nam. (1999). Control of circadian rhythms and photoperiodic flowering by the *Arabidopsis GIGANTEA* gene. *Science* **285**:1579–1582.

Park, S., R. Howden, and D. Twell. (1998). The *Arabidopsis thaliana* gametophytic mutation *gemini pollen 1* disrupts microspore polarity, division asymmetry and pollen cell fate. *Development* **125**:3789–3799.

Pastore, J., A. Limpuangthip, N. Yamaguchi, M. Wu, Y. Sang, S. Han, L. Malaspina, N. Chavdaroff, A. Yamaguchi, and D. Wagner. (2011). LATE MERISTEM IDENTITY2 acts together with LEAFY to activate APETALA1. *Development* **138**:3189–3198.

Pauw, A. (1998). Pollen transfer on birds' tongues. *Nature* **394**:731–732.

Pauw, A. (2006). Floral syndromes accurately predict pollination by a specialised oil-collecting bee (*Rediviva peringueyi*, Melittidae) in a guild of South African orchids (Coryciinae). *American Journal of Botany* **93**:917–926.

Pazhouhandeh, M., J. Molinier, A. Berr, and P. Genschik. (2011). MSI4/FVE interacts with CUL4-DDB1 and a PRC2-like complex to control epigenetic regulation of flowering time in Arabidopsis. *Proceedings of the National Academy of Sciences of the USA* **108**:3430–3435.

Pedersen, M. W. (1967). Cross-pollination studies involving three purple-flowered alfalfas, one white-flowered line, and two pollinator species. *Crop Science* **7**:59–62.

Pelaz, S., G. Ditta, E. Baumann, E. Wisman, and M. Yanofsky. (2000). B and C floral organ identity functions require *SEPALLATA* MADS-box genes. *Nature* **405**:200–202.

Pellegrini, L., T. Song, and T. Richmond. (1995). Structure of serum response factor core bound to DNA. *Nature* **376**:490–498.

Pena, L., M. Martin-Trillo, J. Juarez, J. A. Pina, L. Navarro, and J. M. Martinez-Zapater. (2001). Constitutive expression of Arabidopsis *LEAFY* or *APETALA1* genes in citrus reduces their generation time. *Nature Biotechnology* **19**:263–267.

Perez-Rodriguez, M., F. Jaffe, E. Butelli, B. J. Glover, and C. Martin. (2005). Development of three different cell types is associated with the activity of a specific MYB transcription factor in the ventral petal of *Antirrhinum majus* flowers. *Development* **132**:359–370.

Perl-Treves, R., A. Kahana, N. Rosenman, Y. Xiang, and L. Silberstein. (1998). Expression of multiple *AGAMOUS*-like genes in male and female flowers of cucumber (*Cucumis sativus* L.). *Plant Cell Physiology* **39**:701–710.

Piatelli, M. (1981). The betalains: structure, biosynthesis, and chemical taxonomy. In: E. Conn (ed.), *The Biochemistry of Plants: A comprehensive treatise*, pp. 557–575. Academic Press, New York.

Pin, P., R. Benlloch, D. Bonnet, E. Wremerth-Weich, T. Kraft, J. Gielen, and O. Nilsson. (2010). An antagonistic pair of FT homologs mediates the control of flowering time in sugar beet. *Science* **330**:1397–1400.

Pinyopich, A., G. Ditta, B. Savidge, S. Liljegren, E. Baumann, E. Wisman, and M. Yanofsky. (2003). Assessing the redundancy of MADS-box genes during carpel and ovule development. *Nature* **424**:85–88.

Ponomarenko, A. G. (1995). The geological history of beetles. In: J. Pakaluk and S. A. Slipinski (eds), *Biology, Phylogeny and Classification of Coleoptera*. Muzeum I Instytut Zoologii PAN, Warsaw.

Portereiko, M., L. Sandaklie-Nikolova, A. Lloyd, C. Dever, D. Otsuga, and G. Drews. (2006). *NUCLEAR FUSION DEFECTIVE 1* encodes the Arabidopsis RPL21M protein and is required for karyogamy during female gametophyte development and fertilisation. *Plant Physiology* **141**:957–965.

Preston, J., and E. Kellogg. (2006). Reconstructing the evolutionary history of paralogous *APETALA1/FRUITFULL*-like genes in grasses (Poaceae). *Genetics* **174**:421–437.

Preston, J., and L. Hileman. (2012). Parallel evolution of TCP and B-class genes in Commelinaceae flower bilateral symmetry. *EvoDevo* **3**:6.

Preston, J., C. Martinez, and L. Hileman. (2011). Gradual disintegration of the floral symmetry gene network is implicated in the evolution of a wind-pollination syndrome. *Proceedings of the National Academy of Sciences of the USA* **108**:2343–2348.

Prum, B., R. Seidl, H. Bohn, and T. Speck. (2012). Impact of cell shape in hierarchically structured plant surfaces on the attachment of male Colorado potato beetles (*Leptinotarsa decemlineata*). *Beilstein Journal of Nanotechnology* **3**:57–64.

Putterill, J. (2001). Flowering in time: genes controlling photoperiodic flowering in *Arabidopsis*. *Philosophical Transactions of the Royal Society of London Series B* **356**: 1761–1767.

Putterill, J., F. Robson, K. Lee, R. Simon, and G. Coupland. (1995). The *CONSTANS* gene of Arabidopsis pomotes flowering and encodes a protein showing similarities to zinc finger transcription factors. *Cell* **80**: 847–857.

Putterill, J., R. Laurie, and R. Macknight. (2004). It's time to flower: the genetic control of flowering time. *BioEssays* **26**:363–373.

Puzey, J., S. Gerbode, S. Hodges, E. Kramer, and L. Mahadevan. (2012). Evolution of spur length diversity in *Aquilegia* petals is achieved solely through cell shape anisotropy. *Proceedings of the Royal Society Series B* **279**: 1640–1645.

Qiao, H., F. Weng, L. Zhao, J. Zhou, Z. Lai, Y. Zhang, T. Robbins, and Y. Xue. (2004). The F-box protein AhSLF-S2 controls the pollen function of S-RNase-based self incompatibility. *Plant Cell* **16**:2307–2322.

Qiu, Y., J. Lee, F. Bernasconi-Quadroni, D. E. Soltise, P. Soltis, M. Zanis, E. A. Zimmer, Z. Chen, V. Savolainen, and M. W. Chase. (1999). The earliest angiosperms. *Nature* **402**:404–407.

Quail, P. H. (2002). Phytochrome photosensory signalling networks. *Nature Reviews Molecular Cell Biology* **3**:85–93.

Quattrocchio, F., J. F. Wing, K. van der Woude, J. N. M. Mol, and R. Koes. (1998). Analysis of bHLH and MYB domain proteins: species-specific regulatory differences are caused by divergent evolution of target anthocyanin genes. *Plant Journal* **13**:475–488.

Quattrocchio, F., J. Wing, K. van der Woude, E. Souer, N. de Vetten, J. Mol, and R. Koes. (1999). Molecular analysis of the *anthocyanin2* gene of petunia and its role in the evolution of flower colour. *Plant Cell* **11**:1433–1444.

Quattrocchio, F., W. Verweij, A. Kroon, C. Spelt, J. Mol, and R. Koes. (2006). PH4 of Petunia is an R2R3 MYB protein that activates vacuolar acidification through interactions with basic-helix-loop-helix transcription factors of the anthocyaninin pathway. *Plant Cell* **18**:1274–1291.

Quesada, V., R. Macknight, C. Dean, and G. G. Simpson. (2003). Autoregulation of *FCA* pre-mRNA processing controls *Arabidopsis* flowering time. *EMBO Journal* **22**:3142–3152.

Rahmann, M., K. Uchiyama, M. Kuno, N. Hirashima, K. Suwabe, T. Tsuchiya, Y. Kagaya, I. Kobayashi, K. Kakeda, and Y. Kowyama. (2007). Expression of stigma- and anther-specific genes located in the S locus region of *Ipomoea trifida*. *Sexual Plant Reproduction* **20**:73–85.

Ramsay, N. A., and B. J. Glover. (2005). MYB-bHLH-WD40 protein complex and the evolution of cellular diversity. *Trends in Plant Science* **10**:63–70.

Ramsay, N. A., A. R. Walker, M. Mooney, and J. C. Gray. (2003). Two basic-helix-loop-helix genes (*MYC-146* and *GL3*) from *Arabidopsis* can activate anthocyanin biosynthesis in a white flowered *Matthiola incana* mutant. *Plant Molecular Biology* **52**:679–688.

Ramsey, J., H. D. Bradshaw, and D. Schemske. (2003). Components of reproductive isolation between the monkeyflowers *Mimulus lewisii* and *M. cardinalis* (Phrymaceae). *Evolution* **57**:1520–1534.

Rands, S., B. Glover, and H. Whitney. (2011). Floral epidermal structure and flower orientation: getting to grips with awkward flowers. *Arthropod–Plant Interactions* **5**:279–285.

Rao, N., K. Prasad, P. Kumar, and U. Vijayraghavan. (2008). Distinct regulatory role for RFL, the rice LFY homolog, in determining flowering time and plant architecture. *Proceedings of the National Academy of Sciences of the USA* **105**:3646–3651.

Ratcliffe, O., I. Amaya, C. Vincent, S. Rothstein, R. Carpenter, E. Coen, and D. Bradley. (1998). A common mechanism controls the life cycle and architecture of plants. *Development* **125**:1609–1615.

Ratcliffe, O., D. Bradley, and E. Coen. (1999). Separation of shoot and floral identity in *Arabidopsis*. *Development* **126**:1109–1120.

Rathcke, B. J. (2000). Hurricane causes resource and pollination limitation of fruit set in a bird-pollinated shrub. *Ecology* **81**:1951–1958.

Raven, P. (1972). Why are bird-visited flowers predominantly red? *Evolution* **26**:674.

Razafimandimbison, S., S. Ekman, T. McDowell, and B. Bremer. (2012). Evolution of growth habit, inflorescence architecture, flower size, and fruit type in Rubiaceae: its ecological and evolutionary implications. *PLoS One* **7**:e40851.

Razem, F. A., A. El-Kereamy, S. R. Abrams, and R. D. Hill. (2006). The RNA-binding protein FCA is an abscisic acid receptor. *Nature* **439**:290–294.

Reale, L., A. Porceddu, L. Lanfaloni, C. Moretti, S. Zenoni, M. Pezzotti, B. Romano, and F. Ferranti. (2002). Patterns of cell division and expansion in developing petals of *Petunia hybrida*. *Sexual Plant Reproduction* **15**:123–132.

Redei, G. P. (1973). *Arabidopsis* as a genetic tool. *Annual Reviews of Genetics* **9**:111–127.

Reichmann, J., B. Krizek, and E. Meyerowitz, (1996). Dimerization specificity of Arabidopsis MADS domain homeotic proteins APETALA1, APETALA3, PISTILLATA and AGAMOUS. *Proceedings of the National Academy of Sciences of the USA* **93**:4793–4798.

Richards, A. (1997). *Plant Breeding Systems*, 2nd edn. Chapman and Hall, London.

Riffell, J., H. Lei, L. Abrell, and J. Hildebrand. (2013). Neural basis of a pollinator's buffet: olfactory specialisation and learning in *Manduca sexta*. *Science* **339**:200–204.

Rijpkema, A., J. Zethof, T. Gerats, and M. Vandenbussche. (2009). The petunia *AGL6* gene has a SEPALLATA-like function in floral patterning. *Plant Journal* **60**:1–9.

Robertson, H., and K. Wanner. (2006). The chemoreceptor superfamily in the honeybee, *Apis mellifera*: expansion of the odorant, but not gustatory, receptor family. *Genome Research* **16**:1395–1403.

Rodriguez-Girones, M., and L. Santamaria. (2004). Why are so many bird flowers red? *PLoS Biology* **2**:e306.

Rolland-Lagan, A., A. J. Bangham, and E. Coen. (2003). Growth dynamics underlying petal shape and symmetry. *Nature* **422**:161–163.

Ronse de Craene, L. (2007). Are petals sterile stamens or bracts? The origin and evolution of petals in the core eudicots. *Annals of Botany* **100**:621–630.

Ronse de Craene, L., P. Soltis, and D. Soltis. (2003). Evolution of floral structures in basal angiosperms. *International Journal of Plant Science* **164**:S329–S363.

Rosati, C., A. Cadic, M. Duron, M. Amiot, M. Tacchini, S. Martens, and G. Forkmann. (1998). Flavonoid metabolism in Forsythia flowers. *Plant Science* **139**:133–140.

Rouse, D. T., C. C. Sheldon, D. J. Bagnall, W. J. Peacock, and E. S. Dennis. (2002). FLC, a repressor of flowering, is regulated by genes in different inductive pathways. *Plant Journal* **29**:183–191.

Rubio, V., and X. Deng. (2005). Phy tunes: phosphorylation status and phytochrome-mediated signalling. *Cell* **120**:290–292.

Rudall, P. J. (2006). How many nuclei make an embryo sac in flowering plants? *BioEssays* **28**:1067–1071.

Rudall, P., and R. Bateman. (2007). Developmental bases for key innovations in the seed plant microgametophyte. *Trends in Plant Science* **12**:317–326.

Rudall, P., J. Hilton, F. Vergara-Silva, and R. Bateman. (2011). Recurrent abnormalities in conifer cones and the evolutionary origins of flower-like structures. *Trends in Plant Science* **16**:151–159.

Ruiz-Sola, M.A., and M. Rodríguez-Concepción (2012). Carotenoid biosynthesis in Arabidopsis: a colorful pathway. *The Arabidopsis Book* **10**:e0158.

Rukolainen, S., Y. Ng, V. Albert, P. Elomaa, and T. Teeri. (2010). Large scale interaction analysis predicts that the *Gerbera hybrida* floral E function is provided both by general and specialized proteins. *BMC Plant Biology* **10**:129.

Rutishauser, R., and B. Isler. (2001). Developmental genetics and morphological evolution of flowering plants, especially bladderworts (*Utricularia*): fuzzy Arberian morphology complements classical morphology. *Annals of Botany* **88**:1173–1202.

Rutledge, R., S. Regan, O. Nicolas, P. Fobert, C. Cote, W. Bosnich, C. Kauffeldt, G. Sunohara, A. Seguin, and D. Stewart. (1998). Characterization of an *AGAMOUS* homologue from the conifer black spruce (*Picea mariana*) that produces floral homeotic conversions when expressed in *Arabidopsis*. *Plant Journal* **15**:625–634.

Saddic, L. A., B. Huvermann, S. Bezhani, Y. Su, C. M. Winter, C. S. Kwon, R. P. Collum, and D. Wagner. (2006). The LEAFY target LMI1 is a meristem identity regulator and acts together with LEAFY to regulate expression of *CAULIFLOWER*. *Development* **133**:1673–1682.

Sakai, H., L. Medrano, and E. Meyerowitz. (1995). Role of Superman in maintaining Arabidopsis floral whorl boundaries. *Nature* **378**:199–203.

Samach, A., H. Onouchi, S. Gold, G. Ditta, Z. Schwarz-Sommer, M. Yanofsky, and G. Coupland. (2000). Distinct roles of *CONSTANS* target genes in reproductive development in Arabidopsis. *Science* **288**:1613–1616.

Sanchez, M. (2011). Taste perception in honeybees. *Chemical Senses* **36**:675–692.

Sanda, S., and R. M. Amasino. (1996). Ecotype-specific expression of a flowering mutant phenotype in *Arabidopsis thaliana*. *Plant Physiology* **111**:641–644.

Sanderson, M., J. Thorne, N. Wikström, and K. Bremer. (2004). Molecular evidence on plant divergence times. *American Journal of Botany* **91**:1656–1665.

Sandring, C., and J. Agren. (2009). Pollinator mediated selection on floral display and flowering time in the perennial herb *Arabidopsis lyrata*. *Evolution* **63**:1292–1300.

Sargent, R., and S. Otto. (2004). A phylogenetic analysis of pollination mode and the evolution of dichogamy in angiosperms. *Evolutionary Ecology Research* **6**:1183–1199.

Sargent, R., M. Mandegar, and S. Otto. (2006). A model of the evolution of dichogamy incorporating sex-ratio selection, anther-stigma interference and inbreeding depression. *Evolution* **60**:934–944.

Sargent, R., C. Goodwillie, S. Kalisz, and R. Ree. (2007). Phylogenetic evidence for a flower size and number trade-off. *American Journal of Botany* **94**:2059–2062.

Sato, T., T. Nakayama, S. Kikuchi, Y. Fukui, K. Yonekura-Sakakibara, T. Ueda, T. Nishino, Y. Tanaka, and T. Kusumi. (2001). Enzymatic formation of aurones in the extracts of yellow snapdragon flowers. *Plant Science* **160**:229–236.

Savidge, B., S. Rounsley, and M. Yanofsky. (1995). Temporal relationship between the transcription of two Arabidopsis MADS box genes and the floral organ identity genes. *Plant Cell* **7**:721–733.

Sawa, M., D. Nusinow, S. Kay, and T. Imaizuma. (2007). FKF1 and GIGANTEA complex formation is required for day-length measurement in Arabidopsis. *Science* **318**:261–265.

Schaefer, H., and S. Renner. (2010). A three-genome phylogeny of *Momordica* (Cucurbitaceae) suggests seven returns from dioecy to monoecy and recent long-distance dispersal to Asia. *Molecular Phylogenetics and Evolution* **54**:553–560.

Schemske, D., and H. Bradshaw. (1999). Pollinator preference and the evolution of floral traits in monkeyflowers (*Mimulus*). *Proceedings of the National Academy of Sciences of the USA* **96**:11910–11915.

Schemske, D., and P. Bierzychudek. (2001). Evolution of flower colour in the desert annual *Linanthus parryae*: Wright revisited. *Evolution* **55**:1269–1282.

Schlüter, P., and F. Schiestl. (2008). Molecular mechanisms of floral mimicry in orchids. *Trends in Plant Science* **13**:228–235.

Schoenrock, N., R. Bouveret, O. Leroy, L. Borghi, C. Koehler, W. Gruissem, and L. Hennig. (2006). Polycomb-group proteins repress the floral activator *AGL19* in the *FLC*-independent vernalization pathway. *Genes and Development* **20**:1667–1678.

Schomburg, F. M., D. A. Patton, D. W. Meinke, and R. M. Amasino. (2001). *FPA*, a gene involved in floral induction in Arabidopsis, encodes a protein containing RNA-recognition motifs. *Plant Cell* **13**:1427–1436.

Schopfer, C., M. Nasrallah, and J. Nasrallah. (1999). The male determinant of self-incompatibility in *Brassica*. *Science* **286**:1697–1700.

Schreiber, H., A. Swink, and T. Godsey. (2010). The chemical mechanism for Al^{3+} complexing with delphinidin: a model for the bluing of hydrangea sepals. *Journal of Inorganic Biochemistry* **104**:732–739.

Schruff, M., M. Spielman, S. Tiwari, S. Adams, N. Fenby, and R. Scott. (2006). The *AUXIN RESPONSE FACTOR 2* gene of Arabidopsis links auxin signalling, cell division, and the size of seeds and other organs. *Development* **133**:251–261.

Schwarz-Sommer, Z., I. Hue, P. Huijser, P. Flor, H. Hansen, F. Tetens, W. Lonnig, H. Saedler, and H. Sommer. (1992). Characterisation of the Antirrhinum floral homeotic MADS-box gene deficiens: evidence for DNA binding and autoregulation of its persistent expression throughout flower development. *EMBO Journal* **11**:251–263.

Schwinn, K., J. Venail, Y. Shang, S. Mackay, V. Alm, E. Butelli, R. Oyama, P. Bailey, K. Davies, and C. Martin. (2006). A small family of *MYB*-regulatory genes controls floral pigmentation intensity and patterning in the genus *Antirrhinum*. *Plant Cell* **18**:831–851.

Scora, R. (1964). Dependency of pollination on patterns in *Monarda* (Labiatae). *Nature* **204**:1011–1012.

Searle, I., Y. He, F. Turck, C. Vincent, F. Fornara, S. Kroeber, R. A. Amasino, and G. Coupland. (2006). The transcription factor FLC confers a flowering response to vernalization by repressing meristem competence and systemic signalling in *Arabidopsis*. *Genes and Development* **20**:898–912.

Seitz, C., C. Eder, B. Deiml, S. Kellner, S. Martens, and G. Forkmann. (2006). Cloning, functional identification and sequence analysis of flavonoid 3′-hydroxylase and flavonoid 3′,5′-hydroxylase cDNAs reveals independent evolution of flavonoid 3′,5′-hydroxylase in the Asteraceae family. *Plant Molecular Biology* **61**: 365–381.

Seo, H.S., J. Y. Yang, M. Ishikawa, C. Bolle, M. L. Ballesteros, and N. H. Chua. (2003). LAF1 ubiquitination by COP1 controls photomorphogenesis and is stimulated by SPA1. *Nature* **423**:995–999.

Shalitin, D., H. Yang, T. Mockler, M. Maymon, H. Guo, G. Whitelam, and C. Lin. (2002). Regulation of *Arabidopsis* cryptochrome by blue-light-dependent phosphorylation. *Nature* **417**:763–767.

Shalitin, D., X. Yu, M. Maymon, T. Mockler, and C. Lin. (2003). Blue light-dependent in vivo and in vitro phosphorylation of Arabidopsis cryptochrome 1. *Plant Cell* **15**:2421–2429.

Shang, Y., J. Venail, S. Mackay, P. Bailey, K. Schwinn, P. Jameson, C. Martin, and K. Davies. (2011). The molecular basis for venation patterning of pigmentation and its effect on pollinator attraction in flowers of *Antirrhinum*. *New Phytologist* **189**:602–615.

Sharma, B., and E. Kramer. (2013). Sub- and neo-functionalization of *APETALA3* paralogs have contributed to the evolution of novel floral organ identity in *Aquilegia* (columbine, Ranunculaceae). *New Phytologist* **197**:949–957.

Sharma, B., C. Guo, H. Kong, and E. Kramer. (2011). Petal-specific subfunctionalization of an *APETALA3* paralog in the Ranunculales and its implications for petal evolution. *New Phytologist* **90**:870–883.

Sheldon, C., J. E. Burn, P. P. Perez, J. Metzger, J. A. Edwards, W. J. Peacock, and E. S. Dennis. (1999). The *FLF* MADS box gene: a repressor of flowering in *Arabidopsis* regulated by vernalization and methylation. *Plant Cell* **11**:445–458.

Sheldon, C., E. J. Finnegan, D. T. Rouse, M. Tadege, D. J. Bagnall, C. A. Helliwell, W. J. Peacock, and E. S. Dennis. (2000a). The control of flowering by vernalization. *Current Opinion in Plant Biology* **3**:418–422.

Sheldon, C., D. T. Rouse, E. J. Finnegan, W. J. Peacock, and E. S. Dennis. (2000b). The molecular basis of vernalization: the central role of *FLOWERING LOCUS C (FLC)*. *Proceedings of the National Academy of Sciences of the USA* **97**:3753–3758.

Sheldon, C., A. Conn, E. S. Dennis, and W. J. Peacock. (2002). Different regulatory regions are required for the vernalization-induced repression of *FLOWERING LOCUS C* and for the epigenetic maintenance of repression. *Plant Cell* **14**:2527–2537.

Sheldon, C., E. J. Finnegan, E. S. Dennis, and W. J. Peacock. (2006). Quantitative effects of vernalization on *FLC1* and *SOC1* expression. *Plant Journal* **45**:871–883.

Shiba, H., M. Iwano, K. Entani, K. Ishimoto, H. Shimosato, F. Che, Y. Satta, A. Ito, Y. Takada, M. Watanabe, A. Isogai, and S. Takayama. (2002). The dominance of alleles controlling self-incompatibility in *Brassica* pollen is regulated at the RNA level. *Plant Cell* **14**:491–504.

Shimada, S., Y. Inoue, and M. Sakuta. (2005). Anthocyanidin synthase in non-anthocyanin-producing Caryophyllales species. *Plant Journal* **58**:950–959.

Shinomura, T., K. Uchida, and M. Furuya. (2000). Elementary processes of photoperception by phytochrome A for high-irradiance response of hypocotyl elongation in Arabidopsis. *Plant Physiology* **122**:147–156.

Shindo, S., K. Sakakibara, R. Sano, K. Ueda, and M. Hasebe. (2001). Characterization of a *FLORICAULA/LEAFY* homologue of *Gnetum parvifolium* and its implications for the evolution of reproductive organs in seed plants. *International Journal of Plant Science* **162**:1199–1209.

Shiono, M., N. Matsugaki, and K. Takeda. (2005). Structure of the blue cornflower pigment. *Nature* **436**:791.

Shoji, K., N. Miki, N. Nakajima, K. Momonoi, C. Kato, and K. Yoshida. (2007). Perianth bottom-specific blue color development in tulip cv. Murasakizuisho requires ferric ions. *Plant & Cell Physiology* **48**:243–251.

Shpak, E. D., C. T. Berthiaume, E. J. Hill, and K. U. Torii. (2003). Synergistic interaction of three ERECTA-family receptor-like kinases controls Arabidopsis organ growth and flower development by promoting cell proliferation. *Development* **131**:1491–1501.

Sicard, A., N. Stacey, K. Hermann, J. Dessoly, B. Nueffer, I. Baurle, and M. Lenhard. (2011). Genetics, evolution and adaptive significance of the selfing syndrome in the genus *Capsella*. *Plant Cell* **23**:3156–3171.

Sieber, P., F. Wellmer, J. Gheyselinck, J. Reichmann, and E. Meyerowitz. (2007). Redundancy and specialization among plant microRNAs: role of the miR164 family in developmental robustness. *Development* **134**:1051–1060.

Sieburth, L., and E. Meyerowitz. (1997). Molecular dissection of the *AGAMOUS* control region shows that *cis* elements for spatial regulation are located intragenically. *Plant Cell* **9**:355–365.

Sijacic, P., X. Wang, A. Skirpan, Y. Wang, P. Dowd, A. McCubbin, S. Huang, and T. Kao. (2004). Identification of the pollen determinant of S-RNase-mediated self-incompatibility. *Nature* **429**:302–305.

Simon, R., R. Carpenter, S. Doyle, and E. Coen. (1994). Fimbriata controls flower development by mediating between meristem and organ identity genes. *Cell* **78**: 99–107.

Simon, R., M. Igeno, and G. Coupland. (1996). Activation of floral meristem identity genes in Arabidopsis. *Nature* **84**:59–62.

Simpson, G. G. (2003). Evolution of flowering in response to day length: flipping the *CONSTANS* switch. *BioEssays* **25**:829–832.

Simpson, G. G. (2004). The autonomous pathway: epigenetic and post-transcriptional gene regulation in the control of *Arabidopsis* flowering time. *Current Opinion in Plant Biology* **7**:570–574.

Simpson, G. G., and C. Dean. (2002). Arabidopsis, the rosetta stone of flowering time? *Science* **296**:285–289.

Simpson, G. G., P. P. Dijkwel, V. Quesada, I. Henderson, and C. Dean. (2003). FY is an RNA 3′ end-processing factor that interacts with FCA to control the *Arabidopsis* floral transition. *Cell* **113**:777–787.

Smaczniak, C., R. Immink, J. Muino, R. Blanvillain, M. Busscher, J. Busscher-Lange, Q. Dinh, S. Liu, A. Westphal, S. Boeren, F. Parcy, L. Xu, C. Carles, G. Angenent, and K. Kaufmann. (2012). Characterization of MADS-domain transcription factor complexes in Arabidopsis flower development. *Proceedings of the National Academy of Sciences of the USA* **109**:1560–1565.

Smith, R., and M. Rausher. (2008). Experimental evidence that selection favours character displacement in the ivyleaf morning glory. *American Naturalist* **171**:1–9.

Smith, S., and M. Donoghue. (2008). Rates of molecular evolution are linked to life history in flowering plants. *Science* **322**:86–89.

Smith, S., and M. Rausher. (2011). Gene loss and parallel evolution contribute to species differences in flower colour. *Molecular Biology and Evolution* **28**:2799–2810.

Soltis, D., and P. Soltis. (2004). *Amborella* not a "basal angiosperm"? Not so fast. *American Journal of Botany* **91**:997–1001.

Soltis, D., V. Albert, V. Savolainen, K. Hilu, Y. Qiu, M. Chase, J. Farris, S. Stefanovic, D. Rice, J. Palmer, and P. Soltis. (2004). Genome-scale data, angiosperm relationships, and 'ending incongruence': a cautionary tale in phylogenetics. *Trends in Plant Science* **9**:477–483.

Soltis, D., P. Soltis, P. Endress, and M. Chase. (2005). *Phylogeny and Evolution of Angiosperms*. Sinauer Associates, Sunderland, MA.

Soltis, D., A. Chanderbali, S. Kim, M. Buzgo, and P. Soltis. (2007). The ABC model and its applicability to basal angiosperms. *Annals of Botany* **100**:155–163.

Soltis, P., S. Brockington, M. Yoo, A. Piedrahita, M. Latvis, M. Moore, A. Chanderbali, and D. Soltis. (2009). Floral variation and floral genetics in basal angiosperms. *American Journal of Botany* **96**:110–128.

Somers, D., P. Devlin, and S. A. Kay. (1998). Phytochromes and cryptochromes in the entrainment of the *Arabidopsis* circadian clock. *Science* **282**:1488–1490.

Somers, D., T. Schultz, M. Milnamow, and S. A. Kay. (2000). *ZEITLUPE* encodes a novel clock-associated PAS protein from *Arabidopsis*. *Cell* **101**:319–329.

Sommer, H., P. Beltran, P. Huijser, H. Pape, W. Lonnig, H. Saedler, and Z. Schwarz-Sommer. (1990). Deficiens, a homeotic gene involved in the control of flower morphogenesis in *Antirrhinum majus*: the protein shows homology to transcription factors. *EMBO Journal* **9**:605–613.

Song, Y. H., S. Ito, and T. Imaizumi. (2010). Similarities in the circadian clock and photoperiodism in plants. *Current Opinion in Plant Biology* **13**:594–603.

Sonmez, C., I. Baurle, A. Magusin, R. Dreos, S. Laubinger, D. Weigel, and C. Dean. (2011). RNA 3⊠ processing functions of Arabidopsis FCA and FPA limit intergenic transcription. *Proceedings of the National Academy of Sciences of the USA* **108**:8508–8513.

Sonneveld, T., K. Tobutt, S. Vaughan, and T. Robbins. (2005). Loss of pollen-S function in two self-compatible selections of *Prunus avium* is associated with deletion/mutation of an S haplotype-specific F-box gene. *Plant Cell* **17**:37–51.

Souer, E., A. van der Krol, D. Kloos, C. Spelt, M. Bliek, J. Mol, and R. Koes. (1998). Genetic control of branching pattern and floral identity during Petunia inflorescence development. *Development* **125**:733–742.

Souer, E., A. B. Rebocho, M. Bliek, E. Kusters, R. A. M. de Bruin, and R. Koes. (2008). Patterning of inflorescences and flowers by the F box protein DOUBLE TOP and the LEAFY homolog ABERRANT LEAF AND FLOWER of petunia. *Plant Cell* **20**:2033–2048.

Southwick, S., and T. Davenport. (1986). Characterization of water stress and low temperature effects on flower induction in citrus. *Plant Physiology* **81**:26–29.

Spaethe, J., and A. Briscoe. (2005). Molecular characterisation and expression of the UV opsin in bumblebees: three ommatidial subtypes in the retina and a new photoreceptor organ in the lamina. *Journal of Experimental Biology* **208**:2347–2361.

Spaethe, J., J. Tautz, and L. Chittka. (2001). Visual constraints in foraging bumblebees: flower size and colour affects search time and flight behaviour. *Proceedings of the National Academy of Sciences of the USA* **98**:3898–3903.

Specht, C., and M. Bartlett. (2009). Flower evolution: the origin and subsequent diversification of the angiosperm flower. *Annual Review of Ecology, Evolution and Systematics* **40**:217–243.

Spelt, C., F. Quattrocchio, J. N. M. Mol, and R. Koes. (2000). *anthocyanin1* of Petunia encodes a basic helix-loop-helix protein that directly activates transcription of structural anthocyanin genes. *Plant Cell* **12**:1619–1631.

Spelt, C., F. Quattrocchio, J. N. M. Mol, and R. Koes. (2002). ANTHOCYANIN1 of Petunia controls pigment synthesis, vacuolar pH, and seed coat development by genetically distinct mechanisms. *Plant Cell* **14**:2121–2135.

Srinivasan, M. (2010). Honeybees as a model for vision, perception and cognition. *Annual Review of Entomology* **55**:267–284.

Stacey, M., S. Hicks, and A. von Arnim. (1999). Discrete domains mediate the light-responsive nuclear and cytoplasmic localization of Arabidopsis COP1. *Plant Cell* **11**:349–363.

Stahl, Y., and R. Simon. (2010). Plant primary meristems: shared functions and regulatory mechanisms. *Current Opinion in Plant Biology* **13**:53–58.

Stanton, M., A. Snow, and S. Handel. (1986). Floral evolution: attractiveness to pollinators increases male fitness. *Science* **232**:1625–1627.

Stebbins, G. L. (1970). Adaptive radiation of reproductive characteristics in angiosperms, I: pollination mechanisms. *Annual Reviews in Ecology and Systematics* **1**:307–326.

Stefanović, S., D. Rice, and J. Palmer. (2004). Long branch attraction, taxon sampling, and the earliest angiosperms: *Amborella* or monocots? *BMC Evolutionary Biology* **4**:35.

Stein, J., B. Howlett, D. Boyes, M. Nasrallah, and J. Nasrallah. (1991). Molecular cloning of a putative receptor protein kinase gene encoded at the self-incompatibility locus of *Brassica oleracea*. *Proceedings of the National Academy of Sciences of the USA* **88**:8816–8820.

Stein, J., R. Dixit, M. Nasrallah, and J. Nasrallah. (1996). SRK, the stigma-specific S locus receptor kinase of Brassica, is targeted to the plasma membrane in transgenic tobacco. *Plant Cell* **8**:429–445.

Steiner, U., W. Schliemann, H. Bohm, and D. Strack. (1999). Tyrosinase involved in betalain biosynthesis of higher plants. *Planta* **208**:114–124.

Stellari, G., M. A. Jaramillo, and E. M. Kramer. (2004). Evolution of the *APETALA3* and *PISTILLATA* lineages of MADS-box-containing genes in the basal angiosperms. *Molecular Biology and Evolution* **21**:506–519.

Stewart, R. N., K. H. Norris, and S. Asen. (1975). Microspectrophotometric measurement of pH and pH effect on color of petal epidermal cells. *Phytochemistry* **14**:937–942.

Stinchcombe, J., C. Weinig, M. Ungerer, K. Olsen, C. Mays, S. Halldorsdottir, M. Purugganan, and J. Schmitt. (2004). A latitudinal cline in flowering time in *Arabidopsis thaliana* modulated by the flowering time gene *FRIGIDA*. *Proceedings of the National Academy of Sciences of the USA* **101**:4712–4717.

Stone, G., P. Willmer, and J. A. Rowe. (1998). Partitioning of pollinators during flowering in an African *Acacia* community. *Ecology* **79**:2808–2827.

Stone, S., M. Arnoldo, and D. Goring. (1999). A breakdown of Brassica self-incompatibility in ARC1 antisense transgenic plants. *Science* **286**:1729–1731.

Stone, S., E. Anderson, R. Mullen, and D. Goring. (2003). ARC1 is an E3 ubiquitin ligase and promotes the ubiquitination of proteins during the rejection of self-incompatible *Brassica* pollen. *Plant Cell* **15**:885–898.

Stotz, G., P. de Vlaming, H. Wiering, A. W. Schram, and G. Forkman. (1985). Genetic and biochemical studies on flavonoid 3′-hydroxylation in flowers of *Petunia hybrida*. *Theoretical and Applied Genetics* **70**:300–305.

Strack, D., T. Vogt, and W. Schliemann. (2003). Recent advances in betalain research. *Phytochemistry* **62**:247–269.

Strange, A., P. Li, C. Lister, J. Anderson, N. Warthmann, C. Shindo, J. Irwin, M. Nordborg, and C. Dean. (2011). Major-effect alleles at relatively few loci underlie distinct vernalization and flowering variation in Arabidopsis accessions. *PLoS One* **6**:e199949.

Strayer, C., T. Oyama, T. Schultz, R. Raman, D. Somers, P. Mas, S. Panda, J. Kreps, and S. Kay. (2000). Cloning of the *Arabidopsis* clock gene *TOC1*, an autoregulatory response regulator homolog. *Science* **289**:768–771.

Stromberg, M., and P. Johnsen. (1990). Hummingbird sweetness preferences; taste or viscosity? *The Condor* **92**:606–612.

Suarez-Lopez, P., K. Wheatley, F. Robson, H. Onouchi, F. Valverde, and G. Coupland. (2001). *CONSTANS* mediates between the circadian clock and the control of flowering in *Arabidopsis*. *Nature* **410**:1116–1120.

Subramanian, C., B. Kim, N. Lyssenko, X. Xu, C. Johnson, and A. von Arnim. (2004). The *Arabidopsis* repressor of light signalling, COP1, is regulated by nuclear exclusion: mutational analysis by bioluminescence resonance energy transfer. *Proceedings of the National Academy of Sciences of the USA* **101**:6798–6802.

Sugimoto-Shirasu, K., and K. Roberts. (2003). 'Big it up': endoreduplication and cell cycle control in plants. *Current Opinion in Plant Biology* **6**:544–553.

Sun, G., D. L. Dilcher, S. Zheng, and Z. Zhou. (1998). In search of the first flower: a Jurassic angiosperm, *Archaefructus*, from northeast China. *Science* **282**:1692–1695.

Sun, G., Q. Ji, D. Dilcher, S. Zheng, K. Nixon, and X. Wang. (2002). Archaefructaceae, a new basal angiosperm family. *Science* **296**:899–904.

Sun, G., D. Dilcher, H. Wang, and Z. Chen. (2011). A eudicot from the Early Cretaceous of China. *Nature* **471**: 625–628.

Sundstrom, J., and P. Engstrom. (2002). Conifer reproductive development involves B-type MADS-box genes with distinct and different activities in male organ primordia. *Plant Journal* **31**:161–169.

Sundstrom, J. F., N. Nakayama, K. Glimelius, and V. Irish. (2006). Direct regulation of the floral homeotic *APETALA1* gene by APETALA3 and PISTILLATA in Arabidopsis. *Plant Journal* **46**:593–600.

Sung, S., and R. M. Amasino. (2004a). Vernalization in *Arabidopsis thaliana* is mediated by the PHD finger protein VIN3. *Nature* **427**:159–163.

Sung, S., and R. M. Amasino. (2004b). Vernalization and epigenetics: how plants remember winter. *Current Opinion in Plant Biology* **7**:4–10.

Sung, S., Y. He, T. W. Eshoo, Y. Tamada, L. Johnson, K. Nakahigashi, K. Goto, S. E. Jacobsen, and R. M. Amasino. (2006). Epigenetic maintenance of the vernalized state in *Arabidopsis thaliana* requires LIKE HETEROCHROMATIN PROTEIN 1. *Nature Genetics* **38**:706–710.

Szecsi, J., C. Joly, K. Bordji, E. Varaud, J. Cock, C. Dumas, and M. Bedahmane. (2006). BIGPETALp, a bHLH transcription factor, in involved in the control of Arabidopsis petal size. *EMBO Journal* **25**:3912–3920.

Tadege, M., C. Sheldon, C. Helliwell, N. Upadhyaya, E. Dennis, and W.J. Peacock. (2003). Reciprocal control of flowering time by OsSOC1 in transgenic Arabidopsis and by FLC in transgenic rice. *Plant Biotechnology Journal* **1**:361–369.

Takasaki, T., K. Hatakeyama, G. Suzuki, M. Watanabe, A. Isogai, and K. Hinata. (2000). The S receptor kinase determines self-incompatibility in *Brassica* stigma. *Nature* **403**:913–916.

Takayama, S., H. Shiba, M. Iwano, H. Shimosato, F. Che, N. Kai, M. Watanabe, G. Suzuki, K. Hinata, and A. Isogai. (2000). The pollen determinant of self-incompatibility in

Brassica campestris. Proceedings of the National Academy of Sciences of the USA **97**:1920–1925.

Takayama, S., H. Shimosato, H. Shiba, M. Funato, F. Che, M. Watanabe, M. Iwano, and A. Isogai. (2001). Direct ligand-receptor complex interaction controls *Brassica* self-incompatibility. *Nature* **413**:534–538.

Takebayashi, N., D. Wolf, and L. Delph. (2006). Effect of variation in herkogamy on outcrossing within a population of *Gilia achilleifolia. Heredity* **96**:159–165.

Takeda, S., N. Matsumoto, and K. Okada. (2003). *RABBIT EARS*, encoding a SUPERMAN-like zinc-finger protein, regulates petal development in *Arabidopsis thaliana. Development* **131**:425–434.

Tanabe, Y., M. Hasebe, H. Sekimoto, T. Nishiyama, M. Kitani, K. Henschel, T. Muenster, G. Theissen, H. Nozaki, and M. Ito. (2005). Characterization of MADS-box genes in charophycean green algae and its implication for the evolution of MADS-box genes. *Proceedings of the National Academy of Sciences of the USA* **102**:2436–2441.

Tanahashi, T., N. Sumikawa, M. Ato, and M. Hasebe. (2005). Diversification of gene function: homologs of the floral regulator *FLO/LFY* control the first zygotic cell division in the moss *Physcomitrella patens. Development* **132**:1727–1736.

Tanaka, Y., N. Sasaki, and A. Ohmiya. (2008). Biosynthesis of plant pigments: anthocyanins, betalains and carotenoids. *Plant Journal* **54**:733–749.

Tandre, K., M. Svenson, M. E. Svensson, and P. Engstrom. (1998). Conservation of gene structure and activity in the regulation of reproductive organ development of conifers and angiosperms. *Plant Journal* **15**: 615–623.

Tarutani, Y., H. Shiba, M. Iwano, T. Kakizaki, G. Suzuki, M. Watanabe, A. Isogai, and S. Takayama. (2010). *Trans*-acting small RNA determines dominance relationships in *Brassica* self-incompatibility. *Nature* **466**: 983–986.

Tavares, R., M. Cagnon, I. Negrutiu, and D. Mouchiroud (2010). Testing the recent theories for the origin of the hermaphrodite flower by comparison of the transcriptomes of gymnosperms and angiosperms. *BMC Evolutionary Biology* **10**:240.

Teeri, T., V. Albert, P. Elomaa, J. Hamalainen, M. Kotilainen, E. Pollanen, and A. Uimari. (2002). Involvement of non-ABC MADS-box genes in determining stamen and carpel identity in *Gerbera hybrida* (Asteraceae). In: Q. C. B. Cronk, R. M. Bateman, and J. A. Hawkins (eds), *Developmental Genetics and Plant Evolution*, pp. 173–205. Taylor and Francis, London.

Tepperman, J., T. Zhu, H. Chang, X. Wang, and P. H. Quail. (2001). Multiple transcription factor genes are early targets of phytochrome A signalling. *Proceedings of the National Academy of Sciences of the USA* **98**: 9437–9442.

The Arabidopsis Genome Initiative. (2000). Analysis of the genome of the flowering plant *Arabidopsis thaliana. Nature* **408**:796–815.

Theissen, G., and H. Saedler. (2001). Plant biology—floral quartets. *Nature* **409**:469–471.

Theissen, G., and A. Becker. (2004). Gymnosperm orthologues of class B Floral homeotic genes and their impact on understanding flower origin. *Critical Reviews in Plant Sciences* **23**:129–148.

Theissen, G., A. Becker, A. Di Rosa, A. Kanno, J. T. Kim, T. Münster, K. U. Winter, and H. Saedler. (2000). A short history of MADS-box genes in plants. *Plant Molecular Biology* **42**:115–149.

Theissen, G., A. Becker, K. Winter, T. Munster, C. Kirchner, and H. Saedler. (2002). How the land plants learned their floral ABCs: the role of MADS-box genes in the evolutionary origin of flowers. In: Q. C. B. Cronk, R. M. Bateman, and J. A. Hawkins (eds), *Developmental Genetics and Plant Evolution*, pp. 173–205. Taylor and Francis, London.

Thien, L. B., H. Azuma, and S. Kawano. (2000). New perspectives on the pollination biology of basal angiosperms. *International Journal of Plant Sciences* **161**:S225–S235.

Thompson, B., L. Bartling, C. Whipple, D. Hall, H. Sakai, R. Schmidt, and S. Hake. (2009). *bearded-ear* encodes a MADS box transcription factor critical for maize floral development. *Plant Cell* **21**:2578–2590.

Thornton, T., S. Swain, and N. Olszewski. (1999). Gibberellin signal transduction presents the SPY who O-GlcNAc'd me. *Trends in Plant Science* **4**:424–428.

Tilney-Bassett, R. (1986). *Plant Chimeras*. Edward Arnold, London.

Todesco, M., S. Balasubramanian, T. Hu, B. Traw, M. Horton, P. Epple, C. Kuhns, S. Sureshkumar, C. Schwartz, C. Lanz, R. Laitinen, J. Chory, V. Lipka, J. Borevitz, J. Dangl, J. Bergelson, M. Nordborg, and D. Weigel. (2010). Natural allelic variation underlying a major fitness tradeoff in *Arabidopsis thaliana. Nature* **465**: 632–636.

Torti, S., F. Fornara, C. Vincent, F. Andres, K. Nordström, U. Göbel, D. Knoll, H. Schoof, and G. Coupland. (2012). Analysis of the *Arabidopsis* shoot meristem transcriptome during floral transition identifies distinct regulatory patterns and a leucine-rich repeat protein that promotes flowering. *Plant Cell* **24**:444–462.

Trevaskis, B., D. Bagnall, M. Ellis, W. J. Peacock, and E. Dennis. (2003). MADS box genes control vernalization-induced flowering in cereals. *Proceedings of the Natioanl Academy of Sciences of the USA* **100**:13099–13104.

Trobner, W., L. Ramirez, P. Motte, I. Hue, P. Huijser, W. Lonnig, H. Saedler, H. Sommer, and Z. Schwarz-Sommer. (1992). GLOBOSA—a homeotic gene which interacts with DEFICIENS in the control of Antirrhinum floral organogenesis. *EMBO Journal* **11**:4693–4704.

Tsaftaris, A., K. Pasentsis, P. Madesis, and A. Argiriou. (2012). Sequence characterization and expression analysis of three *APETALA2*-like genes from saffron crocus. *Plant Molecular Biology Reporter* **30**:443–452.

Tsuchimatsu, T., K. Suwabe, R. Shimizu-Inatsuqi, S. Isokawa, P. Pavlidis, T. Stadler, G. Suzuki, S. Takayama, M. Watanabe, and K. Shimizu. (2010). Evolution of self-compatibility in Arabidopsis by a mutation in the male specificity gene. *Nature* **464**:1342–1346.

Tsuji, H., K. Taoka, and K. Shimamoto. (2011). Regulation of flowering in rice: two florigen genes, a complex gene network, and natural variation. *Current Opinion in Plant Biology* **14**:45–52.

Tucker, M., T. Okada, Y. Hu, A. Scholefield, J. Taylor, and A. Koltunow. (2012). Somatic small RNA pathways promote the mitotic events of megagametogenesis during female reproductive development in Arabidopsis. *Development* **139**:1399–1404.

Tuskan, G., S. DiFazio, S. Jansson *et al.* (2006). The genome of black cottonwood, *Populus trichocarpa* (Torr. & Gray). *Science* **313**:1596–1604.

Twell, D. (2011). Male gametogenesis and germline specification in flowering plants. *Sexual Plant Reproduction* **24**:149–160.

Twell, D., S. Park, T. Hawkins, D. Schubert, R. Schmidt, A. Smertenko, and P. Hussey. (2002). Mor1/Gem1 has an essential role in the plant-specific cytokinetic phragmoplast. *Nature Cell Biology* **4**:711–714.

Uehara, S., and H. Morita. (1972). The effect of temperature on the labellar chemoreceptors of the blowfly. *Journal of General Physiology* **59**:213–226.

Uimari, A., and J. Strommer. (1997). Myb26: a MYB-like protein of pea flowers with affinity for promoters of phenylpropanoid genes. *Plant Journal* **12**:1273–1284.

Uimari, A., M. Kotilainen, P. Elomaa, D. Yu, V. A. Albert, and T. H. Teeri. (2004). Integration of reproductive meristem fates by a *SEPALLATA*-like MADS-box gene. *Proceedings of the Natioanl Academy of Sciences of the USA* **101**:15817–15822.

Urzay, J., S. Llewellyn Smith, E. Thompson, and B. Glover. (2009). Wind gusts and plant aeroelasticity effects on the dynamics of pollen shedding: a hypothetical turbulence-initiated wind-pollination mechanism. *Journal of Theoretical Biology* **259**:785–792.

Valverde, F., A. Mouradov, W. Soppe, D. Ravenscroft, A. Samach, and G. Coupland. (2004). Photoreceptor regulation of CONSTANS protein in photoperiodic flowering. *Science* **303**:1003–1006.

Vamosi, J., T. Knight, J. Steets, S. Mazer, M. Burd, and T. Ashman. (2006). Pollination decays in biodiversity hotspots. *Proceedings of the National Academy of Sciences of the USA* **103**:956–961.

Vandenbussche, M., G. Theissen, Y. Van de Peer, and T. Gerats. (2003a). Structural diversification and neo-functionalization during floral MADS box gene evolution by C-terminal frameshift mutations. *Nucleic Acids Research* **31**:4401–4409.

Vandenbussche, M., J. Zethof, E. Souer, R. Koes, G. Tornielli, M. Pezzotti, S. Ferrario, G. Angenent, and T. Gerats. (2003b). Toward the analysis of the petunia MADS box gene family by reverse and forward transposon insertion mutagenesis approaches: B, C, and D floral organ identity functions require SEPALLATA-like MADS box genes in petunia. *Plant Cell* **15**:2680–2693.

Vanoosthuyse, V., G. Tichtinsky, C. Dumas, T. Gaude, and J. M. Cock. (2003). Interaction of calmodulin, a sorting nexin and kinase-associated protein phosphatase with the *Brassica oleracea* S locus receptor kinase. *Plant Physiology* **133**:919–929.

van Tunen, A. J., R. E. Koes, C. E. Spelt, A. R. van der Krol, A. R. Stuitje, and J. N. M. Mol. (1988). Cloning of the two chalcone flavanone isomerase genes from *Petunia hybrida*: coordinate, light regulated and differential expression of flavonoid genes. *EMBO Journal* **7**: 1257–1263.

Varela, F., A. Palacios, and T. Goldsmith. (1993). Colour vision of birds. In: H. Zeigler (ed.), *Vision, Brain and Behaviour in Birds*. MIT Press, Cambridge, MA.

Vazquez, D., N. Bluthgen, L. Cagnolo, and N. Chacoff. (2009). Uniting pattern and process in plant–animal mutualistic networks: a review. *Annals of Botany* **103**: 1445–1457.

Vazquez-Lobo, A., A. Carlsbecker, F. Vergara-Silva, E. Alvarez-Bullya, D. Pinero, and P. Engström. (2007). Characterization of the expression patterns of *LEAFY/ FLORICAULA* and *NEEDLY* orthologs in female and male cones of the conifer genera *Picea*, *Podocarpus*, and *Taxus*: implications for current evo-devo hypotheses for gymnosperms. *Evolution and Development* **9**: 446–459.

Verdonk, J., M. Haring, A. van Tunen, and R. Schuurink. (2005). ODORANT1 regulates fragrance biosynthesis in petunia flowers. *Plant Cell* **17**:1612–1624.

Vert, G., C. Walcher, J. Chory, and J. Nemhauser. (2008). Integration of auxin and brassinosteroid pathways by AUXIN RESPONSE FACTOR 2. *Proceedings of the National Academy of Sciences of the USA* **105**:9829–9834.

Verweij, W., C. Spelt, G. Di Sansebastiano, J. Vermeer, L. Reale, F. Ferranti, R. Koes, and F. Quattrocchio. (2008). An H+ P-ATPase on the tonoplast determines vacuolar pH and flower colour. *Nature Cell Biology* **10**:1456–1462.

Vignolini, S., M. Thomas, M. Kolle, T. Wenzel, P. Rudall, J. Baumberg, B. Glover, and U. Steiner. (2012). Directional scattering from *Ranunculus acris*: how the buttercup lights up your chin. *Journal of the Royal Society Interface* **9**:1295–1301.

Voinnet, O. (2009). Origin, biogenesis and activity of plant microRNAs. *Cell* **136**:669–687.

Vyskot, B., and R. Hobza. (2004). Gender in plants: sex chromosomes are emerging from the fog. *Trends in Genetics* **20**:432–438.

Waelti, M., J. Muhlemann, A. Widmer, and F. Schiestl. (2007). Floral odour and reproductive isolation in two species of *Silene*. *Journal of Evolutionary Biology* **21**:111–121.

Wagner, D., R. Sablowski, and E. Meyerowitz. (1999). Transcriptional activation of APETALA1 by LEAFY. *Science* **285**:582–584.

Wagner, J., J. Brunzelle, K. Forest, and R. Viestra. (2005). A light-sensing knot revealed by the structure of the chromophore-binding domain of phytochrome. *Nature* **438**:325–331.

Waites, R., and A. Hudson. (2001). The Handlebars gene is required with Phantastica for dorsoventral asymmetry of organs and for stem cell activity in Antirrhinum. *Development* **128**:1923–1931.

Walker, A. R., R. A. Davison, A. C. Bolognesi-Winfield, C. M. James, N. Srinivasan, T. Blundell, J. J. Esch, M. D. Marks, and J. C. Gray. (1999). The TRANSPARENT TESTA GLABRA 1 locus which regulates trichome differentiation and anthocyanin biosynthesis in Arabidopsis encodes a WD40 repeat protein. *Plant Cell* **11**:1337–1345.

Wang, H., and X. Deng. (2003). Dissecting the phytochrome A-dependent signalling network in higher plants. *Trends in Plant Science* **8**:172–178.

Wang, H., L. Ma, J. Li, H. Zhao, and X. Deng. (2001). Direct interaction of *Arabidopsis* cryptochromes with COP1 in light control development. *Science* **294**:154–158.

Wang, R., S. Farrona, C. Vincent, A. Joecker, H. Schoof, F. Turck, C. Alonso-Blanco, G. Coupland, and M. Albani. (2009). PEP1 regulates perennial flowering in *Arabis alpina*. *Nature* **459**:423–427.

Wang, X., P. Zhang, Q. Du, H. He, L. Zhao, Y. Ren, and P. Endress. (2012). Heterodichogamy in *Kingdonia* (Circaesteraceae, Ranunculales). *Annals of Botany* **109**:1125–1132.

Wang, Z., Y. Luo, X. Li, L. Wang, S. Xu, J. Yang, L. Weng, S. Sato, S. Tabata, M. Ambrose, C. Rameau, X. Feng, X. He, and D. Luo. (2008). Genetic control of floral zygomorphy in pea (*Pisum sativum*). *Proceedings of the National Academy of Sciences of the USA* **105**:10414–10419.

Warner, K. A., P. J. Rudall, and M. W. Frohlich. (2008). Differentiation of perianth organs in Nymphaeales. *Taxon* **57**:1096–1109.

Waser, N. (1978). Competition for hummingbird pollination and sequential flowering in two Colorado wildflowers. *Ecology* **59**:934–944.

Waser, N. (2006). Specialization and generalization in plant–pollinator interactions: a historical perspective. In: N. Waser and J. Ollerton (eds), *Plant–Pollinator Interactions*, pp. 3–17. University of Chicago Press, Chicago.

Waser, N., and M. Price. (1981). Pollinator choice and stabilizing selection for flower colour in *Delphinium nelsonii*. *Evolution* **35**:376–390.

Waser, N., and J. Ollerton. (2006). *Plant–Pollinator Interactions*. University of Chicago Press, Chicago.

Waser, N., L. Chittka, M. Price, N. Williams, and J. Ollerton. (1996). Generalization in pollination systems, and why it matters. *Ecology* **77**:1043–1060.

Wasserthal, L. (1997). The pollinators of the Malagasy star orchids *Angraecum sesquipedale*, *A. sororium*, and *A. compactum* and the evolution of extremely long spurs by pollinator shift. *Botanica Acta* **110**:343–359.

Watanabe, M., G. Suzuki, S. Takayama, A. Isogai, and K. Hinata. (2000). Genomic organization of the SLG/SRK region of the S locus in *Brassica* species. *Annals of Botany* **85**:155–160.

Weigel, D., and E. Meyerowitz. (1993). Activation of floral homeotic genes in Arabidopsis. *Science* **261**:1723–1726.

Weigel, D., and O. Nilsson. (1995). A developmental switch sufficient for flower initiation in diverse plants. *Nature* **377**:495–500.

Weigel, D., J. Alvarez, D. Smyth, M. Yanofsky, and E. Meyerowitz. (1992). *LEAFY* controls floral meristem identity in Arabidopsis. *Cell* **69**:843–859.

Weiss, D., A. van der Luit, J. Kroon, J. Mol, and J. Kooter. (1993). The petunia homologue of the *Antirrhinum majus candi* and *Zea mays A2* flavonoid genes; homology to flavanone 3-hydroxylase and ethylene-forming enzyme. *Plant Molecular Biology* **22**:893–897.

Weiss, M. R. (1991). Floral colour changes as cues for pollinators. *Nature* **354**:227–229.

Wellensiek, S. J. (1962). Dividing cells as the locus for vernalization. *Nature* **195**:307–308.

Weller, J., J. Reid, S. Taylor, and I. Murfet. (1997). The genetic control of flowering in pea. *Trends in Plant Science* **2**:412–418.

Weller, J., V. Hecht, L. Liew, F. Sussmilch, B. Wenden, C. Knowles, and J. Vander Schoor. (2009). Update on the genetic control of flowering in garden pea. *Journal of Experimental Botany* **60**:2493–2499.

Wellmer, F., J. Reichmann, M. Alves-Ferreira, and E. Meyerowitz. (2004). Genome-wide analysis of spatial gene expression in Arabidopsis flowers. *Plant Cell* **16**:1314–1326.

Wesselingh, R. (2007). Pollen limitation meets resource allocation: towards a comprehensive methodology. *New Phytologist* **174**:26–37.

Wessinger, C., and M. Rausher. (2012). Lessons from flower colour evolution on targets of selection. *Journal of Experimental Botany* **63**:5741–5749.

Wessler, S. R. (2005). Homing into the origin of the AP2 DNA binding domain. *Trends in Plant Science* **10**:54–56.

Weston, E., and K. Pyke. (1999). Developmental ultrastructure of cells and plastids in the petals of wallflower (*Erysimum cheiri*). *Annals of Botany* **84**:763–769.

Weterings, K., and S. Russell. (2004). Experimental analysis of the fertilisation process. *Plant Cell* **16**:S107–S118.

Wheeler, M., B. de Graaf, N. Hadjiosif, R. Perry, N. Poulter, K. Osman, S. Vatovec, A. Harper, F. Franklin, and V. Franklin-Tong. (2009). Identification of the pollen self-incompatibility determinant in *Papaver rhoeas*. *Nature* **459**:992–995.

Wheeler, M., S. Vatovec, and V. Franklin-Tong. (2010). The pollen S-determinant in *Papaver*: comparisons with known plant receptors and protein ligand partners. *Journal of Experimental Botany* **61**:2015–2025.

Whibley, A., N. Langlade, C. Andalo, A. Hanna, A. Bangham, C. Thebaud, and E. Coen. (2006). Evolutionary paths underlying flower color variation in *Antirrhinum*. *Science* **313**:963–966.

Whipple, C., P. Ciceri, C. Padilla, B. Ambrose, S. Bandong, and R. Schmidt. (2004). Conservation of B-class floral homeotic gene function between maize and *Arabidopsis*. *Development* **131**:6083–6091.

Whipple, C., M. Zanis, E. Kellogg, and R. Schmidt. (2007). Conservation of B class gene expression in the second whorl of a basal grass and outgroups links the origin of lodicules and petals. *Proceedings of the National Academy of Sciences of the USA* **104**:1081–1086.

Whitney, H., L. Chittka, T. Bruce, and B. Glover. (2009a). Conical epidermal cells allow bees to grip flowers and increase foraging efficiency. *Current Biology* **19**:1–6.

Whitney, H., M. Kolle, P. Andrew, L. Chittka, U. Steiner, and B. Glover. (2009b). Floral iridescence, produced by diffractive optics, acts as a cue for animal pollinators. *Science* **323**:130–133.

Whitney, H., K.M.V. Bennett, M. Dorling, L. Sandbach, D. Prince, L. Chittka, and B. Glover. (2011). Why do so many petals have conical epidermal cells? *Annals of Botany* **108**:609–616.

Whitney, H., G. Milne, S. Rands, S. Vignolini, C. Martin, and B. Glover. (2013). The influence of pigmentation patterning on bumblebee foraging from flowers of *Antirrhinum majus*. *Naturwissenscaften* **100**:249–256.

Whittall, J., and S. Hodges. (2007). Pollinator shifts drive increasingly long nectar spurs in columbine flowers. *Nature* **447**:706–709.

Whittall, J., C. Voelckel, D. Kliebenstein, and S. Hodges. (2006). Convergence, constraint and the role of gene expression during adaptive radiation: Floral anthocyanins in *Aquilegia*. *Molecular Ecology* **15**:4645–4657.

Wienand, U., H. Sommer, Z. Schwarz, N. Shephard, H. Saedler, F. Kreuzaler, H. Ragg, K. Hahlbrock, R. Harrison, and P. A. Peteson. (1982). A general method to identify plant structural genes among genomic DNA clones using transposable element induced mutations. *Molecular and General Genetics* **187**:195–201.

Wigge, P. A., M. C. Kim, K. E. Jaeger, W. Busch, M. Schmid, J. U. Lohmann, and D. Weigel. (2005). Integration of spatial and temporal information during floral induction in *Arabidopsis*. *Science* **309**:1056–1059.

Wilkinson, M., and G. Haughn. (1995). *UNUSUAL FLORAL ORGANS* controls meristem identity and organ primordia fate. *Plant Cell* **7**:1485–1499.

Willis, K. J., and J. C. McElwain. (2002). *The Evolution of Plants*. Oxford University Press, Oxford.

Willmer, P., D. Stanley, K. Steijven, I. Matthews, and C. Nuttman. (2009). Bidirectional flower color and shape changes allow a second opportunity for pollination. *Current Biology* **19**:919–923.

Wilson, P., M. Castellanos, J. Hogue, J. Thomson, and W. S. Armbruster. (2004). A multivariate search for pollination syndromes among penstemons. *Oikos* **104**:345–361.

Winter, K., C. Weiser, K. Kaufmann, A. Bohne, C. Kirchner, A. Kanno, H. Saedler, and G. Theissen. (2002). Evolution of class B floral homeotic proteins: obligate heterodimerization originated from homodimerization. *Molecular Biology and Evolution* **19**:587–596.

Wolfe, K. H., M. Gouy, Y. Yang, P. Sharp, and W. Li. (1989). Date of the monocot–dicot divergence estimated from chloroplast DNA sequence data. *Proceedings of the National Academy of Sciences of the USA* **86**:6201–6205.

Wollmann, H., E. Mica, M. Todesco, J. Long, and D. Weigel. (2010). On reconciling the interactions between *APETALA2*, miR172 and *AGAMOUS* with the ABC model of flower development. *Development* **137**:3633–3642.

Wood, C. C., M. Robertson, G. Tanner, W. J. Peacock, E. S. Dennis, and C. A. Helliwell. (2006). The *Arabidopsis thaliana* vernalization response requires a polycomb-like protein complex that also includes VERNALIZATION INSENSITIVE 3. *Proceedings of the National Academy of Sciences of the USA* **103**:14631–14636.

Wuerschum, T., R. Gross-Hardt, and T. Laux. (2006). *APETALA2* regulates the stem cell niche in the *Arabidopsis* shoot meristem. *Plant Cell* **18**:295–307.

Wuest, S., D. O'Maoileidigh, L. Rae, K. Kwasniewska, A. Raganelli, K. Hanczaryk, A. Lohan, B. Loftus, E. Graciet, and F. Wellmer. (2012). Molecular basis for the specification of floral organs by APETALA3 and PISTILLATA. *Proceedings of the National Academy of Sciences of the USA* **109**:13452–13457.

Xu, L., Z. Zhao, A. Dong, L. Soubigou-Taconnat, J. Renou, A. Steinmetz, and W. Shen. (2008). Di- and tri- but not monomethylation on histone H3 lysine 36 marks active transcription of genes involved in flowering time regulation and other processes in *Arabidopsis thaliana*. *Molecular and Cellular Biology* **28**:1348–1360.

Xu, Y., L. Teo, J. Zhou, P. Kumar, and Yu, H. (2006). Floral organ identity genes in the orchid *Dendrobium crumenatum*. *Plant Journal* **46**:54–68.

Yadav, V., C. Mallappa, S. Gangappa, S. Bhatia, and S. Chattopadhyay. (2005). A basic helix-loop-helix transcription factor in Arabidopsis, MYC2, acts as a repressor of blue light-mediated photomorphogenic growth. *Plant Cell* **17**:1953–1966.

Yadegari, R., and G. Drews. (2004). Female gametophyte development. *Plant Cell* 16:S133–S141.

Yamaguchi, T., S. Fukada-Tanaka, Y. Inagaki, N. Saito, K. Yonekura-Sakakibara, Y. Tanaka, T. Kusumi, and S. Iida. (2001). Genes encoding the vacuolar Na^+/H^+ exchanger and flower coloration. *Plant and Cell Physiology* 42:451–461.

Yamaguchi, T., D. Lee, A. Kiyao, H. Hirochika, G. An, and H. Hirano. (2006). Functional diversification of the two C-class MADS box genes *OsMADS3* and *OsMADS58* in *Oryza sativa*. *Plant Cell* 18:15–28.

Yamaguchi, N., A. Yamaguchi, M. Abe, D. Wagner, and Y. Komeda. (2012). LEAFY controls Arabidopsis pedicel length and orientation by affecting adaxial-abaxial cell fate. *Plant Journal* 69:844–856.

Yamamoto, Y., M. Matsui, L. Ang, and X. Deng. (1998). Role of a COP1 interactive protein in mediating light-regulated gene expression in Arabidopsis. *Plant Cell* 10:1083–1094.

Yan, L., A. Loukoianov, G. Tranquilli, M. Helguera, T. Fahima, and J. Dubcovsky. (2003). Positional cloning of the wheat vernalization gene *VRN1*. *Proceedings of the National Academy of Sciences of the USA* 100:6263–6268.

Yan, L., A. Loukoianov, A. Blechl, G. Tranquilli, W. Ramakrishna, P. SanMiguel, J. Bennetzen, V. Echenique, and J. Dubcovsky. (2004). The wheat *VRN2* gene is a flowering repressor down-regulated by vernalization. *Science* 303:1640–1644.

Yan, L., D. Fu, C. Li., A. Blechl, G. Tranquilli, M. Bonafede, A. Sanchez, M. Valarik, S. Yasuda, and J. Dubcovsky. (2006). The wheat and barley vernalization gene VRN3 is an orthologue of FT. *Proceedings of the National Academy of Sciences of the USA* 103:19581–19586.

Yang, C., M. Spielman, J. Coles, Y. Li, S. Ghelani, V. Bourdon, R. Brown, B. Lemmon, R. Scott, and H. Dickinson. (2003a). *TETRASPORE* encodes a kinesin required for male meiotic cytokinesis in Arabidopsis. *Plant Journal* 34:220–240.

Yang, H., Y. Wu, R. Tang, D. Liu, Y. Liu, and A. Cashmore. (2000). The C termini of *Arabidopsis* cryptochromes mediate a constitutive light response. *Cell* 103:815–827.

Yang, J., R. Lin, J. Sullivan, U. Hoecker, B. Liu, L. Xu, X. Deng, and H. Wang. (2005). Light regulates COP1-mediated degradation of HFR1, a transcription factor essential for light signalling in Arabidopsis. *Plant Cell* 17:804–821.

Yang, S., L. Xie, H. Mao, C. Puah, W. Yang, L. Jiang, V. Sundaresan, and D. Ye. (2003b). *TAPETUM DETERMINANT 1* is required for cell specialisation in the Arabidopsis anther. *Plant Cell* 15:2792–2804.

Yano, M., Y. Katayose, M. Ashikari, U. Yamanouchi, L. Monna, T. Fuse, T. Baba, K. Yamamoto, Y. Umehara, Y. Nagamura, and T. Sasaki. (2000). *Hd1*, a major photoperiod sensitivity quantitative trait locus in rice, is closely related to the Arabidopsis flowering time gene *CONSTANS*. *Plant Cell* 12:2473–2483.

Yanofsky, M., H. Ma, J. Bowman, G. Drews, K. Feldmann, and E. Meyerowitz. (1990). The protein encoded by the Arabidopsis homeotic gene *agamous* resembles transcription factors. *Nature* 346:35–39.

Yanovsky, M. J., and S. A. Kay. (2003). Living by the calendar: how plants know when to flower. *Nature Reviews Molecular Cell Biology* 4:265–275.

Yant, L., J. Mathieu, T. Dinh Thanh, F. Ott, C. Lanz, H. Wollmann, X. Chen, and M. Schmid. (2010). Orchestration of the floral transition and floral development in Arabidopsis by the bifunctional transcription factor APETALA2. *Plant Cell* 22:2156–2170.

Yoshida, H. (2012). Is the lodicule a petal: Molecular evidence? *Plant Science* 184:121–128.

Yoshida, K., T. Kondo, Y. Okazaki, and K. Katou. (1995). Cause of blue petal colour. *Nature* 373:291.

Yoshida, K., M. Kawachi, M. Mori, M. Maeshima, M. Kondo, M. Nishimura, and T. Kondo. (2005). The involvement of tonoplast proton pumps and $Na^+(K^+)/H^+$ exchangers in the change of petal colour during flower opening of morning glory, *Ipomoea tricolor* cv. Heavenly Blue. *Plant and Cell Physiology* 46:407–415.

Yoshida, K., S. Kitahara, D. Ito, and T. Kondo. (2006). Ferric ions involved in the flower colour development of the Himalayan blue poppy *Meconopsis grandis*. *Phytochemistry* 67:992–998.

Yu, C., X. Liu, M. Luo, C. Chen, X. Lin, G. Tian, Q. Lu, Y Cui, and K. Wu. (2011). HISTONE DEACETYLASE 6 interacts with FLOWERING LOCUS D and regulates flowering in Arabidopsis. *Plant Physiology* 156: 173–184.

Yu, D., M. Kotilainen, E. Pollanen, M. Mehto, P. Elomaa, Y. Helariutta, V. A. Albert, and T. H. Teeri. (1999). Organ identity genes and modified patterns of flower development in *Gerbera hybrida* (Asteraceae). *Plant Journal* 17:51–62.

Yu, H., T. Ito, Y. Zhao, J. Peng, P. Kumar, and E. Meyerowitz. (2004). Floral homeotic genes are targets of gibberellin signaling in flower development. *Proceedings of the National Academy of Sciences of the USA* 101: 7827–7832.

Yun, H., Y. Hyun, M. Kang, Y. Noh, B. Noh, and Y. Choi. (2011). Identification of regulators required for the reactivation of *FLOWERING LOCUS C* during Arabidopsis reproduction. *Planta* 234:1237–1250.

Zachgo, S., E. Andrade Silva, P. Motte, W. Trobner, H. Saedler, and Z. Schwarz-Sommer. (1995). Functional analysis of the *Antirrhinum* floral homeotic DEFICIENS gene in vivo and in vitro by using a temperature-sensitive mutant. *Development* 121:2861–2875.

Zahn, L., J. Leebens-Mack, C. dePamphlis, H. Ma, and G. Theissen. (2005). To B or not to B a flower: the role of *DEFICIENS* and *GLOBOSA* orthologs in the evolution of the angiosperms. *Journal of Heredity* 9:225–240.

Zeevart, J. (1976) Physiology of flower formation. *Annual Review of Plant Physiology and Plant Molecular Biology* **27**:321–348.

Zhang, F., A. Gonzalez, M. Zhao, C. Payne, and A. Lloyd. (2003). A network of redundant bHLH proteins functions in all TTG1-dependent pathways of Arabidopsis. *Development* **130**:4859–4869.

Zhang, H., and S. van Nocker. (2002). The *VERNALIZATION INDEPENDENCE 4* gene encodes a novel regulator of *FLOWERING LOCUS C*. *Plant Journal* **31**:663–673.

Zhang, L., S. C. Barrett, J. Y. Gao, J. Chen, W. W. Cole, Y. Liu, Z. L. Bai, and Q. J. Li. (2005). Predicting mating patterns from pollination syndromes: the case of "sapromyophily" in *Tacca chantrieri* (Taccaceae). *American Journal of Botany* **92**:517–524.

Zhang, P., H. T. W. Tan, K.-H. Pwee, and P. P. Kumar. (2004). Conservation of class C function of floral organ development during 300 million years of evolution from gymnosperms to angiosperms. *Plant Journal* **37**: 566–577.

Zhang, W., E. Kramer, and C. Davis. (2010). Floral symmetry genes and the origin and maintenance of zygomorphy in a plant-pollinator mutualism. *Proceedings of the National Academy of Sciences of the USA* **107**:6388–6393.

Zhang, W., E. Kramer, and C. Davis. (2012). Similar genetic mechanisms underlie the parallel evolution of floral phenotypes. *PLoS One* **7**:e36033.

Zhang, Y., S. Gong, Q. Li, Y. Sang, and H. Yang. (2006). Functional and signaling mechanism analysis of rice CRYPTOCHROME 1. *Plant Journal* **46**:971–983.

Zhao, D.-Z., G.-F. Wang, B. Speal, and H. Ma. (2002). The *EXCESS MICROSPOROCYTES1* gene encodes a putative leucine-rich repeat receptor protein kinase that controls somatic and reproductive cell fates in the *Arabidopsis* anther. *Genes & Development* **16**:2021–2031.

Zhu, Y., J. M. Tepperman, C. D. Fairchild, and P. H. Quail. (2000). Phytochrome B binds with greater apparent affinity than phytochrome A to the basic helix-loop-helix factor PIF3 in a reaction requiring the PAS domain of PIF3. *Proceedings of the National Academy of Sciences of the USA* **97**:13419–13424.

Zik, M., and V. Irish. (2003). Global identification of target genes regulated by APETALA3 and PISTILLATA floral homeotic gene action. *Plant Cell* **15**:207–222.

Zupan, J., T. R. Muth, O. Draper, and P. Zambryski. (2000). The transfer of DNA from *Agrobacterium tumefaciens* into plants: a feast of fundamental insights. *Plant Journal* **23**:11–28.

Index